MODEL-ORIENTED SYSTEMS ENGINEERING SCIENCE

A Unifying Framework for
Traditional and Complex Systems

COMPLEX AND ENTERPRISE SYSTEMS ENGINEERING

Series Editors: Paul R. Garvey and Brian E. White
The MITRE Corporation
www.enterprise-systems-engineering.com

Designing Complex Systems: Foundations of Design in the Functional Domain
Erik W. Aslaksen
ISBN: 1-4200-8753-3
Publication Date: October 17, 2008

Architecture and Principles of Systems Engineering
Charles Dickerson and Dimitri N. Mavris
ISBN: 1-4200-7253-6
Publication Date: January 30, 2009

Model-Oriented Systems Engineering Science: A Unifying Framework for Traditional and Complex Systems
Duane W. Hybertson
ISBN: 1-4200-7251-X
Publication Date: February 15, 2009

Enterprise Systems Engineering: Theory and Practice
George Rebovich, Jr. and Brian E. White
ISBN: 1-4200-7329-X
Publication Date: April 15, 2009

Leadership in Decentralized Organizations
Beverly G. McCarter and Brian E. White
ISBN: 1-4200-7417-2
Publication Date: May 15, 2009

Engineering Mega-Systems: The Challenge of Systems Engineering in the Information Age
Renee Stevens
ISBN: 1-4200-7666-3
Publication Date: June 25, 2009

Complex Enterprise Systems Engineering for Operational Excellence
Kenneth C. Hoffman and Kirkor Bozdogan
ISBN: 1-4200-8256-6
Publication Date: November 16, 2009

Social and Cognitive Aspects of Engineering Practice
Stuart S. Shapiro
ISBN: 1-4200-7333-8
Publication Date: March 30, 2010

RELATED BOOKS

Analytical Methods for Risk Management: A Systems Engineering Perspective
Paul R. Garvey
ISBN: 1-5848-8637-4

Probability Methods for Cost Uncertainty Analysis: A Systems Engineering Perspective
Paul R. Garvey
ISBN: 0-8247-8966-0

MODEL-ORIENTED SYSTEMS ENGINEERING SCIENCE

A Unifying Framework for Traditional and Complex Systems

Duane W. Hybertson

CRC Press
Taylor & Francis Group
Boca Raton London New York

CRC Press is an imprint of the
Taylor & Francis Group, an **informa** business
AN AUERBACH BOOK

Auerbach Publications
Taylor & Francis Group
6000 Broken Sound Parkway NW, Suite 300
Boca Raton, FL 33487-2742

Auerbach Publications is an imprint of Taylor & Francis Group, an Informa business

Printed in the United States of America on acid-free paper
10 9 8 7 6 5 4 3 2 1

International Standard Book Number: 978-1-4200-7251-8 (Hardback)

Library of Congress Cataloging-in-Publication Data

Hybertson, Duane W.
Model-oriented systems engineering science : a unifying framework for traditional and complex systems / author, Duane W. Hybertson.
p. cm. -- (Complex and enterprise systems engineering)
Includes bibliographical references and index.
ISBN 978-1-4200-7251-8 (alk. paper)
1. Systems engineering--Simulation methods. 2. System analysis--Technique. 3. Engineering models. I. Title. II. Series.

TA168.H93 2009
620.001'171--dc22 2008044364

Visit the Taylor & Francis Web site at
http://www.taylorandfrancis.com

and the Auerbach Web site at
http://www.auerbach-publications.com

Contents

Book Summary

A top level overview of the book chapters is presented below. Each chapter name is followed by key concepts in the chapter.

Preface

The field of systems engineering (SE) has achieved a degree of maturity but is currently experiencing a significant expansion of scope beyond its comfort zone. The expansion is predominantly in the area of complex systems (CS); that is, systems that tend to include people or other autonomous agents, cross organization boundaries, change continually, and be less predictable, less deterministic, more chaotic, less centrally controlled, and more self-organizing and adaptive than is true of what we consider traditional systems. The expansion additionally includes scaling — both up and down. The scale-up expansion is reflected in sociotechnical systems such as organizations, enterprises, and systems of systems (SoS). The scale-down expansion is reflected in systems at the level of synthetic biology and nanotechnology.

SE is, therefore, at a juncture in its history where it needs to strengthen and extend its methods in order to support its attempted reach. An important part of the solution is to add complex SE (CSE) to the SE field. Using a machine metaphor for systems of the past and an organism metaphor for systems of the future brings important concepts to the table. However, if that is all we do, we still have not addressed an important reality: Most systems that SE will engineer in the future are combinations of machines and people, not strictly one or the other. In addition, many of them exist at multiple scales, ranging from the nano-level to the macro system-of-systems (SoS) level. This means that CSE does not replace traditional SE (TSE) but must be integrated with it in a unified perspective. The problem is thus one of both expansion and unification.

How should the SE community respond to this evident need? Two difficulties present themselves. First, CSE and TSE are significantly different approaches. Systems engineers would have difficulty applying a conbination of the two to produce a coherent system. Second, the concept of CSE is new and immature; while people are experimenting with certain elements, including nanotechnology, synthetic biology, and CSE-oriented processes, we do not yet really have a body of knowledge on how to engineer CS. Indeed, there is not yet a general understanding of what engineering a CS means, or a consensus that it is even possible. While our view is that it is possible to engineer CS, we acknowledge that the understanding of this area is very limited.

So how can we achieve both expansion and unification? Let's look at what we do have. We have a set of sciences that is relatively mature and available for supporting an expanded and unified SE. These include systems science (SS), which describes characteristics that are common to all systems, and complex SS (CSS), which draws from a family of sciences — such as biology, sociology, cognitive science, organizational theory, management science, computer science, and network science — that have knowledge useful to SE. Some information from these sciences is in the scope of TSE but considerable portions of the available knowledge have yet to be exploited in SE. If we could collect and unify more knowledge from these sciences into a comprehensive and coherent science of SE, that would provide a foundation on which to develop and mature an extended and unified SE.

The goal of this book is to do just that: define an organized and unified science foundation for the extended and unified SE of the future — an SE science (SES) that is the result of collecting and organizing appropriate information from a variety of sciences, including those listed above. The result is intended to be a seamless and tailored science of the class of systems within the extended scope of the new SE. This SES foundation should then provide a good basis for positioning existing SE approaches and for defining and maturing new approaches that address the broader SE spectrum of the future, including the CSE part of the spectrum.

We did not call the book *Systems Engineering Science*; we called it *Model-Oriented Systems Engineering Science (MOSES).* What is model orientation (MO), and where does it fit in this picture? In this book we extend existing modeling approaches into an MO that views all science artifacts (such as theories, laws, and observed patterns) and all engineering artifacts (such as requirements specifications and designs) as models of systems. Based on this orientation, all traditional and complex SE methods, processes, and artifacts and all supporting sciences, are organized into a virtual structured repository called the SE model space. In effect, the model space is envisioned as a container for the unified SES and unified SE body of knowledge of the future.

A brief discussion of the role of SS, CSS, and MO in extension and unification will further clarify the importance of MO in this book. SS provides support for both extension and unification. But the heavy lifting in *extension* is provided by CSS because CSS has knowledge of the class of organic systems that is, to a significant extent, missing from TSE. And the heavy lifting in *unification* is provided by model orientation because MO brings a common perspective to the whole SES/SE enterprise, including all supporting sciences; engineering for the full range of TS, CS, and hybrid systems; and management. Another SE unifying element is software and computation, which by means of agent modeling and other knowledge areas provides a bridge between traditional and complex systems, between mechanistic and autonomous elements. All of these unifying arenas are important for the future of SE and SE science, as we point out in several places throughout this book.

The substance of SES is provided by the bodies of knowledge in the contributing sciences. The integration and organization of this knowledge are supported by MO and the model space. The model space structure also influences the way SES is presented in this book.

The task of defining MOSES is made easier by the fact that most of the solution elements already exist in some form. SS has developed a body of knowledge, sometimes called general systems theory and systems thinking. Considerable knowledge of CS exists in CSS, much of it already in the form of models. An array of modeling approaches exists that extends beyond the mature field of simulation to embrace a broader engineering spectrum. In addition to SE itself, related disciplines such as software engineering and operations research also have extensive knowledge of and experience in modeling. The intent of this book is to organize and integrate all these elements into a common framework, thereby making them more accessible to SE scientists, methodologists, researchers, practitioners, and managers.

The book first describes the need to expand and unify SE, and various opportunities that exist to satisfy this need. The contribution of SS and a view of the relation of SS to SE are presented next, followed by core concepts and a description of some key features of MOSES. Then a series of chapters lays out the model space and the patterns associated with each of its dimensions. The later chapters cover the specification approach, the mapping of traditional SE artifacts to the MOSES structure, and some discussion of the expanded SE process called collective actualization.

Acknowledgments

Three key elements were instrumental in the course of an intermittent journey over the past 20 years that led to this book. I want to express my appreciation to those responsible. First, the Software Productivity Consortium and in particular Fred Van Horne fostered the start of the journey in 1987–1988. Fred stimulated my thinking in numerous conversations we had during that period and in occasional discussions since then, and he pointed me to some seminal books and papers. Second, MITRE funded an internal research project on architecture foundations that I led from 1998 to 2000. Many of the ideas, including the basics of model orientation, reached critical mass from this work. Third, in the past few years, Brian White of MITRE helped expand the ideas and provided the impetus for turning them into a book. Part of Brian's contribution was through connection with an informal group on complex systems at MITRE and participation in the systems science group that Brian led at INCOSE. The other part was through support for my work and through the establishment, with Paul Garvey, of this systems engineering book series.

Many people affected my thinking that went into this book — too many to acknowledge here. But I would like to express my thanks to a number of people who have made particular contributions over the past year or two. The INCOSE Systems Science Group contributed both by providing stimulating discussions and expanding my understanding of this important area, and by reviewing my early chapters and providing helpful comments, especially Chapter 2, "Systems Science." The reviewers included Erik Aslaksen, Alireza Fereidunian, Tim Ferris, Joe Krupa, James Martin, Dan McGrath, and Nicky Sizemore.

Many of my MITRE colleagues made contributions. In the early stages of the book, Joe DeRosa provided feedback on some of my preliminary ideas, Paul Garvey gave feedback on one of the first draft chapters, and Tony DiPietro reviewed early versions of the first several chapters and gave me helpful feedback. In the later stages, Huei-Wan Ang, Paul Barr, Fatma Dandashi, George Rebovich, Michael Samuels, and Mike Tierney reviewed and commented on various chapters.

Several MITRE colleagues provided expertise in specific areas. Leo Obrst made substantial contributions to Chapter 8, to the extent that he became coauthor of the chapter. Eric Blair made significant improvements to the treatment of probability in Chapter 11, while Marc Colosimo provided feedback and validation of biological aspects of Chapter 9. Several people assisted on Chapter 10. John Anderson contributed to the discussion of architecture frameworks as well as other parts of the chapter; Gary Vecellio and Geoff Raines provided feedback and clarification on object-oriented inheritance; and Angel Asencio, Paul Barr, Keven Ring, and Edward Burke provided feedback on aspect-oriented development.

Special acknowledgments go to two individuals who reviewed the entire book and provided comments. They were Mike Nueslein of MITRE and James Martin of Aerospace, with whom I interacted in the INCOSE/Systems Science Group. Finally, Madge Harrison of MITRE edited the entire book and made numerous suggestions for improvements.

I would also like to thank Rich O'Hanley and Ari Silver of Taylor & Francis for their guidance through the publishing process.

Duane Hybertson

About the Author

Duane Hybertson is a researcher and member of the technical staff in the Information and Computing Technologies Center at the MITRE Corporation in McLean, Virginia. He has a broad background in software and systems engineering, both research and practice. He has conducted research in the foundations of architecture and has published papers on a uniform modeling approach to architecture and software engineering. He has supported enterprise architecture development and defined systems engineering approaches for large systems for several civilian, defense, and intelligence agencies of the U.S. Government. He has also conducted research on capturing security patterns and determining how to integrate these patterns into a usable enterprise engineering context. The outcome of this research was a book on security patterns that he co-authored and co-edited. His recent research has been in strengthening the foundation of systems engineering with increased visibility of systems science, and in extending and applying principles of systems engineering to complex systems such as organizations and other systems that include people.

About the Author

Chapter 1

Opportunities and Challenges for Systems Engineering

> "The spiritual temple of the future, while it will be built largely of the old well-proved materials, will require new and ampler foundations in the light of the immense extension of our intellectual horizons."
>
> **—J. Smuts, *Holism and Evolution***

The field of systems engineering (SE) has achieved a degree of maturity and success in its traditional scope of application domains such as aerospace, defense, transportation, and construction. While SE is working to strengthen its methods and increase its degree of success, changes are simultaneously occurring in the SE environment that generate a force for expansion beyond the traditional scope of SE. The primary area of change is the expansion to qualitatively more complex systems. For example, building a power plant is a complex engineering activity involving both technical and social elements but the behavior of a power grid across a region is substantially more complex and less predictable. The SE community is not only beginning to expand from power plants to power grids, but is also beginning to tackle problems even further beyond its traditional horizons. Representative problems include avoiding disease pandemics, creating a more viable healthcare system model, anticipating the effects of alternative energy policies, improving the effectiveness of the criminal justice system, and preventing or dealing with insurgency and terrorism. A secondary force for expansion is the emergence of new technology areas such as nanotechnology and synthetic biology. These technologies are making significant progress but the difficulty of building systems at different scales makes clear the need for extended SE support.

The current state of SE thus contains both opportunities and challenges. The opportunities are the possible areas of expansion in the field. The challenges are to start with an SE that has strengths and weaknesses within its current scope, and to define an SE that is more robust, covers a significantly expanded scope, and constitutes a coherent whole of the old and the new. Fortunately, many of the solution elements for an expanded and robust SE already exist, in some cases within SE but in other cases in related or supporting disciplines. What is needed is to bring the solution elements together into a unified whole. That is what this book aims to do: define a robust and unifying foundation for an expanded SE of the future.

This chapter discusses six topics: (1) the state of affairs of traditional SE, (2) the areas of potential expansion, (3) a bridge between old and new, (4) the challenges of defining a foundation for the future, (5) an introduction to the proposed model-oriented SE science (MOSES) foundation, and (6) the organization of the book.

The State of Affairs of SE

Characterizing Traditional SE

Over the past half-century, SE has developed as a discipline in certain classical areas, most notably aerospace and defense. It has also been applied in transportation and construction; energy; management, and manufacturing; and medical, environmental, and information systems. It was developed to promote a holistic or systemic approach to systems, as opposed to focusing on each separate part and addressing integration, if needed, as an afterthought. The SE that has been in play during this period, TSE, or good old traditional SE (GOTSE), has achieved considerable success in many cases but has also experienced more than a few significant failures.

TSE has a reasonably well-understood stance toward the systems it engineers. The TSE view of engineered systems is that

- They are large-scale machines that have relatively predictable behavior.
- They have components from multiple disciplines, such as power sources, electrical and mechanical systems, computer hardware and software, control systems, safety, security, and communications.
- Their components are machines or mechanistic elements, which can include any physical material used to construct systems such as aircraft, as well as traditional software and electronic items (e.g., computers). This scope was clearly articulated in Goode and Machol (1957, 5), who acknowledged common points of interest with social, biological, and ecological systems, but excluded them from the scope of SE. This limit has been reiterated almost verbatim by Kossiakoff and Sweet (2003, 4), who say that SE excludes complex systems such as social structure, living organisms, and ecosystems.
- Their components exclude people. This is implied by the previous item but is important enough to point out separately. The claim is sometimes made in TSE that systems include people and processes. However, in TSE, people usually play one of three roles, all of which reside *outside* the system and are treated as distinct from the system. The first role is the systems engineer; the second is the project manager (i.e., the manager of an SE project that engineers a system); and the third is the user or operator of a system (addressed in the areas of human-machine interaction, human factors, and ergonomics). Therefore, in the traditional view, people are not system components.
- They are relatively stable. They change only if the owner decides to change them. Left to their own devices, they remain static or they deteriorate through normal use.
- Development and operation are separate phases of the system's life.
- They are designed and organized by an external designer — that is, by the systems engineer.

These characteristics seem reasonable, even obvious, and they indicate the breadth of TSE. Many of these characteristics are implicit in the SE literature. Examples of the literature include Blanchard and Fabrycky (2005), Kossiakoff and Sweet (2003), Sage and Armstrong (2000),

Martin (1997), Buede (2000), DoD (2001), ISO 15288 (ISO/IEC 2008), IEEE (2005a, b), and INCOSE (2007). Books and standards on TSE typically focus on SE life-cycle processes and management, that is, both management and technical engineering aspects of SE. The latter typically include concept formulation, definition and analysis of alternatives, decision support, system design, system integration, and evaluation. Other common processes are requirements definition, technology assessment, modeling and simulation, and other aspects such as human factors, laws and regulations, and security and safety. In some cases, management processes have been extended to reflect more of the whole systems modeling scope discussed in the section entitled "Response: Model-Oriented SE Science." For example, ISO/IEC 15288 (ISO/IEC 2008), in addition to project management processes, defines enterprise processes such as investment management and resource management, and agreement processes such as acquisition.

Our purpose in providing this brief portrayal is to set a context for the focus and scope of what MOSES addresses. From that perspective, the characteristics described above constrain the type of system addressed by SE to what we call a system-as-machine metaphor; they reflect a mechanistic view of systems. We argue in this chapter that this constraint is beginning to make traditional SE less viable in light of the recent tendency of systems engineers to expand into other more complex types of systems. We need to rethink the scope of SE, not only to support current expansion, but also to provide a growth potential for future opportunities.

The Need to Strengthen TSE

Failed Projects

Many systems engineering projects fail. They do not produce a system on schedule, they overrun the budget, the system does not perform adequately, or the system is not even built. Bar-Yam (2003) describes a number of well-known SE projects that failed, including the FAA Advanced Automation System (AAS) that was in development over a decade in the 1980s and early 1990s, and then was abandoned after more than $1 billion was spent.

Why have so many SE projects failed? A significant part of the problem, according to Bar-Yam and other observers, is that these systems have become so large and complex that the traditional big-bang approach — that is, build the entire system and deploy it at one time — has little chance of succeeding. A significant part of the complexity in many of these systems, including the FAA AAS, was the software. One answer is to use an evolutionary approach. Start with something small and manageable; then grow and evolve it over time into the full system.* This type of process is discussed further in Chapter 13.

Dealing with Change and Other General Shortcomings

In its modern form, SE was developed in the 1950s at least in part as a more systematic approach than the *ad hoc* activities undertaken during World War II (although not all World War II SE work was *ad hoc* — the Manhattan project that produced the atomic bomb certainly was not). Some of this history is examined by Ferris (2007a, 2007b, 2008). During the Cold War, SE in the U.S.

* In most of this book, we follow the common engineering usage of the term "evolution," as the gradual change in an individual system over time, even though this is not what biological evolution means. We discuss the distinction in Chapter 9.

military had a great luxury: plenty of money and plenty of time. The process over time became large and cumbersome because there was little or no forcing function to keep it lean and agile. In fact, according to the NRC (2008), the length of time for the DoD to procure a major system in recent years is two or three times as long as it was 40 or 50 years ago. A sample of major programs from 1945 through 1970, including developing the atomic bomb and going to the moon, showed a range from 2½ to 8 years, while a sample over the past three decades shows a range from 11 to 20 years. This trend is occurring even though, since the end of the Cold War, the time pressure has presumably increased because military situations were changing much more rapidly. Both government agencies and practitioners have been slow to move beyond traditional processes and adopt others that support natural problem solving more effectively.

Alternatives to the rigid sequence of the "waterfall" process have emerged, such as incremental, spiral, and "agile" methods, but government — and to some extent industry — is struggling to devise SE processes that are sufficiently adaptable and flexible while still maintaining planning and accountability. Significant progress is needed in this area. The DoD recognizes some of the shortcomings of the current processes and is taking steps to revitalize SE in the DoD community. Some of the efforts are described in the October 2007 issue of the DoD journal *Crosstalk* (see *Crosstalk* 2007). This issue of the journal also emphasizes the expanding role of software in systems engineering, a topic that we discuss in the section entitled "The Bridge: Information, Computation, Software."

SE systems that change rapidly go well beyond military systems. As Braha et al. (2006, especially Chapter 1) argue, TSE is oriented toward well-behaved systems that can be precisely described, with characteristics of stability, reliability, certainty, and optimality. But the field is now dealing with systems, such as the Internet, power grids, markets, multinational enterprises, and transportation networks, that have different characteristics, as we discuss further in the next section. In a sense, many of these systems have a life of their own; while we can influence them, we do not control them. They continually change, and grow, and involve significant uncertainties, yet they seem to work. TSE is not well equipped to handle this type of system.

Other general weaknesses of TSE have been described by INCOSE (2006) and Rouse (2003). INCOSE argues that there is no set of unified principles and models to support SE use over a wide range of domains. The view in this book is that a clear deficit in TSE is the fact that the foundational concepts of systems science (SS) have not been used in a systematic, explicit way to support TSE. The TSE and SS communities, including the literature, have been largely disjoint. Another weakness stated by INCOSE is that TSE does not consistently address and integrate all elements of a system, beyond hardware and software. Facilities, procedures, processes, and the roles of people may also be significant parts of systems.

Rouse (2003) argues that we need an SE that incorporates and integrates a set of views that includes both the traditional SE perspective of hierarchy and system states, and the complex systems perspective of nonlinear mechanisms and autonomous agents. This SE needs to show how all these views fit together.

Unexpected Side Effects

Many of the problems with systems that SE builds are the unexpected side effects in the environment in which the system operates. Dörner (1996) discusses the difficulty of anticipating the side effects of decisions and systems in complex situations. His concept of a system is that it has many interdependent variables, and there are many side effects when a system is changed or a solution is introduced into the environment. Taking a more holistic view of the environment helps anticipate the effects of introducing a system. A similar idea is represented in the concept of intervention system (Martin 2004) and its effects on the context when it is deployed. Those effects are often different from what was intended.

SE now needs (and to some extent always needed) to consider all interconnected variables, implications, and side effects of decisions. Both Dötner (1996) and Martin (2004) imply that we may be able to model this arena and approach it more knowledgeably and systemically, which means that we may be able to bring these issues into the realm of engineering. That view is shared in this book.

Opportunities: Types of Expansion

Several current trends represent extensions to TSE. The dominant general category of the extensions is complex systems (CS). Here are ways in which complex systems extend SE: more types of systems, more people involved in the engineering, and more scaling.

Characteristics of CS

Complex systems have the potential to exhibit one or more of the following characteristics:

- They are thought of as organisms whose behavior is to some degree unpredictable.
- They have a variety of components, and often a very large number of components, that are highly interconnected, and the behavior of the whole system depends in a significant way on the interactions among components.
- The systems are to some degree autonomous, acting on their own.
- Their components are considered to be people or organic or autonomous elements, which can include anything in the realm of biology, psychology, and sociology, and a range of related disciplines such as economics, law, and organizational theory. In addition to humans, examples include cells, organs, and social groups such as organizations and societies. Other examples, such as self-modifying learning software and cellular automata, are not necessarily organic but can be modeled as organic or autonomous.
- They are adaptive and learn, grow, and evolve in response to their environment. Left alone, they grow and improve. They continually change but they also have significant aspects of stability. They also change their environment.
- Following a formation period, their development and operation are conflated into an ongoing evolution process in which change occurs while the system is in operation.
- They are self-organizing, and the whole system emerges over time as a result of cooperation and competition among components.
- Emergent properties and behavior are often considered a key characteristic.
- They have expanding scope: they go beyond technical issues to social, political, and organizational issues.

These characteristics reflect a system-as-organism metaphor, an organic view of systems, an autonomous agent view. Clearly, these characteristics are in contrast to those described earlier from the perspective of TSE. We emphasize the constraint that a complex system potentially exhibits one or more of the characteristics in the list — not all of them. The purpose is not to equate organic, autonomous, adaptive, emergent, and self-organizing. Part of the job of bringing these types of systems into the SE arena is to differentiate the variety of characteristics that the community, and this book, identify under the rubric of "complex systems."

The contrast of the organism or agent perspective with the machine perspective is reflected in certain changes in mindset, where constructs that are viewed negatively in TSE are embraced as positive or at least potentially positive in a CS approach. These constructs include

- *Change:* Change is something to be avoided in TSE, or at least carefully controlled. In CS, change is a natural process. It can be positive, negative, or neutral, but in many cases, change reflects learning, growth, improvement, and adaptation.
- *Risk:* TSE attempts to minimize or avoid risk. A CS approach sees not only risk, but also opportunity, the flip side of risk, as a positive potential for significant gains.
- *Uncertainty and lack of control:* TSE tries to maintain maximum control and to minimize uncertainty. With CS, engineers can yield some of that control to systems that exhibit autonomy and self-organization, which reduces the amount of necessary instructions or explicit specifications. Instead of the engineer having to anticipate everything for the system, let the system work things out for itself as it goes along. In MOSES, this overall process is called collective actualization.
- *Contradiction and paradox:* TSE seeks to achieve consistency and avoid contradiction, paradox, tension, and contrast. Complex systems accept contradiction, tensions, dualities, and contrasts as representing balance and diversity, in the spirit of yin and yang. Contrasting elements are reconciled, in part, with the use of views.

The inclusion of people in systems and the lack of control are clearly connected. People are not simply information processing components of a system; their behavior in response to environmental stimuli is significantly mediated through their internal model or view of the world and their intentions. Developments in cognitive science, and their application in cognitive systems engineering (e.g., Hollnagel and Woods 2005), provide not only models of humans as components of systems, but also go beyond that to provide holistic models of systems that include both people and machines. In cognitive systems engineering, such systems are called joint cognitive systems; in this book we call them hybrid systems.

For more information on the CS view, see representative literature. A sampler: Axelrod and Cohen (2000), Boccara (2004), Holland (1992), Kuras and White (2005), Simon (1996), and Sussman (2002). A significant number of institutes, universities, and other organizations have been conducting research in complex systems and complex SS for decades. Examples include the Santa Fe Institute (SFI 2007), the New England Complex Systems Institute (NECSI 2007), and the University of Michigan Center for the Study of Complex Systems (CSCS 2005) in the United States, as well as many in other parts of the world.

Example CS Models and Systems

The CS community has produced a variety of models that characterize systems across multiple disciplines. Models include chaotic time series; spatio-temporal chaos; dynamical systems models, including logistic maps; quantum theory; genetic regulatory networks; distributed control in network-based systems; chain formation and network self-organization; biological models such as evolution and autopoiesis; percolation behavior and cascades in networks; and bilingualism and linguistic models. A basic set of complex system models is described in Boccara (2004).

System examples include economic markets; population epidemiology; immunology; pandemics; using bacteria for drug transport; weather systems; transportation systems; organizational management and societal governance; ecology and social ecosystems; politics and elections; health care and cultural shifts; air traffic control; emergency preparedness (e.g., for hurricanes); congestion sensing and mitigation in critical infrastructure systems; criminal justice systems; education; distance learning; the Internet; public policy and urban planning; military conflict, insurgency, and terrorism; and autonomous and adaptive information systems, both natural and artificial. The SE community is already beginning to address systems in these categories. For example, see MIT (2007), Rouse (2007), and Blumstein (2007).

In many cases, a CS model applies to a variety of disciplines or domains. An example is the application of biological models to many other domains such as economic markets, traffic patterns, and social organizations. The field of biomimicry examines biological organisms and processes as a basis for improving the design of human-engineered systems. See for example Benyus (2002) and Margaliot (2008).

Changing Perspectives on Complex Systems

From the CS and SS literature, we can interpret three perspectives on complex systems that have evolved over the past century. The first we will call "Trying to engineer complex systems as machines." This perspective, dominant in the early twentieth century but still in existence, includes the "scientific management" of Taylor (1911), management theory X described in 1960 by McGregor (2006), and in general, hierarchical levels, span of control, chain of command, and optimal efficiency.

The second perspective we will call "Trying to understand characteristics of complex systems as a science." This perspective has been active since the 1920s, in the form of cybernetics, systems theory, organizational theory, and the development of natural and artificial complex system models.

The third perspective returns to engineering; we will call it "Trying to engineer complex systems as organisms or agents." This perspective has emerged in the past decade or so. The new thinking is that the century-old concept of engineering CS did not fail because it was inherently a bad idea; it failed because it used inappropriate models. Regarding people as mechanistic systems did not work. Regarding people as organic systems or agents may work better because it is a better fit. We now have several decades of additional scientific knowledge of people and other complex systems to bring to bear on the task of engineering complex systems, so let's have another go at it — this time, with more realistic and useful models.

The three perspectives can be summarized this way: (1) use TSE to engineer CS; (2) study CS to define a science (CSS); and (3) define and use CSE based on CSS to engineer CS.

This book uses the second perspective, CSS, as a key part of a foundation to support the third perspective, CSE, within the larger SE. The expanded scope of systems identified in the previous section represents fertile ground for SE. The belief is that not only can we observe and describe these complex and organic systems, but we can also engineer (in the context of collective actualization) specific systems of these types, that is, systems that to varying degrees control themselves, act autonomously, and change on their own. Some of the examples above illustrated that the field is expanding in this direction. Potentially SE can address systems even at the level of societies, public policy, cultures, religions, war, peace, coalitions, treaties, laws, and forms of government — if we can define adequate models that work.

More People Involved: Mass Collaboration

The story of the Linux operating system is well known. In 1994, Linus Torvalds released the kernel of a variant of the UNIX operating system and opened it up for people to review it and change it. This was a radically different development model from the traditional closed proprietary team approach to software development. The result that has emerged from this approach is that Linux is seen as robust, reliable, and adaptable, and it has grown into widespread use as a consequence. It continues to evolve from the efforts of thousands of volunteers worldwide, and the open-source movement has spread to other types of software.

The online free encyclopedia, *Wikipedia*, is also a well-known story. A small coordinating group oversees a large number of people who have volunteered to write entries and review and edit other entries, for the encyclopedia. The result: in the space of a few years, Wikipedia has accumulated over 11 million articles in 260 languages, including over 2.6 million in English as of December 2008 (Wikipedia 2008). Although there is some debate on the quality, a 2005 analysis suggested that the quality of Wikipedia articles was close to that of *Encyclopedia Britannica* articles (Giles 2005).

How are these examples related? Linux and the open-source movement, and Wikipedia are examples of what Tapscott and Williams (2006) call "prosumers," which is a category of mass collaboration in which consumers become producers. The success of this model has made clear that non-hierarchical, self-selecting groups can manage and control themselves, and can work successfully as a very distributed group — even globally distributed, as with Linux and Wikipedia — through minimal rules and social norms instead of hierarchical control.

Tapscott and Williams also describe several other categories of mass collaboration, under the general term "Wikinomics." One is called the global plant floor, used by companies for engineering and producing physical goods. In this domain, the approach Boeing used for its earlier aircraft, such as the 777, contrasted with its approach on the 787 aircraft (Tapscott and Williams 2006, Chapter 8). Traditionally, Boeing defined all detailed requirements and developed the complete design of the aircraft and its parts. It gave detailed design specifications to suppliers, who were expected to build parts to those specifications.

In the 787 approach, both the requirements and design were defined collaboratively with suppliers and users, whom Boeing brought in as partners early in the process. This contrast is a clear example of the distinction between a traditional SE approach and the movement toward a potentially more collaborative SE approach.

It is important to note that the jury is still out on the result of Boeing's experiment. Whether that experiment succeeds or fails, the pattern of increased collaboration seems to be gaining traction in multiple industries. For example, the motorcycle industry in China has rapidly overtaken that of the Japanese in the world market by exploiting this pattern, according to Tapscott and Williams (2006, 218–224).

Not everyone agrees on the success of mass collaboration. Keen (2007) strongly criticizes the "amateur" nature of resulting content. Many question whether mass collaboration produces a lowest common denominator and excludes works of genius.

Our point in discussing this approach is not to say that it is a panacea or is better than more traditional approaches in all SE situations, but it can be seen as an alternative approach that has achieved enough success to merit further consideration and analysis for SE, and to include it as a potential tool in a larger toolset. The general point is that it is another way in which SE is expanding.

Traditional Mass Collaboration

Of course, mass collaboration is not a new phenomenon. A long-standing example of mass collaboration is the creation, meaning, and usage of words in a natural language such as English. We tend to think that dictionary writers define what is correct meaning and usage, and that dictionaries are normative or prescriptive. But dictionaries are a description of what the population thinks words mean and how they should be used. Dictionary writers are observers and recorders, not prescribers. We illustrate with three words defined and discussed in *Merriam-Webster* (2008): inflammable (versus flammable), irregardless (versus regardless), and comprise. The first two pairs are words that look like antonyms but are synonyms. The third word is bipolar; it has two meanings that are antonyms.

- *Inflammable* means flammable. It is judged by the population to be an acceptable word.
- *Irregardless* means regardless. Although it has been used for nearly a century, it has consistently been judged by the population to be an unacceptable word in published works.
- *Comprise* has two opposite meanings: 1. include, be made up of. (Example: This book comprises 13 chapters.) 2. compose, constitute. (Example: 13 chapters comprise this book, or this book is comprised of 13 chapters.) The two meanings have been around for over a century, and the second meaning has always been criticized as wrong. This second meaning has been used in scientific and technical writing, but not in literary writing. Recently, however, literary usage has shifted, and now the second meaning is used a bit more — but it is not completely accepted yet in that community and still criticized. Nevertheless, the tide is turning. The meaning over a period of decades appears to be reversing itself; it is changing polarity, or perhaps changing from bipolar to polar.

The dictionary does not say what is correct and incorrect; it records what is generally considered (by mass collaboration) to be correct and incorrect. And the population assessment evolves over time. In addition, dictionary writers do not create new words; the population creates its own words. Natural language is organic, not mechanistic. The population is self-governing when it comes to natural language.

Storytelling

An approach that has been developing over the past decade, distinct from but somewhat related to mass collaboration, is storytelling and the use of narrative to communicate and understand technical information. Storytelling is an ancient form of knowledge representation and transmission but now its strengths are being brought into the engineering arena. Two sample references are Manning (1999) and Kumar et al. (2008). Manning describes how storytelling has been used to provide technical information and some of the challenges involved. Kumar et al. illustrate how storytelling is becoming a formalized approach.

Chapters 3 and 8 discuss additional SE contexts of storytelling.

Scaling in All Directions

SE is experiencing extensions of scale in all directions: up, down, and horizontal. Horizontal scaling is reflected in the mass collaboration practices we have discussed, and also in the very large numbers of parts in complex systems, discussed in the span of control portion of Chapter 6. Up and down scaling is addressed here.

Scaling Up SE

SE is addressing larger systems. The example of extending from power plant to power grid reflected this. The term "system of systems" is sometimes used to reflect a higher scale, where a very large system is put together from existing systems. Another area of scale-up is organizations or enterprises. Enterprise can mean organization or mission but can extend up to multi-organization entities or multinational coalition forces. This level of scale extends beyond the scope of management of a single organization and poses its own challenges.

Scaling Down SE

In recent years, small-scale biological and physical systems have been an area of considerable interest, creativity, research, and development. Overlapping topics include micro-engineering, nanotechnology, nanobiology, nanosystems engineering, synthetic biology, systems biology, biological engineering, molecular computing, and quantum computing. These areas have great promise in many application domains, including medicine and computing. University degree programs and journals have been created to focus on many of these individual topic areas.

Systems biology and synthetic biology are particularly interesting and partially overlapping areas. If there is a distinction between the two, it is that the goal of synthetic biology is to engineer biological parts or "BioBricks," and the goal of systems biology is to build systems from these standard parts. MIT has established a Registry of Standard Biological Parts (Registry 2007). The purposes and emerging practice of synthetic biology are discussed in *Nature* (2005). The economics of systems biology, in terms of the prospects of successfully building systems from standard biological parts, is discussed in Henkel and Maurer (2007).

Do these small-scale systems relate to SE? Yes, systems are being engineered at this level, and anywhere systems are engineered — even if they are at a very small scale — would seem to be an appropriate arena for applying SE. This sentiment is reinforced by the current state of the effort described above to build systems from libraries of standard biological parts. Parts have been successfully built, along with some experimental circuits, but the community is struggling to build working systems from these parts. An example is a report that on one systems biology effort, the group spends 95% of its time trying to find and fix unintended interactions between parts (Keasling 2007, reported in Henkel and Maurer 2007). This problem sounds vaguely familiar to systems engineers. These systems biologists might benefit from additional SE assistance and education. Advocates for an SE view of nanotechnology include Das et al. (2005) and Rouse (2007, 31).

Multiscale SE

Many of these systems involve analysis and modeling of different scales in the same system. The presence of emergent properties at higher levels is a desired characteristic of these systems, but it also complicates the analysis and engineering. One example of this is the analysis of the physical, chemical, and biological properties of ionic liquids, to combine them in a way that produces desired characteristics at a higher scale (Rogers 2007). An application of this is the production of liquid mirrors for telescopes. For example, one proposal is to produce a mirror 20 to 100 meters in diameter for a telescope on the Moon that it is believed would give unprecedented views of the early universe (Borra 2007). Another example is the expected future use of carbon nanotubes in commercial aircraft, as described in *Nature* (2007, 1032). The expectation is that composites made from nanotubes will result in aircraft structures that retain the necessary strength but significantly reduce the weight in comparison to polymer-based fibers.

Systems such as these require an SE approach that can analyze and model many different scales in a system and can relate the different scales and levels to each other. The salient SE question for a given system is: What levels or scales does SE need to address? A multiscale modeling and analysis community is examining this issue. An example is the Center for Integrative Multiscale Modeling and Simulation (CIMMS) at the California Institute of Technology (CIMMS 2007). CIMMS is a multidisciplinary center set up to model and analyze complex systems across multiple scales of both time and length. Areas of study include astrodynamics, atmosphere, earthquakes, fluid mechanics, signal processing, imaging, materials, biological systems, and nanoscale systems.

Scale Viewpoints

Many of the examples presented in this chapter reflect changes of scale. One perspective on mass collaboration on a global scale is that it represents a scale-up of engineering, business, and knowledge production. Clearly, biological systems and nanotechnology represent scale-down. Both scale-up and scale-down have occurred over the past few decades. This stretching in both directions presents important challenges for systems engineering.

To adequately understand the role and extent of scales, we need to look at a more general context, namely, the potential range of scales from several viewpoints. These viewpoints typically are treated separately although they overlap significantly. They include

- *Physical:* The potential range of scales extends from sub-atomic particles up to the entire universe as we know it.
- *Social or organizational:* The range extends from individual people up to societies, nations, and multinational coalitions or groups, such as the United Nations.
- *Enterprise:* The range in this viewpoint typically extends from systems to organizations or enterprises. In this context, system typically means something like information system or machine, and enterprise typically means organization or mission, but can go up to multi-organization groups or multinational coalitions. In recent years, the concept of "system of systems" — that is, a large system put together from existing systems — has been discussed, often but not always in the context of this viewpoint.
- *Biological:* The range extends from nanobiology (assemblies of biomolecules) up to ecosystems.
- *Life or living systems:* The range extends from organic molecules or cells up to multinational groups, or approximately the combined range of biological, enterprise, and social/organizational scales. The fullest expression of this viewpoint is James Miller's monumental work *Living Systems* (1995).
- *Machine:* The range in this viewpoint extends from nano-devices up to large machines that are analogous to organisms in the social or biological viewpoint. That is, a car or train or aircraft or ship or computer is often regarded as a machine, but a transportation system or a fleet of ships or a computer network is usually regarded as an interacting group of machines and other elements — that is, a system — and not as a machine.*
- *Information:* The range extends from bits to a large global information system, of which the Internet is the dominant example.
- *Time:* The range of temporal scales extends from nanoseconds or time cycles of atomic behavior to billions of years or the life of the universe.

* Note that because some systems are not regarded as machines does not mean that machines are not regarded as systems. Machines, and in fact most items across the range of scales in all the viewpoints identified here, are regarded as systems, as this book emphasizes throughout.

The Bridge: Information, Computation, Software

Information systems, including computation and especially software, form a potential bridge between traditional systems and complex systems. Recent developments in this domain represent significant steps toward realizing this potential. This section describes the potential of and steps toward realizing it. Computation and software characteristics are divided into three sets: TSE related; CSE related; and bridge related.

TSE-Related Characteristics

Software and computing systems play multiple roles in SE. One role is manipulator of information. Information has become recognized as an important element and asset in all types of SE systems, and in all organizations. Software processes and analyzes the information. Software and computers are used in virtually every application domain, and as a component in virtually every SE system.

Another role of software is as an engineering tool. Software tools are used extensively in SE, and several forces are expected to increase the role of software tools. A prominent example is simulation. Others include analysis tools.

A major role is automation. Software is used to automate many activities and business processes that were previously manual. One area of automation is control functions in machines. Kossiakoff and Sweet (2003) identify automation as the fastest-growing and strongest influence on SE.

The Internet and technologies associated with the Internet have become increasingly important in SE. Software is a key part of Internet processes and technologies.

CSE-Related Characteristics

A major area of software support to CSE is agent-based systems. Much work has been done in this area, exploring characteristics such as autonomy, learning, adaptation, evolution, fault tolerance, and robustness.

Over the past decade or more, the concept of (software) cognitive agents has been pursued. For example, Huhns and Singh (1998) discuss the general concept, and Majid and Ahmed (2008) discuss a specific application. In fact, the idea of ascribing cognitive or mental qualities to machines (software) was proposed 30 years ago by John McCarthy (1979). A common model used in defining and reasoning about cognitive characteristics of software agents — as well as modeling human agents — is the BDI model (beliefs, desires, and intentions) (Rao and Georgeff 1991).

In 2001, IBM proposed a concept called autonomic computing (Kephart and Chess 2003), which they describe as a systemic view of computing modeled after a self-regulating biological system. The four areas were self-configuration, self-healing, self-optimization, and self-protection. The original motivation was largely the fact that many complex information systems in operation are brittle and non-robust against input or environment vagaries, and keeping them operating is becoming a difficult and resource-intensive effort. Since then, the concept has become an active area of research, and the scope has broadened to incorporate more management aspects (Salehie and Tahvildari 2005). For example, models of biological processes are being applied to assist in policy decomposition (Keeney and Wade 2008).

Another area is the increasing emphasis on ontology engineering and machine-manipulable semantics. Software plays a major role in these efforts. This area is discussed further in Chapter 8.

Bridge-Related Characteristics

Software is a prominent segue, a link, between the mechanistic and organic models. It can range from mechanistic straightforward algorithms or data processing to appearing to be an intelligent agent. The point applies to autonomy as well as intelligence: software and computer systems can exist anywhere in the autonomy spectrum between the controlled machine and the autonomous human.

Software is inherently model oriented. The software engineering field has investigated modeling in great depth and has produced much modeling machinery for design and analysis of all domains. Modeling is something that applies to the engineering of both traditional and complex systems.

Social and biological sciences are a source of much information for the new, expanded SE. The software community has already given the new SE a big head start toward preparing this information for SE by developing and executing models. Much of this work is in agent modeling.

Another perspective on the software role focuses on the challenges of software itself, especially large-scale systems. The Software Engineering Institute at Carnegie Mellon produced a report on ultra-large-scale systems of the future and their challenges, implications, and areas of needed research (Feiler et al. 2006). These large software systems are characterized as different in kind; they need to be viewed not as systems or systems of systems, but as sociotechnical ecosystems. They exhibit many of the characteristics of large complex systems, including continual change and evolution, conflicting goals, and decentralized control.

Software plays a key role in both the extension of SE (via CS models and agents) and in unification (spectrum of autonomy, bridge, TSE and CSE aspects). Extension and unification are challenge areas discussed in the next section.

Challenges: A Foundation Needed for the Future SE

Based on the state of TSE, the opportunities for SE expansion, and the software bridge, this section presents the challenges of defining the SE of the future. Two elements represent the SE challenge and are discussed in this section. The first is the general need to support an expanded SE. The second is the general need for unification of traditional and complex systems, including a unified taxonomy of systems, and a unified foundation. The challenges are then captured as a set of requirements that an SE approach needs to satisfy to adequately support the expanded and unified SE.

General Challenge: Support Expanded Scope

If SE is going to expand in the directions indicated in the section entitled "Opportunities: Types of Expansion," it will need to define methods that support the associated CS characteristics — CS engineering, or CSE. Such support includes a larger variety of system types and their associated disciplines, mass collaboration, and more scaling, both up and down.

Although some in the SE community have begun to examine how to engineer them — for example Braha et al. (2006) and Kuras and White (2005)* — the concept of CSE is still relatively immature. The difficulties have been variously described as wicked problems (Rittel and Webber 1973), messes (Ackoff 1999, 117–118), and a swamp (Schön 1995). In their books on system

* A set of potential CSE methods is identified in Chapter 13.

architecting, Rechtin (1991) and Maier and Rechtin (2000) have extended beyond TSE into architecting of complex systems such as economic systems, organizations, competing or adversarial systems such as warfare, and the category of sociotechnical systems. The focus is on heuristics and general discussion of issues, as well as the use of models.

Significant research, experimentation, and practice in the engineering of CS are needed to develop the requisite knowledge for CSE. Indeed, there is not yet a general understanding of what engineering a complex system means, or any consensus that it is even possible. While developments in related areas — such as nanotechnology, synthetic biology, cognitive SE, design of agent-based systems, and large enterprises — seem to suggest that it is possible to engineer certain types of complex systems, it is fair to characterize the CSE body of knowledge as being in an early stage of development.

General Challenge: Unify

Unify TSE and CSE

Even if we meet the challenge of defining a CSE method to support the CS expansion of SE, we still have a problem. The new scope is so large and heterogeneous that SE will not have a coherent method that covers it all — TSE and CSE have significantly different assumptions. Conceptualizing the systems of the past as machines and the systems of the future as organisms or agents brings useful and needed concepts to the table but leaves out an important reality: Most systems that SE will engineer now and in the future are a combination of machines and people, not strictly one or the other. We cannot simply replace TSE with CSE. And a disjoint SE consisting of the TSE book and the CSE book will not suffice when most SE systems reflect aspects of both. Our position is that they must be integrated into a unified SE.

Unify Multiple Disciplines

TSE is considered multidisciplinary. The new fields of synthetic biology, nanobiology, and others at that scale are multidisciplinary in themselves, so they add to the multidisciplinary scope of SE. In addition, complex systems are broadly multidisciplinary because their models apply to many different application areas. The result is that the extended SE reinforces and extends the multidisciplinary scope of SE. In fact, there are several ways in which SE is multidisciplinary, including

- SE systems are built in many different application domains — defense, aerospace, transportation, construction, energy plants, medicine, organizations, etc.
- Any given SE system typically represents multiple disciplines in its components.
- SE models, including mathematical and complex system models, cut across or support multiple disciplines.
- The SE of the future can be expected to partner with, or at least use knowledge provided by, the various supporting disciplines of SE, including SS, CSS, and others that we list below and identify more fully in Chapter 2. If we are serious about using holistic approaches to solve problems in the new extended SE, we need to move beyond pigeonholed and stovepiped disciplines.

To make this multitude of disciplines work together in a coherent fashion requires significant unification.

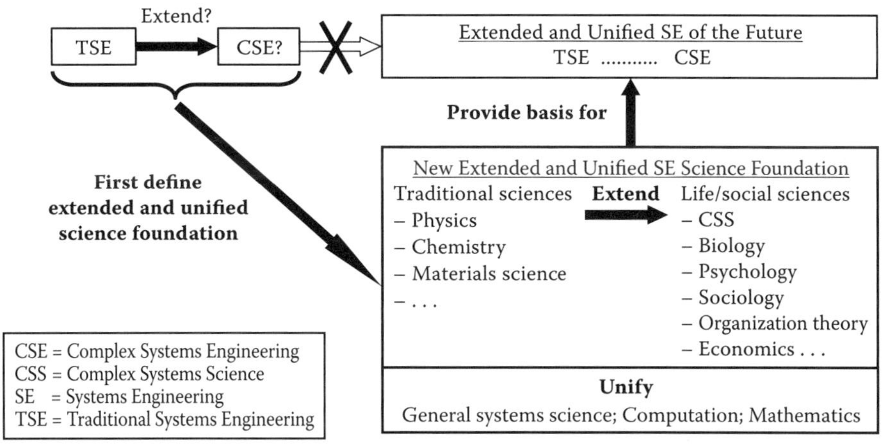

Figure 1.1 Need for extended and unified SE science foundation.

How does the SE community respond to this evident need? Our answer is illustrated in Figure 1.1. It reflects the view that simply adding CSE to TSE is not adequate, as we indicated above.

So what is a better way to proceed? Let's look at what we do have. In addition to traditional disciplines such as physics and mathematics, we have a set of sciences and other supporting disciplines that are relatively mature and available for supporting an expanded and unified SE. SS describes characteristics that are common to all systems, and along with computation and mathematics forms the unifying foundation. Organic or life/social sciences, such as CSS, biology, psychology, sociology, organization theory, and economics, provide knowledge of complex systems that is useful to SE. If we could collect and unify knowledge from these sciences into a coherent science of SE, or more specifically a science of all potential SE systems, that would provide a good foundation for evolving an extended and unified SE.

Specific Requirements List for New SE Foundation

Based on the general challenges and the issues raised in the chapter thus far, we can define a list of requirements for a foundation of the future SE. An enabling foundation for a unified SE should satisfy the requirements listed in Table 1.1. Column 2 indicates a judgment of which requirements TSE satisfies substantially (S) or partially (P). For the majority of requirements not satisfied by TSE, the reason is the use of the mechanistic model of systems.

TSE appears to at least partially satisfy most of these requirements. But the holes are significant: organic, hybrid, change, and unified approach are large gaps.

Response: Model-Oriented SE Science

The Smuts quote given at the start of this chapter was targeted to "the main problems of science and philosophy, ethics, art, and allied subjects" (Smuts 1926, x). However, the quote can also be interpreted as a good summary of the current situation facing SE depicted in this chapter: the systems engineering temple of the future, while it will be built largely of the old proven materials, will require new and ampler foundations in the light of the immense extension of our intellectual horizons.

Table 1.1 Requirements of a Future SE Foundation

No.	Requirement	TSE Satisfies?
1	Support mechanistic characteristics of systems	S
2	Support organic and agent characteristics of systems	
3	Support a hybrid mix of traditional and complex or sociotechnical systems in multiple disciplines (people and machines) in a seamless way, including characteristics in common to both TS and CS, and characteristics in which TS and CS differ	
4	Support a mix of matter, energy, and information elements in systems (i.e., a mix of conceptual and physical elements in systems) in a seamless way, including the dominant role of information and computation	S
5	Support scale-up and scale-down of systems, and multiple scales in a single system, to include a variety of scale viewpoints	P
6	Support arbitrary, fuzzy, and uncertain system boundaries	
7	Support the modeling of the interaction of a system with its environment, to anticipate mutual adaptation and side effects	P
8	Facilitate precision, minimalism, and simplicity	P
9	Support formal and informal approaches	S
10	Be flexible, adaptable, and extensible, and evolve as needed	
11	Facilitate incorporation and use of knowledge and experience	S
12	Support appropriate mass collaboration in systems engineering, including innovation and mechanisms for intellectual property	
13	Support multiple views and perspectives of a given system or class of systems	S
14	Support tensions, dualities, contradictions, contrasts, and paradoxes in a unified way that includes both their separation and their integration; in the manner of yin-yang principles	
15	Support balance of simplicity and complexity: everything should be as simple as possible, but no simpler; recognize imperative of complexity	P
16	Support balance of commonality and variability: Find everything in common but do not force one-size-fits-all; recognize imperative and benefits of diversity	S
17	Support balance of stability and change: find everything that is stable but do not force false stability; recognize imperative and benefits of change	
18	Support clear communication among all SE stakeholders but recognize that not all clear communication is explicit; support both explicit and implicit communication	P
19	Support clear articulation of problems and solutions	P
20	Support analysis of alternatives	S
21	Support uncertainty in specifications	P
22	Support SE management by providing clear communications and analysis via appropriate management models	P
23	Support the engineering modeling process; automate as much of it as possible	P
24	Provide a unified approach to organizing and harnessing the knowledge from the array of disciplines supporting SE	

We cast this task in terms of defining an SE science (SES). The approach we take is what we call model orientation (MO), so the name of our proposed foundation is model-oriented systems engineering science, or MOSES. It is intended to provide the "new and ampler" foundations. MOSES depicts a future state of a science of SE (SES), where SE has the expanded scope described in this chapter, and both the SES and the SE are mature. The purpose of MOSES is to satisfy the requirements identified in the previous section.

The characteristics or aspects of MOSES are introduced and briefly discussed in this section; they are addressed more fully in the remaining chapters of the book. Also described in this section is the rationale for MOSES and potential audiences of MOSES.

Scope and Basic Aspects

MOSES is depicted here in two ways. First, the scope of the composite MOSES is shown in Figure 1.2. Second, the items in the figure are elaborated and described as a set of MOSES aspects.

Scope of MOSES

This chapter has discussed the extension of scope in the types of systems that SE is attempting to engineer. However, the total scope of SE, and the corresponding scope of an SE science, go beyond the systems that are engineered. Figure 1.2 summarizes the scope of MOSES. The systems

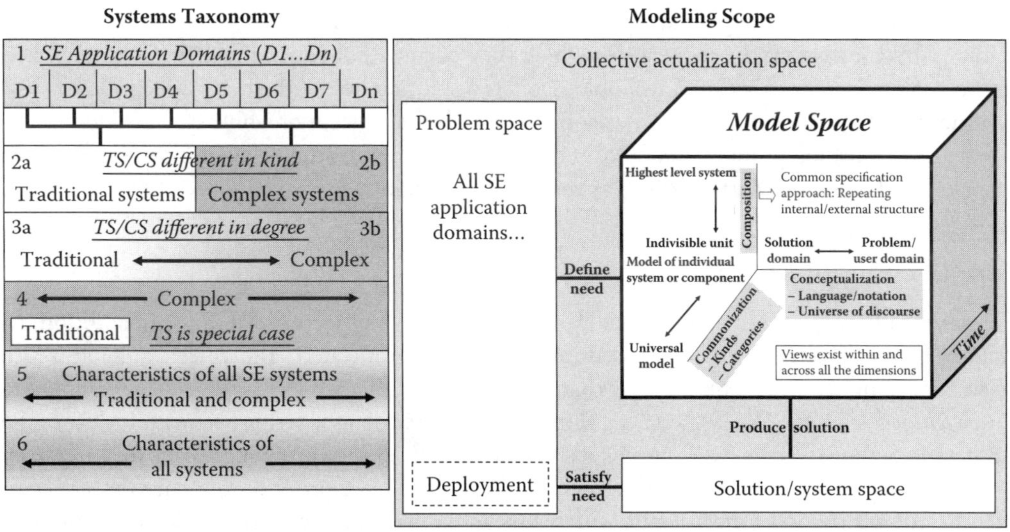

- Systems taxonomy partitions SE systems and shows traditional and complex systems relations
- Collective actualization space includes all SE and management processes, artifacts, contexts
- Problem space includes all application domains SE supports and in which systems are deployed
- Model Space contains all engineering and management artifacts (as models)
- Solution/system space includes all solutions/systems produced by actualization processes

Figure 1.2 Composite MOSES map.

taxonomy* on the left shows the scope of systems addressed, while the modeling scope on the right shows the scope of models and the modeling context. A specific link between the two is that the systems taxonomy shows the variety of systems that show up in the system space on the right. The systems taxonomy incorporates all systems of interest to SE, both traditional and complex systems. The taxonomy also shows categories of systems ranging from specific application domains, such as transportation or defense, to the most general category of all systems defined by SS.

The modeling scope comprises several spaces. One space is the overall collection of entities required to engineer the desired system. Martin (2004) calls this space the realization system; we call it the collective actualization space. The entities can include organizations and people, facilities, equipment, materials, policies, standards, processes, methods, tools, and information. Second is the problem space, the environment in which a problem appears. SE often engineers a system in response to a problem in some application domain and in some operational environment or class of environments.

Third is the model space, an organized repository of all engineering and management artifacts or documents (in MOSES, models) that feed into or result from the actualization process. The model space does not include the problem space or solution space, but it does include models of the problem space and models of solutions in the solution space, as well as models of the collective actualization space and models of other models in the model space. Fourth is the solution or system space, which contains a solution (typically a system) deployed to satisfy a need. The fifth space is the environment in which the system is deployed, which is closely related to the problem space because the system is deployed to solve the problem. However, the deployment space also includes all entities that interact with the deployed system and the entities required to operate, manage, sustain, and evolve the system. The system, when deployed, changes the problem environment.

Note that the lines designated "Define need," "Produce solution," and "Satisfy need" are not arrows. This means that the sequence in which they occur is not constrained. Sometimes, a need drives a solution but other times, technology offers a solution before the need is identified.

The modeling scope in Figure 1.2 is consistent with the concept of whole systems modeling as defined by a working group at INCOSE (Ring 2008), and with the collection of systems in the scope of SE defined by Martin (2004) as the seven samurai.

Aspect Descriptions

The items in Figure 1.2 are now further elaborated as a set of aspects of MOSES. In addition, the chapter(s) where each aspect is discussed in more detail is indicated. The aspects are itemized in Table 1.2, along with the requirements that each satisfies.

Expanded concepts of science and engineering: A number of supporting sciences, including SS, CSS, and computing science, cut across natural and artificial systems, which leads to the concept of artificial science as well as natural science. CSE attempts to engineer or influence those systems that include people and other agents, which cuts across natural and artificial elements and leads to the concept of engineering natural and artificial components into mixed or hybrid systems. This topic is discussed further in Chapter 2. The expanded scope is addressed by another MOSES aspect, collective actualization, which is discussed below.

Model orientation, described more fully in Chapter 4, is a prominent feature of MOSES, and consists of several components:

* The systems taxonomy is developed and described in Chapter 2.

Table 1.2 Basic Aspects of MOSES

Aspect of MOSES	*Reqmts Satisfied*
Expanded and unified taxonomy of systems, including TS and CS	1, 2, 3, 14, 15
Expanded and unified supporting disciplines	11, 24
Expanded scope of science and engineering, especially relative to natural vs. artificial systems, driven by SS, CSS, and CSE	2, 3, 24
Model orientation:	
–Artifacts of SE and all supporting disciplines are models	1, 2, 3, 4, 11, 20, 24
–Whole systems modeling: problem, model, solution, actualization spaces	7, 24
–Model space: dimensions: self-similar repeating patterns; scope includes models in all spaces, across full systems taxonomy, in all supporting disciplines, and in SE	5, 10, 11, 13, 16, 17, 18, 20, 24
–Modular decoupled concepts	8, 10
–Unified specification approach: external, internal; common info types	1, 2, 3, 4, 6, 8, 9, 12, 13, 14, 16, 17, 18, 19, 21, 24
–Exploit software and computation as unifying bridge	3, 4, 7, 23, 24
Collective actualization:	
–Actualization as a technical and social process: model vs. specification	2, 8, 12, 18
–Management as part of actualization: Management as engineering	22
–Actualization process: opportunistic; support for change	12, 23

1. The basic concept is that artifacts of all engineering, science, math, computing, and other supporting disciplines throughout the systems taxonomy are considered to be models.
2. Whole systems modeling: a complete treatment of SE consists of modeling the Problem space, Model space, Solution space, and Actualization space, which is where modeling, engineering, management, and self-organizing processes exist. These spaces and their basic relationships are shown in Figure 1.2.
3. The Model space is the central focus of MOSES. It is structured by the dimensions of composition (whole–part), commonization (general–specific), conceptualization (language and concepts), and time (stability and change), along with the cross-cutting concept of view. It contains models in all spaces, across the full systems taxonomy, in all SE and supporting disciplines. It represents a knowledge base of SE and all supporting disciplines, identified in Chapter 2. Its structure is discussed in Chapter 5, and aspects of it are detailed in Chapters 6 through 10. Ultimately, the vision of the model space is that it will be substantially autonomic or self-managing — self-learning, self adapting, self-correcting.
4. Modular decoupled concepts: MOSES is highly modular. Concepts that are often coupled, such as hierarchy and control, are decoupled in MOSES. These are discussed in Chapter 4.
5. Unified specification approach: MOSES defines a common approach to specification of models that applies throughout the system taxonomy and throughout the model space. There are two primary model types: external and internal. The approach and the common set of information for each type are described in Chapter 11. The spec approach is also

Table 1.3 Elements of Collective Actualization

	Engineering	*Autonomy*
Enterprise	EEM enterprise engineering/management	EA enterprise autonomy
Produced System	SE system engineering	SA system autonomy

identified in the Model space in Figure 1.2. The arrow indicates that the internal-external specification types arise from the structure of the Composition dimension.

6. Exploit software and computation: MOSES exploits the roles, extension, bridging, and available models (including agent models) in the software and computation area, as discussed in the section entitled "The Bridge: Information, Computation, Software." This is a theme that appears in numerous places in the book, including the role of software in partially autonomous systems.

Collective actualization is discussed here, first in terms of structure and then in terms of process. The structure of collective actualization consists of the elements in Table 1.3. Collective actualization addresses both enterprise and produced system, both engineering and autonomy. The table reveals ways in which MOSES extends traditional management and systems engineering, and how they relate to each other.

Enterprise is a generalized term for managed entity at any scale. It is intended to include not only organizations and projects, but also many other types such as social groups, societies, social networks, multi-organizational partnerships or alliances, or multinational coalitions. The term "managed entity," and therefore "enterprise," is not restricted to a management style of hierarchical control. It also admits a style of complete autonomy or self-organization, or anywhere between the two. A simple and common use of enterprise is that it is a project to produce a system, and the project is associated with a larger enterprise called an organization. For example, Boeing as an organization sets up a project to design and produce the aircraft system 787. We use this example to illustrate the definitions of the table elements.

- *Produced system* in the table means a system that is being engineered or produced by the enterprise. The 787 aircraft is a produced system.
- *Engineering,* in general (as defined in Chapter 2), is the systematic practice of agents taking purposeful and constrained action to achieve a system of a certain class with certain desired characteristics. In Table 1.3, two classes of system are being engineered: enterprise and produced system — 787 project and 787 aircraft.
- *Enterprise engineering/management* (EEM) is the application of science-based management knowledge to an enterprise. The name of the role performing EEM can be enterprise engineer or manager. In the Boeing 787 project, EEM is the portion of the project managed or influenced by the Boeing project management team.
- *System engineering* (SE) is engineering applied to the produced system. Note that this SE box in Table 1.3 is a subset of the larger systems engineering role that is the scope of this

book and that includes all the elements of Table 1.3. In the Boeing 787 project, the SE box is the portion of the 787 aircraft engineering performed or influenced by the system engineering team.

- *Autonomy* is characterized by self-defined goals and purposeful action. *Enterprise autonomy* (EA) is autonomy practiced by components of the enterprise. In the Boeing 787 project, EA is the autonomous management-related actions and suggestions of the worldwide partners that helped shape the project.
- *System autonomy* (SA) is autonomy practiced by components of the system. In the 787 aircraft, we can consider two levels of autonomy: design and operations. At the design level, the internal designs of parts by partners would be considered autonomous from the viewpoint of the overall system SE, that is, the autonomous design agents in this case are the partners. Another design level variety of SA is that perhaps parts of the design were automated and performed by software agents. At the operations level, there are no doubt many self-monitoring, self-adapting, self-correcting, self-managing, autonomic elements in the 787 aircraft.

Collectively, EEM and EA reflect a change to traditional management, namely, the addition of autonomous elements of enterprise structure and behavior. Instead of a manager or enterprise engineer making all enterprise management decisions, the decisions are now shared or split. The enterprise components themselves make some decisions regarding their behavior, their role in the enterprise, and how the enterprise is organized and operates. The enterprise engineer or manager now shapes and influences rather than dictates. In addition, the EE part of EEM is intended to emphasize more systematic management based on science, including management science, organizational theory, and community-vetted best practice patterns. This does not mean that intuition and personal experience are not used in EEM (or in SE). It means that the intuition and personal experience are better supported and grounded. See also the discussion of heuristics below.

Collectively, SE and SA reflect a change to TSE analogous to the change in traditional management, namely, the addition of autonomous elements of system design or behavior. Instead of a system engineer making all system design decisions, the decisions are now shared or split. The system components themselves make some decisions regarding their behavior, their role in the system, and how the system is organized and operates. The system engineer now shapes and influences rather than dictates.

Both engineering and autonomy involve purposeful activity. The reason the autonomy activity is not considered engineering of the produced system is this: the purposeful activity of engineering includes the holistic and emergent behavior and properties of the produced system. The purposeful and autonomous activity of the SA, on the other hand, is typically directed to its own behavior and its interaction with other components in the system. This distinction implies that an entirely self-organizing system emerges or is produced with no engineering activity; it is all from autonomous behavior (SA). At the other extreme, if no autonomous behavior is present, it is equivalent to traditional engineering. The same argument applies to the distinction between EEM and EA.

Another important aspect of both EEM and SE is the use of heuristics. These are general rules of thumb based on insight and experience that have not been or cannot be reduced to specific engineering rules. But they can be shared across a community. Heuristics are an integral part of systems architecting in Rechtin (1991) and Maier and Rechtin (2000). In MOSES, some heuristics can be captured in models and specifications. An example of a heuristic in Rechtin (1991, 34) is "Success is defined by the beholder, not by the architect." This is at least partially captured in the Quality of Experience item in the external specification described in Chapter 11.

Collective actualization as process: MOSES addresses these additional supporting concepts of actualization:

1. Actualization as a social and a technical process; the actualization process involves many stakeholders or agents of different types, including both human and software agents. Achieving a common or shared understanding of a model among agents is important. But in a mature domain, much of the shared understanding between knowledgeable agents is implicit and does not need to be written down. In MOSES, the complete shared meaning of a model is called a specification, and the explicit part of that specification is the model. In Figure 1.2, models are contained in the Model space; specifications exist in the larger Collective actualization space. More complete definitions of model and specification, and a discussion of their relation, are given in Chapter 3.
2. The actualization process defined in MOSES has two primary differences from TSE engineering processes. First, the MOSES process relaxes most of the constraints of the traditional linear waterfall, and instead promotes an opportunistic process that works on any aspect of the problem that can get traction. Second, the MOSES process accommodates change and evolution.

Although most of the subject matter of MOSES in this book is intended to apply to any system in the purview of extended SE, Chapter 13 specifically addresses SE actualization processes and explores the above concepts in more detail.

Most if not all the elements of this comprehensive MOSES solution already exist in some form. However, some assembly is required; integration and unification of the parts is needed.

Why MOSES?

Why do we suggest that MOSES is a good answer to the needs of SE? What value does it have? A preliminary argument is given here, and the remainder of the book is intended to make a more detailed case.

The bottom line is the unification of traditional and complex systems SE. Others have noted their differences, and some have recently begun to define CSE methods, but no one to our knowledge has defined a way to bring TSE and CSE together into a common SE framework.

This book claims that leaving the two parts of SE separate and disjoint is not acceptable. As argued in the subsection entitled "General Challenge: Unify," the reason is that most systems addressed by systems engineers in the foreseeable future are a hybrid combination of traditional and complex elements woven together. Defining CSE methods is an important and necessary step in supporting the expanded SE — but it is only the first step. The second step is to reconcile and bring together TSE and CSE into a single whole SE, while not losing the strengths and uniqueness of each. That is the objective of this book.

One might assume that this is a daunting job but the job is made easier by the fact that the solution pieces for such a foundation already exist. We simply have to recognize the appropriate pieces, and organize them properly into a coherent foundation. The appropriate solution pieces we identify include a few basic principles, modeling dimensions, the body of knowledge of systems science and complex systems, approaches to uncertainty and change, knowledge from other contributing disciplines, and approaches to specification.

Because of the quality and pedigree of the solution pieces that already exist, and because of the flexibility and extensibility of the model-oriented foundation resulting from the way we have organized the pieces, we believe that the foundation we describe in this book will be able to serve the future SE for an extensive period of time, and certainly in a much more robust way and for a longer period of time than would the approach of simply adding to the current SE.

There is an aspect of the job that is daunting. The volume of material representing knowledge in the various disciplines that support SE means that one book cannot begin to cover it all. However, even if we could somehow capture all that knowledge in one large book, we would not choose to do it. Why? Because that composite body of knowledge should be captured and organized in the proposed online repository that we call the model space. Therefore, this book does not attempt to define the contents of the large body of knowledge. Rather, it indicates the overall scope and then focuses on the elements of model orientation and model space that unify and bring everything together. Much of the book is on the structuring variables and dimensions that organize the model space. It describes illustrative examples but does not populate the model space. It is a framework for unifying the engineering of the full range of traditional and complex systems in this space.

Potential Audiences

Who could reasonably use the MOSES approach? There are several possible audiences. One is practicing systems engineers who want to incorporate more systems thinking into their practice, or who want to extend their work to include more complex/organic systems and to understand how their engineering can be integrated with more familiar practice. The approach does not require all new methods and tools; rather, it is intended to help the practicing systems engineer see more clearly how different elements of SE tie together, and to use their tools and methods in a way that is more extensible, adaptable, and sustainable.

However, the approach does introduce some important changes for this audience. First, it extends the repertoire of systems engineers to include people, organizations, and social or political entities. Second, it provides improvements in current practice, in areas where current practice is generally inadequate — especially in how change is understood and addressed. Third, it promotes and facilitates the movement from reliance on documents as the primary artifacts of SE to the use of models as the primary artifacts, and the capture of models as both an SE and SES body of knowledge (BoK) in a machine processable repository. These changes will take some time to materialize but they will affect systems engineers who adopt a MOSES-based approach.

A second audience is managers, decision makers, policy makers, operations researchers, and industrial engineers who want to understand how the SE of the future can provide more support, including automated support, for managing projects and organizations. The book supports this audience by promoting the use of models that represent management knowledge, management patterns, best practice, and metrics. It extends SE support to higher levels, to leaders of enterprises that include cities, military forces, and even societies, by bringing engineering practice to bear on social and organizational issues. It also supports the concept of self-managing, which reduces the management load.

A third audience is SE methodologists, who can use the approach as a framework to define next-generation SE methods and processes, based on the foundations, extensions, orientation, and unification of MOSES. They have a common orientation and a small number of unifying concepts, artifacts, and principles on which to build, and their methods will be well grounded.

A fourth audience is professors and students in undergraduate or graduate SE courses, who could use this book as a supplement or as a primary text for a graduate seminar on the science foundations of SE or the future of SE.

A fifth audience is researchers, who can use the approach as a source of research topics. There are many areas where research and experimentation are needed to better understand the full scope of the new SE, including how to engineer CS, common specification information, variable specification information, how to exploit the basic external-internal patterns, different dimensions, relations among dimensions, a set of common view types, domain specific models and views, a composite body of knowledge, languages/representations for multiple audiences, tools, actualization process, applying ontology, semantic Web, all current technologies and how they relate to the MOSES approach. Some of these areas are further described in the book.

A sixth audience is all who are interested in software SE, the interplay between software engineering and SE, and the numerous and prominent roles software plays in the SE of the future.

A seventh audience is those interested in model-based engineering and especially extending it to the full scope of SE.

Additional potential audiences include

- System of systems engineers who want to scale up SE and nanotechnologists who want to scale down SE to support their work
- Systems scientists who are interested in supporting SE and making systems science a more visible element of SE education and practice
- SE ontologists, including those who want to specialize the general ontology of MOSES into specific domain ontologies, and those who want to formalize the ontology of MOSES
- SE standards bodies who want to firm up the foundations of existing standards, or who want to orient SE standards more toward artifacts/models and decrease the constraints on processes
- SE associations and leaders who are interested in compiling a digital BoK that includes the scope of all SE disciplines and is in the form of models, and who have considerable patience, because this will take a while
- Repository specialists interested in the challenge of building a working SE repository; they could support those interested in compiling the SE BoK described above.

Organization of the Book

This chapter has described the challenges and opportunities in SE, and the requirements for a science foundation for an extended and unified SE. The remaining chapters spell out the MOSES approach in some detail. The subject of each chapter is indicated in Figure 1.3, which is the composite map from Figure 1.2 with chapter numbers overlaid. Chapter 2 describes the relation between science and engineering in MOSES, including the roles of natural and artificial science. It also describes the systems science universal element of the foundation. Chapter 3, after giving a flavor of MOSES in an extended example, introduces some of the core concepts and definitions. Chapter 4 follows up on this with a description of the key features of MOSES, including model orientation. Chapter 5 goes into the context and structure of the model space, the structured repository of knowledge of all SE-related disciplines. Succeeding Chapters 6 through 10 present more details of each dimension or structuring variable of the model space: composition, commonization, conceptualization, time, and views. Chapter 11 presents the specification approach, which unifies all models into two types: internal and external. Chapter 12 maps traditional artifacts to MOSES and discusses elements of MOSES unification. Chapter 13 describes the unified MOSES collective actualization processes.

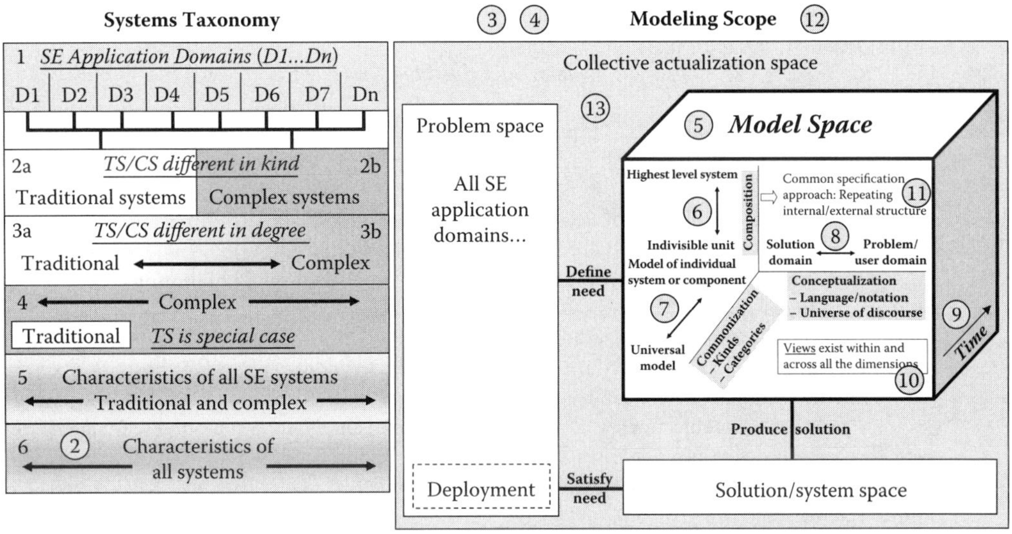

- Systems taxonomy partitions SE systems and shows traditional and complex systems relations
- Collective actualization space includes all SE and management processes, artifacts, contexts
- Problem space includes all application domains SE supports and in which systems are deployed
- Model Space contains all engineering and management artifacts (as models)
- Solution/system space includes all solutions/systems produced by actualization processes

Figure 1.3 Composite map elements allocated to chapters.

References

Ackoff, R. 1999. *Ackoff's Best: His Classic Writings on Management.* New York: John Wiley & Sons.

Axelrod, R. and M. Cohen. 2000. *Harnessing Complexity: Organizational Implications of a Scientific Frontier.* New York: Basic Books.

Bar-Yam, Y. 2003. When Systems Engineering Fails — Toward Complex Systems Engineering, *Int. Conf. Systems, Man & Cybernetics,* 2:2021–2028, IEEE Press, Piscataway, NJ.

Benyus, J. 2002. *Biomimicry: Innovation Inspired by Nature.* New York: HarperCollins.

Blanchard, B. and W. Fabrycky. 2005. *Systems Engineering and Analysis,* 4th ed., Englewood Cliffs, NJ: Prentice Hall.

Blumstein, A. 2007. An OR missionary's visits to the criminal justice system. *Operations Res.,* 55(1):14–23.

Boccara, N. 2004. *Modeling Complex Systems.* New York: Springer-Verlag.

Borra, E. 2007. Making the paper: how an infrared-spectrum telescope could be built on the Moon. *Nature,* 447(7147): xv.

Braha, D., A.A. Minai, and Y. Bar-Yam (Eds.). 2006. *Complex Engineered Systems: Science Meets Technology.* New York: Springer.

Buede, D. 2000. *The Engineering Design of Systems: Models and Methods.* New York: John Wiley & Sons.

CIMMS. 2007. Center for Integrative Multiscale Modeling and Simulation (CIMMS), California Institute of Technology. http://www.cimms.caltech.edu/whatiscimms.html. (Accessed January 11, 2009.)

CSCS. 2005. Univ. of Michigan Center for the Study of Complex Systems, "About the Science of Complexity." http://cscs.umich.edu/about/complexity.html. (Accessed January 11, 2009.)

Das, S., G. Rose, M. Ziegler, C. Picconatto, and J. Ellenbogen. 2005. Architectures and Simulations for Nanoprocessor Systems Integrated on the Molecular Scale. Chapter in *Introducing Molecular Electronics,* Heidelberg: Springer-Verlag.

DoD (Department of Defense). 2001. *Systems Engineering Fundamentals.* Fort Belvoir, VA: Defense Acquisition University Press.

Dörner, D. 1996. *The Logic of Failure: Recognizing and Avoiding Failure in Complex Situations.* New York: Metropolitan Books.

Feiler, P., R. Gabriel, J. Goodenough, R. Linger, T. Longstaff, R. Kazman, M. Klein, L. Northrop, D. Schmidt, K. Sullivan, and K. Wallnau. 2006. *Ultra-Large-Scale Systems*: The Software Challenge of the Future. CMU/SEI.

Ferris, T.L.J. 2007a. Some Early History of Systems Engineering — 1950's in IRE Publications (part 1) — The Problem. 17th Int. Symp. INCOSE. 24–28 June, San Diego, CA.

Ferris, T.L.J. 2007b. Some Early History of Systems Engineering — 1950s in IRE Publications (part 2) - The Solution. 17th Int. Symp. INCOSE. 24–28 June, San Diego, CA.

Ferris, T.L.J. 2008. Early History of Systems Engineering (Part 3) — 1950's in Various Engineering Sources. 18th Int. Symp. INCOSE. 15–19 June, The Netherlands.

Giles, J. 2005. Internet encyclopaedias go head to head. *Nature,* 438(7070):900–901. See also Nature's response to Britannica's objections: Encyclopaedia Britannica and Nature: a response. 23 March 2006. http://www.nature.com/press_releases/Britannica_response.pdf. (Accessed January 11, 2009.)

Goode, H. and R. Machol, 1957. *System Engineering: An Introduction to the Design of Large-Scale Systems,* New York: McGraw-Hill.

Henkel, J. and S. Maurer. 2007. The economics of synthetic biology. *Molecular Systems Biol.,* 3(117):1–4. http://www.nature.com/msb/journal/v3/n1/pdf/msb4100161.pdf. (Accessed January 11, 2009.)

Holland, J. 1992. *Adaptation in Natural and Artificial Systems: An Introductory Analysis with Applications to Biology, Control, and Artificial Intelligence (Complex Adaptive Systems).* Cambridge, MA: MIT Press.

Hollnagel, E. and D. Woods. 2005. *Joint Cognitive Systems: Foundations of Cognitive Systems Engineering.* Boca Raton, FL: CRC Press.

Huhns, M. and M. Singh. 1998. Cognitive agents. *Internet Computing, IEEE,* 2(6):87–89.

IEEE. 2005a. Adoption of ISO/IEC 15288:2002. Systems Engineering — System Life Cycle Processes. IEEE 15288-2004. New York: IEEE Computer Society.

IEEE. 2005b. IEEE Standard for Application and Management of the Systems Engineering Process, IEEE Std 1220-2005, New York: IEEE Computer Society.

INCOSE. 2006. International Council on Systems Engineering (INCOSE), Systems Engineering Vision 2020 v2.0, 6 October 2006.

INCOSE. 2007. INCOSE Systems Engineering Handbook, v3.1. Seattle, WA: INCOSE (International Council on Systems Engineering).

ISO/IEC. 2008. Systems and software engineering — System life cycle processes. ISO/IEC 15288:2008, 2nd ed. 18 March.

Keasling, J. 2007. Video lecture on synthetic biology, March 8, 2007.

Keen, A. 2007. *The Cult of the Amateur: How Today's Internet Is Killing Our Culture.* New York: Doubleday.

Keeney, J. and V. Wade. 2008. Towards Policy Decomposition for Autonomic Systems Governance by Applying Biologically Inspired Techniques. *Network Operations and Management Symposium Workshops,* 7–11 April. IEEE, p. 309–313.

Kephart, J. and D. Chess. 2003. The vision of autonomic computing. *IEEE Computer,* 36(1):41–50.

Kossiakoff, A. and W.N. Sweet. 2003. *Systems Engineering: Principles and Practice.* Hoboken, NJ: Wiley.

Kumar, D., N. Ramakrishnan, R. Helm, and M. Potts. 2008. Algorithms for storytelling. *IEEE Trans. Knowledge and Data Eng.,* 20(6):736–751.

Kuras, M. and B. White. 2005. Engineering Enterprises Using Complex-System Engineering. *Proc. INCOSE Int. Symp.,* Rochester, NY, July 2005.

Maier, M. and E. Rechtin. 2000. *The Art of Systems Architecting, 2nd ed.,* Boca Raton, FL: CRC Press.

Majid, S. and K. Ahmed. 2008. Post-Disaster Communications: A Cognitive Agent Approach. *Seventh Int. Conf. on Networking.* ICN 2008. 13–18 April, p. 645–650.

Manning, A. 1999. Storytelling as a Tool of Technical Explanation — Improvisation Risks and Benefits. *Proc. 1999 IEEE Int. Professional Communication Conf.* 7–10 September 1999, p. 345–351.

Margaliot, M. 2008. Biomimicry and Fuzzy Modeling: A Match Made in Heaven. *IEEE Computational Intelligence Magazine,* 3(3):38-4.

Martin, J. 1997. *Systems Engineering Guidebook: A Process for Developing Systems and Products.* Boca Raton, FL: CRC Press.

Martin, J. 2004. The Seven Samurai of Systems Engineering: Dealing with the Complexity of 7 Interrelated Systems. Presented at the *14th Annu. Int. Symp. Int. Council on Systems Engineering (INCOSE).* Toulouse, France. 21–24 June.

McCarthy, J. 1979. Ascribing Mental Qualities to Machines. In *Philosophical Perspectives in Artificial Intelligence*, M. Ringle, Ed. Brighton, UK: Harvester Press.

McGregor, D. 2006. *The Human Side of Enterprise: Annotated Edition.* New York: McGraw-Hill. (Original edition published 1960, McGraw-Hill.)

Merriam-Webster. 2008. *Merriam-Webster Online: Dictionary and Thesaurus.* http://www.merriam-webster.com/. (Accessed January 11, 2009.)

Miller, J. 1995. *Living Systems.* Winot, CO: University Press of Colorado. (Originally published 1978.)

MIT. 2007. Center for Engineering Systems Fundamentals: Research. http://cesf.mit.edu/research.html. (Accessed January 11, 2009.)

Nature. 2005. Life is what you make it, Editorial, *Nature,* 438(7067):396.

Nature. 2007. Editorial: Electric skies? How to navigate a flight path to greener air travel. *Nature,* 447(7148):1032.

NECSI. 2007. New England Complex Systems Institute. http://www.necsi.org/. (Accessed January 11, 2009.)

NRC. 2008. *Pre-Milestone A and Early-Phase Systems Engineering: A Retrospective Review and Benefits for Future Air Force Acquisition.* Committee on Pre-Milestone A Systems Engineering, National Research Council. Washington, D.C.: The National Academies Press. http://www.nap.edu/catalog/12065.html. (Accessed January 11, 2009.)

Rao, A. S. and M. P. Georgeff. 1991. "Modeling rational agents within a BDI-architecture," in *Proc. 2nd Int. Conf. on Principles of Knowledge Representation and Reasoning.* San Mateo, CA: Morgan Kaufmann, pp. 473-484.

Rechtin, E. 1991. *Systems Architecting: Creating & Building Complex Systems.* Englewood Cliffs, NJ: Prentice Hall.

Registry. 2007. Registry of Standard Biological Parts. MIT. http://parts.mit.edu/registry/index.php/Main_Page. (Accessed January 11, 2009.)

Ring, J. 2008. Discovering the strategy for whole system modeling. *Presented at 6th Annual Conf. Systems Engineering Research (CSER 2008),* Redondo Beach, CA. 4–5 April.

Rittel, H. and M. Webber. 1973. Dilemmas in a general theory of planning. *Policy Sciences,* 4:155–169. http://www.uctc.net/mwebber/Rittel+Webber+Dilemmas+General_Theory_of_Planning.pdf. (Accessed January 11, 2009.)

Rogers, R. 2007. Reflections on ionic liquids. *Nature,* 447, 21 June 2007, p. 917–918.

Rouse, W. 2003. Engineering complex systems: implications for research in systems engineering, *IEEE Trans. Systems, Man and Cybernetics, Part C: Applications and Reviews,* 33(2):154–156.

Rouse, W. 2007. Complex Engineered, Organizational & Natural Systems: Issues Underlying the Complexity of Systems and Fundamental Research Needed to Address These Issues, Georgia Institute of Technology, March 2007.

Sage, A.P. and J.E. Armstrong, Jr. 2000. *Introduction to Systems Engineering.* New York: Wiley.

Salehie, M. and L. Tahvildari. 2005. Autonomic Computing: Emerging Trends and Open Problems. *Proc. Design and Evolution of Autonomic Application Software (DEAS 2005).* St. Louis, MO, 21 May. New York: ACM Press, p. 82–88.

Schön, D.A. 1995. Knowing in action: The new scholarship requires a new epistemology. *Change,* November/December, p. 27–34. http://www.encyclopedia.com/printable.aspx?id=1G1:17839288. (Accessed January 11, 2009.)

SFI. 2007. Santa Fe Institute. http://www.santafe.edu/. (Accessed January 11, 2009.)

Simon, H. 1996. *The Sciences of the Artificial, 3rd ed.* Cambridge, MA: MIT Press.(Originally published 1969.)

Smuts, J. 1926. *Holism and Evolution.* London: Macmillan.

Sussman, J. 2002. Collected Views on Complexity in Systems. MIT ESD Working Papers Series, ESD-WP-2003-01.06, May 2002. *Proc. Engineering Systems Division Internal Symposium*. Cambridge, MA: Engineering Systems Division, MIT, p. 1–25. http://esd.mit.edu/WPS/ESD%20Internal%20Symposium%20Docs/ESD-WP-2003-01.06-ESD%20Internal%20Symposium.pdf. (Accessed January 11, 2009.)

Tapscott, D. and A. Williams. 2006. *Wikinomics: How Mass Collaboration Changes Everything*. Portfolio (Penguin Books).

Taylor, F. 1911. *The Principles of Scientific Management*, published in Norton Library, 1967 by arrangement with Harper & Row, Publishers, by W. W. Norton & Company, New York.

Wikipedia. 2008. Wikipedia. http://en.wikipedia.org/wiki/Wikipedia. (Accessed January 11, 2009.)

Chapter 2

Systems Science Context

> Scientists study the world as it is; engineers create the world that has never been.
>
> **—Theodore von Karman**

> Adapting von Karman's observations, it may be said that *engineering* science is the study of that part of the world which has been created by man.
>
> **—Michael Griffin, 2007**

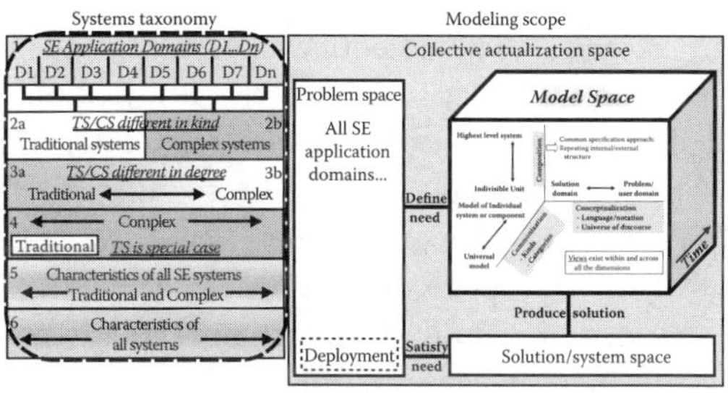

Since the intent of this book is to define a science of SE (although most or all of the pieces already exist), we start laying out MOSES with a chapter on science and its relation to engineering, and their role in a science of SE. Such a discussion is necessitated by the expansion of SE to include organic elements such as humans. This entails some reorientation in how we think about science and engineering, especially with regard to natural and artificial systems. This chapter addresses the reorientation in the expanded SE context, including the relation between systems science (SS) and complex systems science (CSS). It also introduces the foundation portion of the systems taxonomy shown in dashed outline in the left part of the reduced graphic above.

Conceptualization of Science

This section presents the MOSES view of science in general and introduces a set of prerequisite laws for a science.

What Is Science?

We start with some common definitions of science.

- Random House (2008):
  1. A branch of knowledge or study dealing with a body of facts or truths systematically arranged and showing the operation of general laws: the mathematical sciences.
  2. Systematic knowledge of the physical or material world gained through observation and experimentation.
- Morris (1992):
  1. The systematic observation of natural events and conditions in order to discover facts about them and to formulate laws and principles based on these facts.
  2. The organized body of knowledge that is derived from such observations and that can be verified or tested by further investigation.
- Dorland (2007):
  1. The systematic observation of natural phenomena for the purpose of discovering laws governing those phenomena.
  2. The body of knowledge accumulated by such means.
- American Heritage (2007):
  1. The observation, identification, description, experimental investigation, and theoretical explanation of phenomena.
  2. Such activities restricted to explaining a limited class of natural phenomena.
  3. Such activities applied to an object of inquiry or study.
- Simon (1996, 3–4):

  Natural science is knowledge about natural objects and phenomena. We ask whether there cannot also be "artificial" science — knowledge about artificial objects and phenomena [where artificial means] man-made as opposed to natural.

Science is defined both as a process and as the knowledge that results from that process. Some definitions restrict science to the study of natural phenomena while others do not make such a restriction. Simon explicitly poses two categories of science: (1) natural science is knowledge about natural objects and phenomena, and (2) artificial science is knowledge about artificial objects and phenomena.

For the purposes of this book, science is the systematic practice of observing patterns in the characteristics of existing phenomena or systems, and sharing those observations in such a way that they can be independently examined or replicated. Knowledge in a scientific field, at a given point in time, is the set of patterns that most of the specialists in that field believe to be true at that point in time.

This definition deliberately emphasizes four aspects of science:

1. *Objective:* One aspect of science is that phenomena in the world have an objective existence independent of what people observe or perceive.
2. *Perceptive:* Another aspect is that human knowledge of what exists in the world is shaded or colored by what is perceived via the five senses and augmented by various sensors and instruments. Central to science is that humans or observers perceive patterns — that is, the same characteristics occurring in multiple cases. Observing the characteristics of one single system at one point in time or even multiple points in time, for example, would not be sufficient to establish a pattern. Patterns in traditional science are typically called theories or natural laws. In this book, a science pattern is a general model of some part or aspect of the world. However, in each case, a pattern is what observers see or perceive, which may differ from

what actually exists in the phenomena being observed.* Those differences are often due in part to preexisting conceptions or beliefs about the phenomena, which may shape and limit the resulting observations, as discussed by Ferris (1997). This shaping and limiting may be interpreted as negative or positive. On the negative side, a preconceived model of what is there can blind the observer to, or distort the perception of, what is really there, and the observations will not be valid or cannot be used to validate the model. On the positive side, the shaping and limits may reflect a deliberate and explicit modeling choice (described in Chapter 4) whose purpose is to take a certain stance toward or view of the phenomena.

3. *Social:* The social aspect of science has two related components. First, scientific knowledge is a corpus of agreement or belief at a given point in time among the scientists in a field such as physics, or a subfield such as particle physics. Not every scientist in the field must believe, but a significant portion of the community must believe in a theory or model for it to be accepted as scientific knowledge. This means that whether a theory is considered scientific knowledge is not a binary yes/no, but rather a continuum that ranges from universally rejected to universally accepted. The social process of arriving at these beliefs includes the usual human elements of social interaction, such as personal conflicts, competition, reputation, organizational affiliation, and cultural differences. These elements have some degree of influence on scientists' attitudes toward theories. The second component of the social aspect is that scientists or agents other than the proposer of a theory or model can independently examine the claim and find evidence that either strengthens or weakens the claim.
4. *Change:* The view of a scientific community toward a scientific theory changes over time, and sometimes one theory is replaced by another that the community believes is a better model of what is being observed. A well-known model of change in scientific theory is Kuhn's depiction of scientific revolutions (Kuhn 1970). Kuhn asserted a common pattern in which a scientific theory, in the form of a general model or paradigm, is dominant over a period of time during which it is fine-tuned, and then at some point is replaced with another theory that represents a different paradigm. This repeating pattern of normal science periods interrupted periodically by revolutions is analogous to the concept of punctuated equilibrium in evolutionary biology. Eldridge and Gould (1972) argued that instead of gradual and steady change, the pattern is one of long periods of equilibrium in species, disturbed periodically by relatively rapid speciation events. In engineering or technology, a comparable pattern might be technology revolutions, or disruptive technologies as described in Christensen (2003). We note these models because change and evolution are dominant elements of MOSES, both in engineering and in science.

The view of scientific knowledge or theories as models has been expressed by some scientists and philosophers of science over several decades, including Stephen Hawking (1988, 9): "a

* Observation can take at least two forms: *direct* observation or measurement of the pattern, and *indirect* observation or measurement by identifying and observing *implications* of the pattern. Implications are in the form of implied observable conditions: If <pattern> holds, then <observable condition> will be observed. The trend in the past century in physics, for example, has been away from direct observation and toward examination of implications. As theories have become more subtle, probabilistic, and multidimensional, direct observation of predicted properties has become less feasible.

[scientific] theory is just a model of the universe, or a restricted part of it, and a set of rules that relate quantities in the model to observations that we make." Ronald Giere has written extensively about a philosophy of science that is consistent with the models concept of science. See (Giere 1999; Giere 2004). In addition, *The Oxford Companion to Philosophy* (Honderich 1995, 448), in its discussion of knowledge and science, is also consistent with this view.

An example from biology is reported by Mayr (1982, 660) in discussing the length of time required to gain an understanding of the phenomenon of fertilization in the nineteenth century. Two opposing arguments were made. Embryologists argued that the spermatozoon played a purely mechanical role in fertilizing the egg, while those focusing on heredity saw fertilization as the mixing of paternal and maternal hereditary endowment. Mayr states that these two diverging interpretations of fertilization were based on two entirely different explanatory models. The resolution came later: Fertilization has a dual significance; the two opposing views are both correct. Fertilization belongs to both functional and evolutionary biology. The more complete model embraces the duality.

The model concept has also been extended beyond the traditional natural sciences. In 1972, a seminar at the University of Edinburgh brought together scholars in mathematics, computer science, engineering, philosophy, and social sciences to discuss the use of models in the social sciences (Collins 1976). While the overall focus was on social sciences, two of the presented papers (Hesse 1976; Harré 1976) addressed the prominent role of models that had emerged in the natural sciences and how that relates to their role in the social sciences.

The view of this book is that SE science encompasses both natural and artificial sciences. A list of supporting sciences and related disciplines appears later in this chapter.

Prerequisite Laws

John Warfield has made several contributions to the areas of systems science and complex systems. One contribution is the identification of several "laws" that must be followed when developing a science (Warfield 1994). We list the laws here that are most important to MOSES and briefly describe how MOSES addresses them.

Law of Requisite Variety

Ashby (1956) defined this law in the context of cybernetics and the regulation of a system. The law quantifies the variety needed in a regulator, where variety can be measured by the number of potential states. To successfully regulate a system, the variety of the regulator must be at least as large as the variety of the regulated system. Later Conant and Ashby (1970) identified an implication of this law, which they expressed in the "Good Regulator" theorem: The simplest optimal regulator of a system is a model of that system, in the sense that the regulator's actions are a mapping of the system's actions. This theorem applies not only to regulators of separate systems, but also to self-regulating or homeostatic systems.

We apply this law and the associated theorem in two ways. First, it motivates the scope of MOSES. The scope of any SE science must be equal to or greater than the scope of the SE that it supports. Because we argued in Chapter 1 that SE is expanding, care must be given to define an adequate scope of MOSES as a supporting science. We claim that the scope of MOSES defined in Chapters 1 and 4 is sufficient.

Second, the law and its associated theorem apply to models. Specifically, a condition for achieving adequate fidelity of a model to a system is that the model must represent a variety as great as the variety of the system it is modeling, but only on the characteristics of interest to the model. Chapter 3 discusses the latter restriction further.

Law of Gradation

This law, defined by Warfield (1994), states that a science must acknowledge that the objects it studies — in our case, systems — occur at varying degrees of complexity and scale, from simplest to most comprehensive, and must support understanding at all levels and scales.

This law is an important requirement for a science of the expanded SE because it represents the range of complexity and scale discussed in Chapter 1. MOSES addresses this law in terms of systems by defining a taxonomy of systems that covers the full range of traditional and complex systems at multiple scales, and allows this range to be a continuum. It addresses the law in terms of models by defining a range for each structuring variable in the model space, as described in later chapters.

Law of Requisite Parsimony

This law, defined by Warfield (1994), states that a science must present information in a way that prevents human information overload. This law derives from Miller's (1956) findings on the limits of human information processing capacity.

MOSES addresses this law using abstraction as a tool to help focus on important issues and ignore the rest, using views to address selected concerns, supporting implicit elements of system specifications, and exploiting autonomy in systems. All these conceptual tools are discussed in subsequent chapters.

Science and Engineering

The expansion of SE and the corresponding supporting sciences introduce some different perspectives on science, engineering, and the relation between them. Therefore, it is important to discuss these perspectives, including some nontraditional roles of both science and engineering, and we do so in this section. We start with a preview that summarizes the section, and then discuss an extension of classic science into a nontraditional role, that of science of artificial systems. Subsequently, we present a brief conceptualization of engineering, which is followed by an extension of classic engineering into a nontraditional role: engineering of natural or hybrid systems. These perspectives provide the basis for a discussion of the relation between science and engineering in general, between SS and CSS, and between SES and SE.

Preview: Scope of Science and Engineering

The discussion of science and engineering in the context of an expanded SE of the future reflects a conceptualization or view that calls for reexamining the definitions of science and engineering. The view espoused in this chapter and this book retains the essence, the core, of science and engineering, while removing nonessential restrictions that may exist, to allow the necessary breadth of application of both. A preliminary extension of science was discussed in the previous section. The MOSES view of the extension of both science and engineering is discussed further in this section. Table 2.1 summarizes the extensions and how they complement each other. Science is fully

Table 2.1 Extended Scope of Science and Engineering

	Natural systems	Artificial systems
Science	Traditional	Full extension
Engineering	Partial extension	Traditional

extended from the traditional scope of natural systems to artificial systems. That is, the definition of science as observing patterns can be applied to all types of systems, natural and artificial. Engineering, however, is only partially extended from its traditional scope of artificial systems to natural systems. The extent to which engineering produces natural systems, such as humans or groups, is noticeably reduced from the extent for artificial systems. That is reflected in terms such as "influencing" and "shaping" rather than "building" or "creating." That is why we use the broader term "actualization" in Chapter 13 rather than "engineering" to reflect this partial extent into the natural systems arena.

The partial extension is another view of the distinctions in Table 1.3 in Chapter 1: The shaded portion in Table 2.1 that reflects the partial extension is the SE box in Table 1.3, the systems engineering work that can be done. The white or unshaded portion in Table 2.1 is the SA box in Table 1.3, the natural system autonomy.

Natural and Artificial Science: Distinctions and Parallels

When we think of natural versus artificial, a classic distinction that comes to mind is science versus engineering. We tend to think of science as observing natural things and engineering as building artificial things, as reflected in the "Traditional" cells in Table 2.1. Classic examples of science are physics, chemistry, and biology. Engineers build artificial things such as radios, cars, buildings, bridges, aircraft, ships, and computers. But in the definitions above, we discussed the possibility of science observing artificial things. Our definition of science was targeted to existing phenomena or systems,* which allows the study of both natural and artificial phenomena, reflecting natural science and artificial science, respectively. We now discuss some parallels between natural and artificial science.

Natural science works because natural systems exhibit common patterns. If nature were *ad hoc* and everything was unique, natural science would get little traction and, in fact, would probably not exist. But natural things have many characteristics in common; that is, they exhibit patterns, and those patterns allow natural science to say much that is meaningful about many of those

* In this context, the term "existing" is used somewhat loosely in a temporal sense. The real restriction is that appropriate *information* about systems must exist. Science, for example, can study members of animal or plant species today that have been extinct for thousands or even millions of years, if adequate information is available today, from fossils, for example.

systems and to group those systems into categories. All the atoms in the known universe fit into about a hundred categories, organized into the periodic table. Biology categorizes all known living things into categories called kingdoms, species, etc. All members of a species exhibit the same basic architecture pattern. For example, a book on human anatomy describes an architecture pattern that applies to all individual humans. Although each human is unique in certain respects, the anatomy book focuses primarily on what is common — which is considerable.

The reason for stating the obvious with natural systems is to set up a parallel with artificial systems. It turns out that artificial systems, including those that engineers build, also have many characteristics in common and they exhibit patterns that allow artificial science to say much that is meaningful about many of those systems and to group those systems into categories. We regard as artificial science the work that identifies and captures the key categories, properties, functions, measures, and conventional designs of systems in those categories. Within given engineering domains such as bridges, cars, aircraft, or computers, the domain, as it matures, converges on one or a small number of basic architectures for that domain of systems. Examples are the suspension bridge, the internal combustion engine, fixed-wing aircraft, and von Neumann computer architecture. We have listed classical engineering domains but the same holds true for expanded SE — for example, organizations or nano-systems. All members of a domain exhibit the same basic architecture pattern, or one of a small set of patterns. For example, an engineering handbook on bridges describes a suspension bridge architecture pattern that applies to all individual suspension bridges. Although each suspension bridge is unique in certain respects, the handbook focuses primarily on what is common — which is considerable.

In the quote at the start of the chapter, Griffin uses the term "engineering science" for what Simon calls "artificial science." We use Simon's term in this book.

Missile Guidance as a TSE Artificial Science

An example of an artificial science is the domain of compiled knowledge of missile guidance and control. One compilation of this knowledge is (Siouris 2004). It includes mathematical models, taxonomies of missiles and other entities, and terms (referring to the universe of discourse). Chapter 4 of the Siouris book presents laws of tactical missile guidance. Two samples are selected from the book. The first is a taxonomy of missiles (Figure 2.1). The common characteristics and the differences between the various types are described. The second sample is the dominant guidance law, called the proportional navigation guidance law for homing missiles. It is expressed as follows:

$$(4.24)\quad a_n = NV_c\,(d\lambda/dt)$$

where

a_n = Commanded normal (or lateral) acceleration, in [ft/sec^2] or [m/sec^2]
N = *Navigation constant* (also known as *navigation ratio, effective navigation ratio,* or *navigation gain*), a positive real number, [dimensionless]
V_c = Closing velocity, in [ft/sec] or [m/sec]
$d\lambda/dt$ = LOS (line of sight) rate measured by the missile seeker, in [rad/sec]

The argument that this domain of knowledge represents an artificial science is this: it is artificial because its subject matter is a domain of engineered

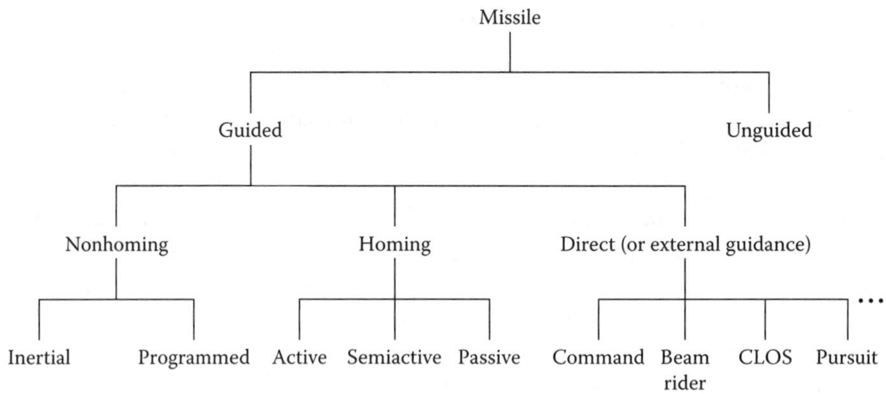

Figure 2.1 Missile types and classification. (*Source:* Siouris, G. *Missile Guidance and Control Systems*. New York: Springer, 2004. Used with permission of Springer Science + Business Media.)

(i.e., artificial) systems. It is a science because it fits our definition of science given above: "the systematic practice of observing patterns in the characteristics of existing phenomena or systems…" In the Siouris book, the field is defined in the title: *Missile Guidance and Control Systems*. A large number of patterns of the characteristics of existing missile guidance and control systems are defined and shared in this book, so clearly they can be independently examined. If other specialists disagree with these patterns or models, they can challenge them.

Another parallel is that for both natural and artificial systems, the architectures evolve across generations of systems, and in some cases the changes are significant enough that the architecture is judged to be a new one, not just a modification of the old one. Science, natural and artificial, tracks this evolution.

We regard many sciences as a mix of both natural and artificial if they study both natural and artificial phenomena, or if the systems they study represent a fuzzy or ambiguous boundary between natural and artificial. The laws of physics apply equally to natural and artificial physical systems. Both an apple and a baseball are composed of elementary particles and atoms. The fall of a baseball to the Earth is no less — and no more — instructive to the theory of gravitation than is the fall of an apple. Nanotechnology and synthetic elements in chemistry and biotechnology suggest that even the most natural of sciences — physics, chemistry, and biology — may now be a mix of both natural and artificial. Perhaps we are moving into an era in which no science is strictly natural or strictly artificial. Psychology, sociology, economics, linguistics, and organizational theory… are they natural or artificial, or some of both? Is a social group, such as a family or church or mob, a natural or man-made system, or both?

The possibility of artificial science and the increasing interest in systems that seem to blur the boundary between natural and artificial are key elements in MOSES as a unifying foundation for SE. Our interest here is not to allocate systems and sciences to a strict taxonomy of natural, artificial, hybrid, or fuzzy. It is rather to make the point that science must be able to support all these types of systems, the full range of systems, in a seamless way. A significant driver of MOSES is the need for an approach to SES and to SE that can comfortably cross the boundary between artificial and natural systems without loss of continuity, and in fact in many respects can be agnostic to this distinction. The use of models and model orientation is the major factor in achieving such a capability.

Conceptualization of Engineering

Engineering is the systematic practice of agents taking purposeful and constrained action to achieve a system of a certain type or category with certain desired characteristics. Constrained action means that the actions take place under a variety of constraints, including resource constraints. Typically we think of agents as humans. If humans build a dam, we think of it as engineering. If beavers build a dam, we do not usually think of it as engineering, although it satisfies all aspects of the definition of engineering as stated here. The purpose of making this point is not so much to argue that beavers do engineering, but rather to raise the possibility that agents other than humans (e.g., software agents) could be viewed as performing engineering, in some circumstances. Category means that engineers do not produce all types or categories of systems; they produce systems in their field, such as civil engineering, electrical engineering, software engineering, or systems engineering.

Engineering: What Does It Mean for Complex or Hybrid Systems?

The expanded SE challenges traditional notions of engineering. Large-scale complex systems (CS) such as enterprises and organizations, small-scale systems in nanotechnology and synthetic biology, and hybrid systems in general, exhibit the phenomenon of bringing natural and artificial components together and blurring the distinction between what is artificial and what is natural. What does it mean to engineer a natural component or a hybrid system? Or to engineer a person, a group, or an organization? Is it really engineering?

Complex systems, as described in Chapter 1, are characterized as having the potential to exhibit behavior that is autonomous, unpredictable, difficult to understand, likely to appear chaotic, or continually changing. In addition, there are many nontechnical aspects, such as social, organizational, economic, and legal. These characteristics seemingly do not lend themselves to engineering as we traditionally conceive it.

The MOSES approach is that engineering will adequately support the new expanded SE, including CSE, under the broader umbrella of collective actualization, introduced in Chapter 1. In this section we briefly recapitulate the engineering aspects of this process.

The basic orientation is that SE systems will range from fully TS to fully CS but most will be a hybrid, a combination. This means that a systems engineer needs to perform some combination of engineering planning and design along with facilitating , shaping, and taking advantage of autonomous behavior in the system. The systems engineer has less control than in a traditional system but still has some shaping influence. The systems engineer does not have to decide everything. Systems engineers will need to learn how to apply influence and exploit help from the system itself.

This combined activity of engineering and autonomous action applies also to enterprises. A project, agency, or company is shaped and managed through a combination of science-based enterprise engineering and management (EEM), and enterprise self-organizing activity.

Chapter 13 describes the overall actualization process further.

Relationship between Science and Engineering

Shaw and Finch Historical Pattern

The general relationship between science and engineering is that science "supports" engineering, that is, it provides analytic tools and an understanding of the properties that engineering can use in predicting qualities of what it produces. More specifically, science supports engineering in two ways: (1) science (natural and artificial) studies properties of materials that go into the products or systems that engineering constructs, and (2) artificial science studies the properties, including emergent properties, of the products or systems that are engineered.

The evolution of this relationship is more complex. Historically, science and engineering were two separate cultures that did not interact. Only in the past two or three centuries has a scientific foundation been explicitly built and used for engineering. A description of how the relationship between science and engineering develops as part of the maturation process is given for engineering disciplines such as mechanical and chemical engineering, by Shaw (1990) and Finch (1951), and summarized here.

The initial phase is *ad hoc* practice, in which problems are solved by talented amateurs and virtuosos. The solutions of these craftsmen then gradually move into the folklore, become routine, and begin limited production. This gradually leads to commercial-scale production. The problems of large-scale production then stimulate the development of a supporting science that defines the principles that explain the behavior that practitioners have already observed and exploited. As the science matures, it finally gets out ahead of current practice in the sense that it formulates principles that can be used to improve current practice: more predictability, fewer errors, more precise safety margins that allow less over-design and therefore lower cost. At this point, people can be educated in both the theory and practice of the field, and the field is now a mature discipline.

As the science and engineering mature, they contribute to each other. Engineering builds things, the properties of which science needs to analyze and understand in order to enable more efficient and economical engineering.

MOSES Conceptualization

The relation between science and engineering is that science uses systems (or phenomena) to build models, and engineering uses models to build systems. Science observes existing systems such as a cell or a solar system and defines general models that define what scientists believe about the corresponding domain (e.g., the physical world of moving bodies and forces, or the biological world of cells and organisms) at a given point in time. There are two tendencies related to this distinction. One is that in science the systems tend to precede the models, while in engineering the models tend to precede the systems. The other tendency relates to which type of system is used as the standard or assumed to be correct when testing correspondence between a model and a system. Science tends to use the system as the standard, assuming that the model is incorrect if it does not correspond to the system.* Engineering tends to use the model as the standard, assuming that the system is incorrect if it does not correspond to the model. For example, engineers tend to believe that if a system is tested against a set of requirements, and it fails the test, then the requirements are correct and the system is wrong.

---

* But that is not always the case; sometimes scientists believe a theory (model) is correct and their experimental data (of the system) are flawed if the two do not agree.

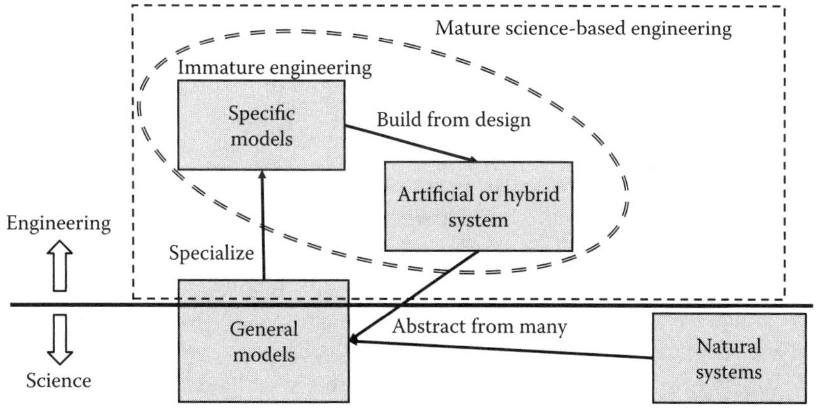

Figure 2.2 General relationship between mature science and engineering.

Figure 2.2 represents our model of the relationship between mature science and engineering. We first discuss this in general terms and then illustrate it with three examples.

The degree of maturity varies for both science and engineering. In Figure 2.2, the variable of engineering maturity is explicit while that of science maturity is implicit. Mature science-based engineering exhibits the following characteristics:

- Both the engineering discipline and the supporting science discipline(s) have some considerable degree of maturity.
- The engineering discipline may draw on multiple science disciplines; it is not limited to one science discipline per engineering discipline.
- A given science abstracts general models or patterns from multiple natural systems if it is strictly a natural science; or from multiple artificial systems if it is strictly an artificial science; or from some combination of multiple natural, artificial, or hybrid systems if it is a mix of natural and artificial science.

Specializing the general models it obtains from science, engineering produces specific models (e.g., requirements, architecture, and design) to build one specific system, which may be artificial or hybrid. Immature engineering is at a stage where specific models are defined, such as design, to guide the building of the system. Immature engineering does not have or use general models from science.

Maturity of science is reflected in the number of general models available and the fidelity or correspondence of those models to the systems. In general, a mature science has more models of higher fidelity than an immature science does.

Three examples are presented. The first two are taken from Shaw (1990) and are interpreted and described in the context of Figure 2.2.

- *Example 1: Civil engineering.* Civil engineering is the discipline of large-scale construction efforts such as buildings, bridges, dams, tunnels, highways, and railroads. It has been practiced for millennia — aqueducts in Rome, irrigation in Mesopotamia, pyramids in Egypt. Like most traditional engineering disciplines, it is based on both natural and artificial parts of the science spectrum but it was based on artificial science long before it was based on natural science. At the natural end are models of statics and strength of materials — both of which were developed around 1700 — and other properties of materials, such as elasticity, which were developed later. At the artificial end are properties, conventional designs, and methods of analysis for particular categories of artificial systems, such as bridges. Civil engineering uses the general models of both natural and artificial systems to define specialized models and build artificial systems of each type listed above.

- *Example 2: Chemical engineering.* Chemical engineering is concerned with the industrial-scale production of chemical goods such as solvents, pharmaceuticals, and synthetic fibers. The natural science supporting chemical engineering is, of course, chemistry. But in the 1880s a chemical plant inspector, G.E. Davis, observed a common pattern among the wide variety of goods being produced. The pattern was that all plants, producing all types of chemical goods, used the same set of steps: filtration and clarification, heat exchange, distillation, screening, magnetic separation, flotation, etc. He wrote down these processes, which later became known as unit operations. We call the set of common processes a general model, and it forms the basis of chemical engineering as it is practiced today. Shaw quotes the 1922 Chemical Engineers Committee on Education as saying that chemical engineering is a science in its own right, whose basis is unit operations and associated laws, materials, and equipment. In our view, they were on the right track but we interpret chemical engineering not as a science but as an engineering discipline, based on natural and artificial sciences, that illustrates the pattern in Figure 2.2. The natural science produced general models in the form of "natural laws" of chemistry. The artificial science, as performed by Davis, produced general models in the form of "artificial laws" of unit operations. Chemical engineering specializes these general models for a wide variety of chemical goods. For each type of chemical product, it defines specific models and "builds" a system, which is a manufacturing plant (with the aid of mechanical and civil engineering) and the unit operations process to use. Shaw said that civil engineering has a basis in theory, and chemical engineering has a basis in practice. We interpret the distinction this way: both disciplines draw on natural and artificial systems but civil engineering relies more visibly on natural science, which defines the statics and strength of materials, while chemical engineering relies more visibly on artificial science, which defines unit operations.
- *Example 3: Software engineering.* This example proposes a natural and artificial science of software. The role of "natural" science is played by computer science, with its focus on formal languages, structures, and computability. The artificial science may be called software science. We identify three samples of (artificial) software science. The first is the work on software evolution laws or patterns by Lehmann and Belady (1985). This work examined a set of engineered software systems and tracked their change over time, and observed the evolution patterns described in the book. The second sample is work on identifying and codifying software patterns, such as design patterns and architecture patterns — for example, Gamma et al. (1995) and Buschmann et al. (1996). The assumption in naming these efforts as science is that the patterns they identified were abstracted from examination of multiple existing systems. In both these samples, the results were general models that apply to a population or class of systems. The third sample is of a different form. Poore (2004) critiques the state and direction of software engineering in comparison with how other disciplines have matured. He notes the

> increasing gap between software education and practice, despite the valiant efforts of Vic Basili, Fred Brooks, Barry Boehm, Harlan Mills, and David Parnas to close the gap and align education and practice. Our purpose here is not to comment on the gap, but rather to make this observation. If one were to compile a list of members of the software engineering community who have made significant contributions in observing and publishing patterns or models abstracted from software systems, these individuals would be on the list. The knowledge they have contributed to the repository of general models, while informed by computer science, has come directly from artificial software systems. Thus, in terms of Figure 2.2, they form a core part of the artificial software science community. The general models that software engineers use come from both (natural) computer science and (artificial) software science, or, alternatively, from both computer science theory and software practice.

Another good treatment of the relation between scientific and engineering knowledge is Vincenti's (1990) book on what engineers know, using examples taken from the history of aeronautical engineering. The patterns and models presented in Vincenti's book are the result of the natural science of physics and the artificial science of aircraft.

Systems Science and CSS

This section discusses the essence of SS and its relation with CSS. Together they form a core part of the basis of an SES, described in the section entitled "Disciplines Contributing to SES."

Why Systems Science Is Important for SE

We have shown that a major component of the science foundation of SE is SS. The notion that SS should provide a foundation for SE has a *prima facie* validity but there are more substantive reasons for explicitly including SS. One reason is that, despite the potential benefits of SS for traditional SE (TSE), SS and TSE have not had a close relationship up to this point. To a large extent, TSE practitioners have not incorporated the concepts of SS into their practice.

A second reason, at least as important, is that the nature of the extended SE brings natural and artificial systems into closer proximity, and in some cases abstracts away the differences. SS supports this abstraction; it is interested in what characteristics hold for all systems of any size, shape, or type, just because they are systems. SS is not only multidisciplinary; it is universal in the sense that it supports all disciplines that deal with systems, which in our view include at least all engineering, science, and mathematical disciplines.

A third reason is that SS is sufficiently mature to be used as a foundation. Although it continues to evolve, it has existed in something like its present form for more than a half century.

Themes

Three themes seem to dominate SS, in both historical and contemporary treatments: (1) holism, (2) multidisciplinarity, and (3) close intertwining with CSS. In the literature, the themes are mostly implicit, even tacit. They simply recur in the course of the discussion in SS or systems theory publications across several decades. We do not take these themes to directly constitute the essence of SS, but we derive the essence from them and then map them to the MOSES view of the world.

Holism

A key part of systems thinking is holism: Consider a system as a whole with its own properties that are separate from the properties of its parts. This was in response to atomism or reductionist thinking that emphasized understanding a system by analyzing its parts. Historically, holism appeared in multiple disciplines around the same time nearly a century ago, such as psychology (gestalt), physics, and biology (Klir 2001, 37–42; Koestler 1967, 47–51). Holism is a way to manage complexity, not by dividing and conquering, but by combining and conquering. Smuts (1926) was one of the first systems theorists to discuss and advocate holism, followed by von Bertalanffy (1933). Many other systems theorists, such as Klir (2001, 37–44), have since emphasized holism as an aspect of systems science. Koestler (1967) took this concept a step further based on the notion that many systems are themselves part of larger wholes. He defined a system or element that is both a stand-alone whole to its parts, and a part of a larger whole, as a holon. A hierarchy of such holons is a holarchy.

Koestler made two significant and related contributions to holism: First, he extended beyond just one level of a system and its parts, to a more scalable model of multiple levels in which a given system plays both the role of a whole and the role of a part. Second, he reconciled holism and atomism by recognizing that both are inadequate by themselves. We need to examine both wholes and parts.

Multidisciplinarity or Generality

A prime motivation for the development of SS was the recognition that similar relations and models are found in widely differing domains of systems. This theme was a reaction against the increasing specialization of science, which resulted in a disintegrated stovepipe structure of knowledge. Cross-cutting and cross-discipline knowledge was getting lost between all the stovepipe specialties. This has arguably been the dominant theme in SS, emphasized by many SS contributors, including Boulding (1956), Wiener (1961), von Bertalanffy (1968), and Klir (2001). The formation of the Society for General Systems Research (SGSR) (now called the International Society for Systems Sciences [ISSS]) in 1954, focused heavily on this theme. Cybernetics, a sub-branch of SS concerned with information and self-regulation, recognizes information and feedback models that hold across multiple system types and disciplines (Ashby 1956; Wiener 1961). Klir (2001) describes the multidiscipline theme in terms of isomorphic models across disciplines, where the isomorphic or common aspects are based on common *relations* ("systemhood"), not on common *things* (thinghood). That is, in the conception of system as a whole and as a relation on parts, SS abstracts away differences in the types of system parts, and focuses on common types of relations on parts. Different types of things or parts yield different disciplines and different specialized sciences.

Close Intertwining with CSS in the Literature

Much of the systems literature over its history has intertwined and even conflated SS and CSS. A discussion of holism, which is typically portrayed as a theme of SS, often includes a discussion of emergent properties, which is typically portrayed as a theme of CSS. Both SS and CSS are multidisciplinary. A fair amount of systems literature connects SS with organic models in contrast to traditional mechanistic models, and describes how the organic models apply across multiple disciplines.

Many, if not most, primary SS authors exhibit all three of these themes but few if any make the distinctions explicit. They simply go back and forth between holism and multidisciplinarity without indicating that they are changing topics. The literature has neither made a clear distinction between SS and CSS, nor has it noted their historical and contemporary intertwining. Some recent publications, however, have begun distinguishing SS from CSS. Gharajedaghi (1999) implicitly distinguishes between holism and CSS, covering the two themes as dual paradigm shifts: one is a shift from analytic to holistic systems thinking and the other is a shift from a machine model to a biological model to a social model of organizations. He does not present these dual shifts as a distinction between systems science and complex systems science. That is our interpretation of his dual shifts.

Essence of SS and Its Relationship with CSS

From these themes, we consider the essence of SS to include at least the following elements:

1. Wholes and parts are both essential to SS, and they form repeating structures or holarchies. MOSES supports this relation in the composition dimension of the model space, discussed in more detail in Chapter 6.
2. Multidisciplinarity is an essential element of SS. More specifically, SS captures characteristics of systems across all disciplines: what they have in common solely by virtue of the fact that they are systems. This information is captured in general models. SS is a generalization of all other sciences, and a general support for all engineering disciplines. This relationship is supported in the commonization dimension of the model space, discussed in more detail in Chapter 7.
3. The generality of SS informs the issue, implicitly raised in the literature, of the relation between SS and CSS — their intertwining and even conflation. Our position is that the two are distinct. While both SS and CSS are multidisciplinary, the scope of CSS is a subset of SS. CSS is concerned with complex/organic systems across multiple disciplines while SS is concerned with all systems, which includes — for the purposes of SE — both complex/organic and traditional/mechanistic. Complex systems engineering (CSE), supported by CSS, forms the dominant area of the *extension* from TSE to the new SE, and SS forms the dominant *foundation* for the full extended range of SE.

Disciplines Contributing to SES

We identify in Table 2.2 a set of disciplines that contributes to an SES and in particular to MOSES.* The disciplines are listed alphabetically within four somewhat loose categories: sciences, mathematics, computation and informatics, and other disciplines. Significant overlap exists among many of these disciplines. The work of integrating these disciplines, especially from a modeling perspective, is already being done to a significant extent by CSS and to a lesser extent by SS. SS restricts itself to characteristics and taxonomies of all systems, which means it is very broad, but not deep. We use SS as the anchor for a systems taxonomy defined in the section entitled "Systems Taxonomy and the MOSES Foundation," while CS are contrasted with TS in that taxonomy.

* These disciplines contribute to the *science* of SE, in contrast to the multiple application disciplines or domains to which SE is *applied*, such as defense and transportation.

Table 2.2 Disciplines That Contribute to SES

Sciences	*Mathematics*	*Computation, Informatics*	*Other Supporting Disciplines*
Anthropology	Algebra	Artificial intelligence	Artificial life
Bioinformatics	Category theory	Computer science	Conceptual modeling
Biology	Complex functions theory	Computing science	Decision theory
Catastrophe theory	Continuous math (calculus, ...)	Formal languages and methods	Epistemology
Chaos theory	Discrete math (logics, ...)	Formal semantics	Forecasting
Chemistry	Dynamical systems	Software science	Game theory
Cognitive science	Factor analysis		Information theory
Complex systems science	Geometry		Knowledge management
Cybernetics	Graph theory		Knowledge representation
Ecology	Nonlinear systems		Law
Economics	Numerical analysis		Mechatronics
Epidemiology	Probability and statistical theory		Mereology
Geology	Topology		Public policy
Linguistics			Ontology
Medical/health science			Operations research, management science
Military science			Simulation
Network science			Small worlds
Organization theory/science			Urban planning
Physics-classical, theory of relativity, quantum theory			
Political science			
Psychology			
Sociology			
Systems science			

These disciplines support SE in a variety of ways. Examples throughout the book illustrate the support of a number of these disciplines.

It was noted earlier that SS and CSS models have been available for a long time but have not been extensively used by TSE. This illustrates that the existence of a science discipline, with models that would be helpful to an engineering discipline, does not necessarily mean that the engineering discipline exploits those models. SE is finally looking at CSS models (and to some extent SS models) in a serious way, and MOSES is designed to facilitate the use of these models.

We conjecture that many, if not most, systems engineered in the future SE will be hybrid systems that include both people and machines. The primary cause of this will not be that the mosaic, the context in which systems are designated, will change. Rather, the big changes will be that (1) SE designates systems in a more holistic way, and (2) SE will view systems differently. We have had a significant blind spot, trying to treat our systems as mechanistic, and the future embrace of a broader SES basis for SE will remove the blind spot and enable us to use more appropriate models, including organic CSS models, for the actual systems we attempt to produce.

In the context of the supporting disciplines, a slightly different and more formal perspective on systems science has been offered under the term "Rosetta Stone" (Baez and Stay, forthcoming). They discuss at some length the analogies among physics, topology, category theory, logic, and computation or computer science, and suggest that research in the common areas of overlap among these fields is beginning to build a new general science of systems and processes. The commonalities among these areas are made more difficult to discern by differences in terminology and notation. The common concepts among all these fields are, in general terms, things and ways to go between things — or systems and processes on systems. The respective terms in category theory are object and morphism; in physics, system and process; in topology, manifold and cobordism; in logic, proposition and proof; and in computation, data type and program.

Unity and Unification

There has been debate in the past century about the unity of science, both its desirability and its achievability. Historically, the unity of science has often been discussed in the context of reducing all science to physics. This has sparked resistance from the CS community, which has noted the existence of emergent properties and has argued that not everything can be reduced to the lowest-level parts, or atoms. Some systems theorists, such as Von Bertalanffy (1968) and Klir (2001), have discussed a role for SS in promoting unity of science but they make a different kind of argument; that is, unity of science is achieved not by reduction, but by abstraction or generalization. SS unifies by abstracting away the differences in types of systems and defining what is common to all.

Klir (2001) says that unity is promoted by focusing on common relations and abstracting away the "thing" or part differences. Von Bertalanffy (1968, 87) puts it this way: "Unity of Science is granted, not by a utopian reduction of all sciences to physics and chemistry, but by the structural uniformities of the different levels of reality." That is, unity of science is achieved not at the lowest level parts but at the broadest level concepts. In terms of the model space dimensions described in Chapter 5, the reductionist position was that unity of science is achieved at the detailed bottom of the *composition* spectrum, while the systems theorists argue that it is achieved at the top of the *generalization* spectrum.

MOSES aligns more with the Klir–Von Bertalanffy generalization argument. However, our position is not about unity of science, *per se*, but about unification of all areas of the expanded SE and SES, through model orientation and the model space. Unification does not mean everything is forced to be the same. It means everything is made commensurate by a common framework: treating all engineering artifacts, scientific theories and laws, and mathematical relations as models of systems. The framework facilitates capturing and expressing in models all essential similarities in systems, all essential differences between systems, and all essential stabilities and all essential changes.

Systems Taxonomy and the MOSES Foundation

System is the most general construct, the pinnacle of generality or commonality, in the world of systems. However, SS does not stop there. It also identifies and distinguishes between system types. How do we start categorizing systems, descending from that pinnacle? What categories would be useful? Because we are defining a science of SE, we proceed as follows:

- The first category within SS is the set of systems of interest to SE. This includes all systems that SE engineers, environments of those systems, and components of those systems; it also includes all models of those systems, environments, and components.

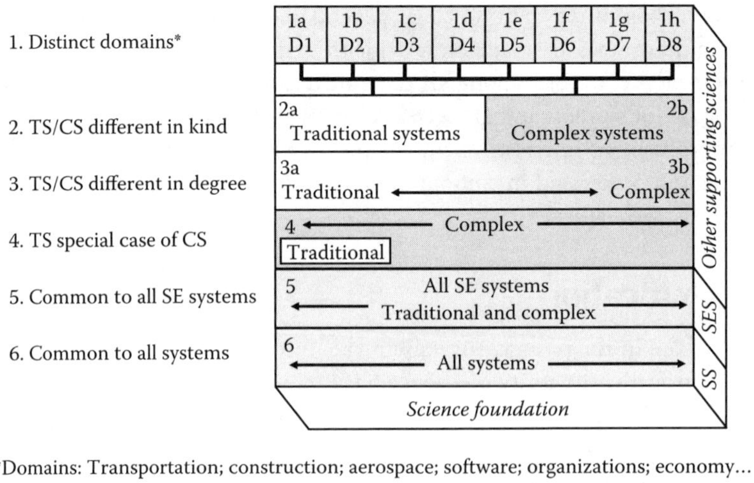

Figure 2.3 SE taxonomy and the range of supporting sciences.

- The next step in the taxonomy differentiates among SE systems. Possible categories include open versus closed systems, target versus model systems, small-scale versus large-scale systems, traditional versus complex systems, physical versus conceptual systems, natural versus artificial systems, or stable versus dynamic systems. Because of the nature of the expansion of SE, we choose the traditional (mechanistic) versus complex (organic/agent) distinction as the next category level.
- Finally, each engineered system is associated with an application domain such as aerospace or transportation. That represents the final level of categories in our systems taxonomy.

The resulting taxonomy, shown in Figure 2.3, is responsive to the need for an extended foundation supporting an extended and unified SE, as was shown in Figure 1.1. The scope of the contributing disciplines in the foundation (spelled out in Table 2.2), ranges from SS, which supports all systems, to disciplines that support specific domains such as transportation or construction. The separation and lines between row 1 and the other rows means that most if not all SE domains include a mix of traditional and complex systems.

Some characteristics are common to all systems and are defined in systems science (row 6). Other characteristics are common to all SE systems and are defined in SES (row 5). The portion of the taxonomy distinguishing TS from CS reflects three types of relations, shown in rows 2, 3, and 4 of Figure 2.3. For some characteristics, TSs are a special case of the more general CS case. An example of this is the presumed traditional characteristic of a deterministic system, while a complex system is probabilistic. Deterministic means a probability of 1.0, which is a special case or a point on the probability spectrum from 1.0 to 0. Many characteristics represent a difference in degree between TS and CS (row 3). These reflect a spectrum from most traditional to most complex. Examples are degree of uncertainty, with traditional being the most certain; and degree of autonomy, with the traditional being the least autonomous. Some characteristics represent a difference in kind (row 2). For example, a complex system can learn while a traditional system cannot learn.

It is important to point out that the scope of SES and MOSES is not just row 5 in Figure 2.3. MOSES is scoped to be a synthesis of knowledge from all contributing sciences and related disciplines, as delineated in Table 2.2, that support all six rows in the taxonomy. This knowledge includes what is common to all systems and to all SE systems, and it includes distinctions between traditional and complex systems, and between domains.

The category of all systems in row 6 is clearly beyond the scope of SE. However, these characteristics will be incorporated in the model space — after all, if they are true of all systems, then they are true of SE systems and can be used in SE. From this point forward, starting with Chapter 3, the scope of the discussion will be SE systems — in terms of the taxonomy, rows 1 through 5.

Another important interpretation of the taxonomy of systems in Figure 2.3, especially the TS versus CS portion, is that the figure does not say that certain systems are inherently on the TS side and others are inherently on the CS side. The correct interpretation is that a systems engineer or a community chooses to model a designated system as mechanistic, as organic, or somewhere in between. In particular, we expect that many machines engineered in SE will usefully be modeled as organic systems or agents.

The discussion in this book does not uniformly cover all the supporting disciplines. For example, it discusses how CSS supports SE more than it does how mathematics supports SE. This is not because mathematics is less important to SE, but rather because its support for SE is currently better understood and applied more extensively than is the support of CSS. An exception is that this book does address the important area of probability and statistics, especially its role in modeling and specifying uncertainty. We also note that continuous math and discrete math map approximately to quantitative and qualitative models, although the qualitative category goes beyond mathematics.

This chapter addressed the science-engineering axis, and the supporting systems science component, of MOSES. This information, along with the systems taxonomy, will serve as a context for the remainder of the book, which discusses model orientation and the model space in greater detail.

References

Ashby, W.R. 1956. Requisite variety and its implications for the control of complex systems, *Cybernetica,* 1(2):1–17. (Requisite variety also discussed in Ashby, W.R. 1956. *An Introduction to Cybernetics.* New York: John Wiley & Sons.)

Baez, J. and M. Stay. Forthcoming. Physics, Topology, Logic and Computation: A Rosetta Stone. To appear in B. Coecke (Ed.), *New Structures for Physics.* Springer. http://math.ucr.edu/home/baez/rosetta.pdf (Accessed January 11, 2009.)

Boulding, K. 1956. General systems theory — The skeleton of science. *Management Science,* 2:197–208.

Buschmann, F., R. Meunier, H. Rohnert, P. Sommerlad, and M. Stal. 1996. *Pattern-Oriented Software Architecture: A System of Patterns.* Chichester, U.K.: John Wiley.

Christensen, C. 2003. *The Innovator's Dilemma: The Revolutionary Book that Will Change the Way You Do Business.* New York: Collins Business.

Collins, L. 1976. *Use of Models in the Social Sciences.* New York: Westview Press.

Conant, R. and W.R. Ashby. 1970. Every good regulator of a system must be a model of that system. *Int. J. Systems Sci.,* 1(2): 89–97.

Dorland, W. 2007. *Dorland's Illustrated Medical Dictionary.* Philadelphia, PA: Saunders.

Eldridge, N. and S. Gould. 1972. Punctuated equilibria: an alternative to phyletic gradualism. In Schopf, T. (Ed.), *Models in Paleontology.* San Francisco: Freeman Cooper.

Finch, J. 1951. *Engineering and Western Civilization.* New York: McGraw-Hill.

Gamma, E., R. Helm, R. Johnson, and J. Vlissides. 1995. *Design Patterns: Elements of Reusable Object-Oriented Software*. Reading, MA: Addison-Wesley.

Gharajedaghi, J. 1999. *Systems Thinking: Managing Chaos and Complexity.* Boston: Butterworth Heinemann.

Giere, R. 1999. *Science without Laws.* Chicago: University of Chicago Press.

Giere, R. 2004. How models are used to represent reality. *Philosophy of Science,* 71(5):742–752.

Griffin, M. 2007. System Engineering and the "Two Cultures" of Engineering. Boeing Lecture, Purdue University. 28 March. https://engineering.purdue.edu/AAE/AboutUs/News/Events/documents/2007GriffinBoeingLecture.pdf. (Accessed January 11, 2009.)

Harré, R. 1976. Constructive role of models. In Collins, L. 1976. *Use of Models in the Social Sciences.* New York: Westview Press, p. 16-43.

Hawking, S. 1988. *A Brief History of Time: From the Big Bang to Black Holes.* New York: Bantam Books.

Hesse, M. 1976. Models versus paradigms in the natural sciences. In Collins, L. 1976. *Use of Models in the Social Sciences.* New York: Westview Press, p. 1–15.

Honderich. T. 1995. *The Oxford Companion to Philosophy.* New York: Oxford University Press.

Klir, G. 2001. *Facets of Systems Science.* New York: Kluwer Academic/Plenum Publishers.

Koestler, A. 1967. *The Ghost in the Machine.* New York: Macmillan.

Kuhn, T. 1970. *The Structure of Scientific Revolutions,* 2nd ed. Chicago: Univ. of Chicago Press.

Lehmann, M. and L. Belady. 1985. *Program Evolution,* London: Academic Press.

Mayr, E. 1982. *The Growth of Biological Thought: Diversity, Evolution, and Inheritance.* Cambridge, MA: Harvard University Press.

Miller, G. 1956. The magical number seven, plus or minus two: some limits on our capacity for processing information. *The Psychological Review, 63, 81–97.*

Poore, J. 2004. A tale of three disciplines … and a revolution. *Computer,* 37(1):30–36.

Shaw, M. 1990. Prospects for an engineering discipline of software. *IEEE Software,* 7(6):15–24.

Simon, H. 1996. *The Sciences of the Artificial,* 3rd ed. Cambridge, MA: MIT Press.(Originally published 1969.)

Siouris, G. 2004. *Missile Guidance and Control Systems.* New York: Springer.

Smuts, J. 1926. *Holism and Evolution.* London: Macmillan. (Reprinted 1999. Sherman Oaks, CA: Sierra Sunrise Publishing, © Sanford Holst.)

Vincenti, W. 1990. *What Engineers Know and How They Know It: Analytical Studies from Aeronautical History.* Baltimore, MD: The Johns Hopkins University Press.

Von Bertalanffy, L. 1933. *Modern Theories of Development: An Introduction to Theoretical Biology,* Oxford: Oxford University Press.

Von Bertalanffy, L. 1968. *General Systems Theory: Foundations, Development, Applications.* New York: George Braziller.

Warfield, J.N. 1994. *A Science of Generic Design* Ames, IA: Iowa State University Press.

Wiener, N. 1961. *Cybernetics, or Control and Communication in the Animal and Machine* (2nd ed.). Cambridge, MA: MIT Press.

Chapter 3

Core Concepts of Model-Oriented SE Science (MOSES)

> "To see what is general in what is particular and what is permanent in what is transitory is the aim of scientific thought."
>
> —**A. Whitehead**, *An Introduction to Mathematics*

This chapter introduces some of the basic concepts of model orientation and related features of model-oriented SE science (MOSES). The main topics are, first, an example of how MOSES can be applied, and then core concepts of MOSES in two parts: system-related concepts and connection-related concepts.

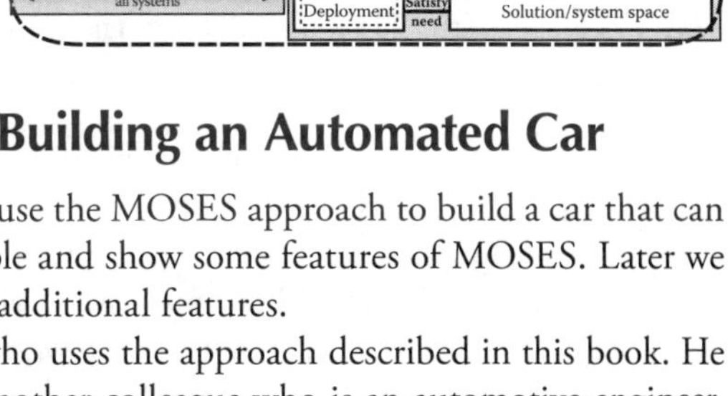

MOSES in Action: A Story about Building an Automated Car

This is a story about Bob and his colleagues who use the MOSES approach to build a car that can drive itself. In this section we develop the example and show some features of MOSES. Later we refer back to this example and use it to illustrate additional features.

The general goal: Bob is a systems engineer who uses the approach described in this book. He has a colleague who is a robotics designer, and another colleague who is an automotive engineer. They decide that they will design and build an automated car that can drive from any current location to any desired location, such as from Bob's home to his office. They will use MOSES in this project. What does that mean? The story that follows narrates how MOSES is applied to engineer a solution. It covers the topics of automation issues, using and tailoring models, approaches to learning, language and precision of models, engineering process and tools, verification and testing, automation features, and documenting. A sidebar outlining multiple types of models for this problem is also provided.

Setting the Stage

MOSES

One of the factors in the approach to engineering a system is the degree to which the general form of a given problem is known and understood, and the general form of solutions are known, understood, and available. If the general problem and solution are known, then the systems engineer can use that general knowledge and tailor it for a specific problem. The solution may be in the form of a general design, which can be tailored for this problem, or in the form of commercially available products, which can be bought. In MOSES, known problems and solutions are characterized as being captured in a model space and made available to a community. The model space is a virtual repository of the body of knowledge (BoK) of a domain. For systems engineering, the scope of the model space covers all domains that SE supports (the application domains), the domain of SE itself, and all domains whose knowledge is required to perform SE (the supporting domains, including mathematics and science, as delineated in Chapter 2). The knowledge is represented in the form of models, which can variously be interpreted as simulation models, requirements sets, concepts of operation, architectures, designs, and any other engineering artifact. For mature domains, the basic process is to find general models that fit the engineering task at hand, and tailor these models to fit the specific problem being addressed and the specific environment in which the problem is to be solved.

On the other hand, if the problem is new, it resembles a research and development situation, and that requires more experimenting and creative activity. Few if any models will be in the model space, and new models will have to be generated. If it is felt that this new problem and solution may be of interest in solving other related problems in the future, then the models resulting from the engineering effort for this experimental system will be put into the model space for others to use and evolve.* More details of the parts of the model space appear in Chapter 5.

Problem Situation

The problem of automating car drivers has characteristics of autonomous agents and robotics, specialized for a traffic environment. The characteristics described here are adapted from Zeller and Mock-Hecker (1993), who defined them in the context of simulating multiple agents driving in traffic. The problem includes the following characteristics:

- An automated autonomous agent (hardware and software) is our system of interest, and it is dealing with an environment that includes machines (cars) and human autonomous agents (drivers). One view of this overall picture is a community of autonomous interacting agents that exhibit a combination of cooperation and competition.
- This community represents a self-organizing, self-adapting system of elements whose interactions and conflicts must be solved in real-time without a central supervisor.

* Note that the issue of intellectual property rights and claims is not ignored in MOSES. That issue puts constraints on what is made available to the larger community. The point of MOSES in this regard is not to change intellectual property rules, but rather to capture and synthesize knowledge currently available in the public forum from books, papers, and online sources and organize it in the modeling space. In addition, each company or organization or person can extend this common base with their own specific knowledge, including their unshared intellectual property.

- It is a dynamic community. From the perspective of our automated driver, this means that the environment — other drivers, the road, etc. — changes continuously.
- Every agent has its own goals.
- Every agent has its own sensor and information processing component, so it can plan the actions to reach its goal independently.
- The agents must cooperate to some extent to reach their goals but they cannot communicate and negotiate explicitly.
- The combination of our automated driver and car is a hybrid system; that is, it includes both machines and autonomous agents

If we consider that our automated agent is the information processor and decision maker in driving the car, then the larger car driver problem involves three basic elements: *sensors* to detect the environment, a *processor* to process the information and decide what action to take, and *actuators* to effect the behavior of the car, such as maintain or change speed or direction, etc.

Automation Issues

In this example case of an automated car, the situation is that there are two relatively mature elements but they must be brought together in perhaps a new way. Clearly, cars represent a very mature engineering domain, and robotics is a fairly mature domain. We assume that robotics includes both the hardware and software agent elements, both sensors and actuators. Because this example is intended to show what MOSES is and how it would be used if adopted, we will assume that there are many models of cars and robots available in the model space. In this context, the task of Bob and his colleagues is to use the existing knowledge and models of cars and robots, and somehow produce a variant of a car that has its own built-in robotic driver. This combination may cause some changes to a "normal car" and to a "normal robot."

Using and Tailoring Models

How do they use and tailor the models? The models of cars that Bob retrieves from the model space repository assume that the driver is located in one place, behind the steering wheel and other controls, and has forward-facing visual sensors. If we are going to automate the driver function, we are no longer constrained to have the robot in one location, or to have the set of controls that we have now. Bob runs some simulations indicating that a more optimum driver would be distributed. Sensors would be positioned in the places with the best visual access in all directions from the car, so they can sense all adjacent vehicles and objects as well as their proximity and relative movement. Actuators would be positioned close (in some sense) to what they are actuating — engine, transmission, and, ultimately, the wheels and tires for accelerating, turning, and braking. For the foreseeable future, the driver controls now in place (steering wheel, accelerator, brake, mirrors, etc.) would likely remain to allow manual override if needed. If at some point in the future the automated technology became sufficiently reliable and trusted, the human driver controls might be removed and the driver's seat would become just another passenger seat.

In any case, the passenger controls would remain. They include temperature control, seat position, window control, and controls for radio, CD, or other entertainment medium. Windshield wipers or ice scrapers would not be needed by the sensors but would likely be retained for backup human driving or for passengers to see the environment.

The robotics models would be in the form of hardware and software models. The hardware would include the sensors and would actuate the turning, acceleration, and braking elements of the car. The software (and data) would include all the rules, patterns, and models of driving; it would receive information from the sensors and would send information to the actuators. The response time would have to be very short because driving requires real-time feedback and response to a continually changing environment. An additional need is to have the driver, if it fails, fail in a graceful and safe way.

Here is an example of a driving vignette as a slice of a continual sequence of {environment input >> applying rules >> actuator/car response}

- *Sensor data:* Red stop light observed 200 ft ahead.
- *Self-monitor data:* Current speed is 35 mph.
- *Software:*
  - Rule 1: If red stop light ahead X ft, then stop in X ft.
  - Rule 2: If stop in X ft and current speed is S mph, then apply brake with force F (computed as a function of X and S; for example, F = kS/X, based on initial X and S at time red light observed) until car stops.
- *Actuator:* Brake is applied to wheels to slow and stop car.

One of the approaches Bob and his colleagues will consider for teaching the robot to drive is to enable it to learn. Therefore, models of machine learning will also be retrieved from the model space.

The reason for identifying these changes and requirements is that this is part of the process Bob and his colleagues go through to find and tailor models to fit the type of system they want to create.

The result is a combined car-robot system that is a modified car with a modified distributed robot driving it based on directions from the passenger(s).

Approaches to Learning

One of the most challenging elements of building this system is effectively teaching the robot how to drive. There are at least three possible approaches, which we call (1) predefined rules based, (2) simulation learning based, and (3) observation learning based.

Approach 1: Predefined Rules Based

The engineers define all the rules and concepts and driving patterns for the robot, load these as instructions for the robot, have the robot operate the car (initially via simulator, later in field testing), refine the rules, and repeat loop until rules are sufficient to enable robot to operate car with a competence similar to an adult human driver.

The rules and patterns include:

- *Concepts* such as external vehicle, pedestrian, road, lane, direction, speed, and self-awareness of state of the car
- *Primitive driving patterns* such as move forward; avoid hitting other vehicles
- *Intermediate patterns* such as turn left at the next stoplight; turn right at the next intersection; find a parking spot

- *Route patterns,* which are routes commonly taken between given locations, such as Bob's home to office
- *Route planning pattern,* which finds an appropriate route to get from current location to desired location, based on current road and traffic conditions

See the sidebar entitled "Example Driving Models or Patterns" for a more extended list of these categories of patterns.

Approach 2: Simulation Learning Based

Specify a minimal set of starter rules; for example,

1. Move forward toward destination.
2. Stay on road.
3. Avoid hitting other objects.

Then select a learning model from categories such as artificial neural networks or Markov chain Monte Carlo methods, and simulate driving situations that enable the robot to learn how to drive — essentially building up the patterns that Bob and his colleagues would need to specify in Approach 1.

Approach 3: Observation Learning Based

The robot, or the engineers, or a combination, learn the rules through analysis of traffic behavior, followed up by simulator feedback to the robot. A precondition of this is the availability of a significant sample of video recordings of traffic movement and flow that shows the gamut of traffic activity and conditions. This video sample would make visible information about the behavior of a given car that includes spacing and movement of other nearby vehicles, and the given car's behavior relative to stop signs, stop lights, road signs, speed limit signs, visibility, weather conditions, road conditions, and traffic conditions (e.g., heavy, light). Then either the robot or the engineers extract and abstract from the video models, rules, and patterns of observed car behavior in the traffic environment. The following are sample rules that could be extracted:

- In normal conditions, at speed S, a car on average leaves a distance D1 between it and the car in front.
- When it is raining or snowing, the distance is D2.
- When the road is icy, the distance is D3, etc.
- If distance < D1 in normal conditions [or < D2 or < D3 in their conditions], the probability of a car being in an accident rises from Px to Py.
- When approaching a stop sign, a car at speed S begins to slow down at distance D4 prior to the stop sign.
- When approaching a stoplight that is red, a car begins to slow down at distance D5.
- When approaching a stop light that is yellow, a car stops if distance > D6 or speeds up if distance < D6.
- On average, prior to turning, a car turns on turn indicator distance D7 before the turn intersection.
- If speed limit is S1, a car with no close traffic on average moves at speed S2.

- If S2 – S1 > X1, the probability of being stopped by a police car rises from 0.0001 to 0.01.
- If S2 – S1 > X2, the probability of a car being in an accident rises from a probability of Pa to Pb.

This method could generate many of the driving patterns but would probably need to be supplemented by Approach 1 if the engineers are doing the rules extraction, or by Approach 2 if the robot extracts the rules. The engineers, assuming they have driving experience in traffic, have a significant advantage because they already know much of the information and many of the rules through their own experience, whereas the robot is starting with a blank slate. The observed traffic would reflect a range of driving habits from poor to good, a distinction that is hinted at in the sample rules above. This distinction, and what the robot learns from it, are further discussed below under "Defensive Driving."

Observe that Approach 3 illustrates a science of driving, by our definition of science in Chapter 2, because it defines patterns based on observations of existing systems. A related and somewhat more subtle characteristic that distinguishes Approaches 1 and 2 from Approach 3 is that the rules defined in the first two approaches are prescriptive (drive this way; do not drive that way, etc.), while the rules defined in Approach 3 are descriptive (if a car behaves this way, X will happen with probability Px; if that way, Y will happen with probability Py, etc.). The descriptive rules in Approach 3 can then be used as a basis for prescriptive rules similar to those in the first two approaches. The reason they can be made prescriptive is that Bob and his colleagues place a positive value on the robot's car behaving in a way that avoids contact with other vehicles, and place a negative value on the car behaving in a way that results in collisions (or getting speeding tickets, or other driving "incidents"). The discussion here relates to two features of MOSES that are discussed in Chapter 4: modality and valuation of behavior. It also hints at the topic of specifications (including probabilistic specifications), which is the subject of Chapter 11.

Example Driving Models or Patterns

Examples of models of driving a car, in the form of various types of patterns:

- *Basic object or concept patterns:* Self (dimensions, mass, engine power, turning radius, braking effectiveness, fuel level, many general health status indicators, direction of movement, speed, etc.); external: car; vehicle; pedestrian; animal; obstacle; bicycle; road; lane; surface condition; speed; destination; direction of movement; adjacent vehicle (front, rear, left, right); clearance between vehicles, curve in road, straight ahead, right turn, left turn, right of way, speed limit, traffic signs, rain, snow, ice, slippery, wind speed and direction, temperature, amount of visibility, etc.
- *Basic primitive driving patterns:* Avoid hitting other vehicles and pedestrians; change from stopped to moving forward, from moving forward to stopped, from stopped to moving backward, from moving backward to stopped; when moving forward, leave space in front = 5 ft + (5 ft * speed in mph); drive in middle of lane; keep speed at posted speed limit; slow down if road is slippery; turn left turn signal on 10 sec and

150 ft prior to turning left; likewise right signal prior to turning right; give right of way to oncoming traffic when turning left; turn left signal on prior to moving one lane to the left, and right signal on prior to moving one lane to the right; go around an obstacle.
  - Stop light pattern: stop at stop light if red, proceed if green, right turn after stop allowed on red if clear.
  - Stop sign pattern: stop, then proceed when clear right and clear left and clear ahead.
- *Basic intermediate patterns:* Keep driving on current road unless otherwise instructed; turn left at the next stop light; turn right at the next intersection; find a parking spot; parallel park; keep speed low enough to maintain traction and keep car on road; override posted speed limit if adjacent cars are all exceeding the speed limit, and adjust to their speed.
  - Legal driving pattern: several primitives, including avoid hitting other cars, comply with traffic signs and stop lights, use turn signals, etc.
  - Defensive driving pattern: all patterns listed so far + drive no slower than 10 mph under the speed limit, conditions permitting; if major obstacles appear, go around them if possible, otherwise stop immediately; if minor obstacles appear (pothole, small to medium size rock, small box) and there is heavy traffic following close behind the car, do not stop, and if traffic is on both sides of car, do not go around the obstacles; at all times, assume the possibility that:
    - Surrounding vehicles may not follow same rules (e.g., they may not stop at stop lights or stop signs).
    - Traffic may suddenly stop.
    - The road may suddenly become slippery.
    - Obstacles may suddenly appear in road.
    - Pedestrians on the side of the road may suddenly walk on road in front of car.
    - Animals may suddenly run onto road in front of car.
  - Emergency driving pattern: avoid hitting other cars and pedestrians and major obstacles, keep speed low enough to maintain traction and keep car on road; proceed as fast as possible within those constraints; slow down when approaching stop sign or red stop light until verifying that cross road is clear; use turn indicators; all other rules are suspended.
- *Route patterns:* The basic form is Drive from location A to location B (e.g., from Bob's home to his office). Each route consists of a connected series of intermediate or primitive patterns, similar to the directions of a typical Internet direction finding service (e.g., turn right out of driveway, go straight 3 blocks, turn left on 11th Avenue, etc.). These patterns may be produced on-the-fly, rather than stored in the repository. But the repository might make it appear to the user that the route was stored; for example, the user asks for a route from A to B; the repository query processor requests an Internet service to determine this and passes the result along as if it were stored in the repository. For the purposes of this automated car, the most common routes between commonly used locations may be loaded into the robot data store and reused each day.

- *Route planning pattern:* There is one basic pattern: It finds an appropriate route to get from current location to desired location, based on current road and traffic conditions (e.g., traffic backups, road construction, ice conditions, etc.). It can work with or without a route pattern.

Example of using the planning pattern *with* a route pattern: Suppose there are two route patterns for getting from Bob's home to work, called Primary and Backup. Each day just before Bob gets into the car, the route planning pattern follows an algorithm something like this: Check Primary route current conditions using online traffic service; if OK, then today's route = Primary; if not OK, then check Backup route current conditions; if OK, then today's route = Backup; if not OK, then go to online route-finding service and get another route, to exclude routes equivalent to Primary and Backup.

Algorithm for using the planning pattern *without* a route pattern: First, go to online service and find a route from location A to location B. Then check current conditions of the route. If OK, use that route. If not OK, go back to the route finding service and get a different route; check current conditions of that; if OK, use that route.

Language and Precision of Models

We used informal language in our statements of the patterns but the actual models in MOSES would likely be more precise. For example, the route pattern is stated as Drive from location A to location B, or in Bob's case from H to W. The MOSES model might look like this:

- **Precondition**: Bob is at home, location H, at time T1.
- **Post condition:** Bob is at his office, location W, safe, uninjured, at time T2 = 8:00 a.m. in time zone where Bob works, where T2 – T1 < 1 hour, with probability P = .999 of Bob reaching W safe and uninjured, and probability P = .9 of Bob reaching W by time T2.
- **Constraint**: Mode of travel is automated car.
- **Dependencies** of the automated car (partial list):
  1. Appropriate air is available for the engine.
  2. Fuel is available.
  3. A continuous surface appropriate for driving a car exists between H and W.
  4. A force exists to push the car against the surface to enable traction to emerge.

All dependencies except (2) would likely be implicit; they can be assumed.

Engineering Process and Tools

What is the sequence for the engineering activities of designing and building an automated car? We briefly describe two aspects: *process* and *tools*. Regarding the process, Bob and his colleagues do not take a sequential waterfall approach. Instead, they follow the MOSES approach of an opportunistic process: that is, they start with the information and knowledge that they have, which is what they already know from their training and experience, and what they found in a search of the model space for car designs and robot designs. They work on areas where they can get traction, rather than adhering to a fixed sequence in developing the concept of operations (CONOPS), requirements, architecture, and design. They go back and forth between the interaction of the automated car with its environment and the robot software and hardware. The goal is to fill in all the pieces, like a jigsaw puzzle, so it will be complete at the end. Reasons for this include (1) all parts of the engineering process are to some degree interdependent, and (2) they want to take advantage of preexisting pieces. For example, they do not need to develop a parallel park pattern; that piece of the puzzle already exists.

The model space is equipped with a variety of associated tools that operate on an array of models. Included are tools for searching, retrieving, analyzing, comparing (including model matching), translating, optimizing, tailoring, modifying, executing, and storing models.

Verification and Testing

A key part of the test is to devise a car simulator (analogous to a flight simulator) that is driven by a robot instead of a human. If car simulator models are available in the model space, then these can be used. If no such models exist, Bob and his colleagues will have to design and build a car simulator, which is probably as difficult as designing and building the robot driver and integrating it into the car. However, in this example we do not explore building the simulator; we simply assume it exists and that Bob will use it.

Bob and his colleagues decide to build and test the automated car in two stages or prototypes. The first prototype knows the basic and intermediate patterns but not the route or route planning patterns. The operation of the first prototype is partially automated: A user (whether called an operator or a passenger) is needed to direct the robot using intermediate patterns: turn left at this intersection, go straight, turn right at the next stop light, take the next exit, etc.

The second prototype knows all the robot planning patterns, and is therefore the full system. The operation of the full system is automated; no operator is needed. The passenger tells the car the desired destination, and the car does route planning, as described in the route planning pattern. When the route is selected, the car then follows the intermediate patterns that constitute the route to reach the desired destination.

This is followed by field testing, starting on a test track with no other traffic, then adding limited traffic, etc. Finally, full road testing is done. Each step involves feedback, iteration, and learning — on the part of both the engineers and the robot.

Automation Features

The passenger has override or interrupt capability. For example, if Bob is part way to the office and remembers something he needs but left at home, he can say, "Interrupt; turn around and go home" or "Change destination to home." Or if he remembers that he promised to pick up a colleague on the way, he can say, "Interrupt; change destination to 123 Oak Street." He can also change to semi-automated mode as in the first prototype. For example, he decides part way to work that he will take the day off and go explore some mountain roads nearby. He can say, "Interrupt; take the next exit and head west on Highway 211." From that point, no route is in effect, and the passenger becomes an operator using the intermediate patterns to direct the car.

Note that this automated car could operate without any occupant/passenger. For example, Bob could define a route from his home to the car maintenance shop, and a route from the maintenance shop to his office. He could program it or tell it to drive itself to the shop, and then when the maintenance was done, to drive itself to his office to pick him up after work.

Documenting

At the end, when they are done testing and the automated car is accepted as successful, someone tells Bob that they need to document their requirements in a detailed requirements spec, and they should have done this up-front. After thinking about this a bit, Bob gives a response. Because the response is different for the three approaches, and we have not specified which approach they took, we provide the response for all three approaches.

Response for Approach 1/Predefined Rules

The requirements model consists of the complete set of models in the form of the various patterns because those models describe exactly what the system does in its interaction with the environment. The route and route planning patterns are the top-level requirements, the intermediate patterns are the next level breakout, and the primitive patterns are the most detailed requirements. The concept patterns are the "ontology," or at least the universe of discourse, of this car-driving domain, and the basis for defining the requirements. We could not have defined them up-front but we developed them and iterated them as we interacted with the simulator or the physical environment. More specifically, the requirements spec we present consists of this virtual document or composite model:

1. The statement: the automated car shall have the ability to perform all the patterns that follow in (2).
2. All the route, route planning, intermediate, and primitive patterns; these represent the substance of the requirements.
3. The concept patterns and definitions; these represent the glossary.

Response for Approach 2/Simulation Learning

We define the set of patterns as the initial minimal set we gave to the learning robot, plus the whole set of patterns and rules that the robot learned. Using that definition, the answer is the same as for Approach 1. That is, the automated car shall have the ability to perform all the patterns in this set. *In this case, the system wrote most of its own requirements.* Because the system recorded in its pattern database each thing it learned when it learned it, we could say that the completion of the requirements model coincided exactly with the completion of the development of the system — because when the system was "done" learning, its development was complete.*

Response for Approach 3/Observation Learning

Under the assumption that the engineers extracted the rules, the documentation of this approach is consistent with that of Approach 1. Under the assumption that the robot extracted the rules, the documentation is consistent with that of Approach 2.

We discuss the features of MOSES demonstrated in this example throughout the book.

---

* The terms "done" and "complete" are used somewhat loosely here. Because this robot can learn, it will continue to learn and evolve its pattern base (and therefore its requirements) while being used operationally. In that sense, the system will never be done learning, and the requirements will never be complete. But it is legitimate to use "done" and "complete" in the sense that a point is reached when the automated car is good enough to meet its acceptance criteria and can go into operational use. The system has learned enough to go to work as an autonomous vehicle.

Core Concepts I: System, Model, Correspondence

This section introduces the first set of basic concepts used in MOSES. These concepts include system, engineering, characteristic, model, correspondence, and specification, which in turn provide the basis for the further depiction in the remainder of the book.

The concepts defined in this section and the next section track to some degree with parts of the PICARD theory of systems (Martin and Ferris 2008). This theory includes the following elements: Parts (or products), Interactions, Context, Actions, Relationships, and Destiny. In a larger sense, the concepts presented here build on many ideas from SS, SE, and related disciplines. Two examples are the foundations part of RM-ODP (1996) and the modeling approach of Selic, Gullekson, and Ward (1994).

Key System Terms

A reasonable starting point for any approach to systems engineering is the concept of *system*. This subsection lays out the MOSES conceptualization of system and its related concepts of world, mosaic, entities, environment, whole, and part.

Preliminary informal concepts:

- A system is an entity that is of interest as a whole and as a set of two or more connected parts, where connection can be due to structural relations or dynamic interactions.
- An open system is a system that is connected with one or more external entities.

In the traditional engineering world of machines, the set of entities connected with an open system is called its *environment*. In the biological world of organisms, the set of entities connected with an open system is called its *ecology*. In the mathematical world of network or graph theory, the set of entities or nodes connected with an open system is called its *neighborhood*. The meaning of these terms can also include the system itself and the connections. In the case of neighborhood, open neighborhood excludes the system, while closed neighborhood includes the system. In MOSES, the three terms can be used interchangeably, but in this book we primarily use the term "environment."

Nominally, a system that is not open is closed; that is, it has no environment. Most systems of interest to systems engineering are conceptualized as open but sometimes it is useful to model certain systems as closed.

While the above informal concepts provide a reasonable starting point, a slightly more formal set of concepts is needed for the MOSES foundation.

- **World**: Space-time region, which may be physical or conceptual space, and may be past, present, or future time.
  - Note 1: Examples of worlds: The current known physical universe; planet Earth over the full time period of its existence as a planet (past and future); all human beings currently alive on Earth; my office last Tuesday at 10:00 a.m. Eastern Standard Time; a specific parking lot during the course of one day; the North American continent 1,000 years from now; the problem space of physics as currently defined; the universe of discourse of systems engineering over the next 10 years; the world of all possible worlds.
  - Note 2: The world of primary interest to SE is the collective actualization space as depicted in Figure 1-2.

- Note 3: Conceptual here means anything not comprising physical atoms. For example, information system modeling includes conceptual, logical, and physical data models. In the context of SE and this definition of world, data models of all three of these types exist in conceptual space, not physical space. All software, data, information, and knowledge are conceptual, even though they are represented and stored in physical media.
- **Mosaic**: Interconnected set of entities that exist in some world, where exist means either physically exists or exists in the shared conceptualization of a community in that world.
  - Note: A prominent mosaic of interest to SE is the set of entities in the problem space (as depicted in Figure 1-2), both before and after solution systems are deployed.
- **System**: A designation by an observer of a group of entities, which entails these commitments:
  - Commitment to a boundary that logically contains the entities in the system and separates it from the entities outside the system, if any.
    - Note: The boundary may be distinct and known, or it may be fuzzy or uncertain.
  - Commitment to a classification of the bounded entities as a single whole system.
  - Commitment to conceptualizing the system as comprising at least two related or interacting parts.
    - Note: Commitment to any constraint on the parts — such as a specific identification, number, or configuration of parts — is not required, but is allowed; if such a commitment is made, it is part of the system designation.
  - If open system: commitment to conceptualizing the designated system as related to or interacting with entities in an environment, without committing to a specific environment.
    - Note: Commitment to a specific environment is allowed; but if such a commitment is made, it is *not* part of the system designation.
  - If closed system: commitment to conceptualizing the designated system as not related to or interacting with entities in any environment.

Based on these concepts, a system requires essentially two things to exist: (1) related or interacting entities in a mosaic, and (2) a designation of the entities as a system.

The Relationship between System and Mosaic

A general view of *mosaic* is that it is the group of all entities, some of which are interconnected, in a defined world, and which form the canvas on which we overlay systems concepts and especially designations of systems. Why do we say that a mosaic exists while a system is merely a designation? The preliminary answer is that a mosaic is a context that exists in some sense separately from an observer's decision to designate it or a subset of it as a system. The meaning is similar to situation, although situation maps more to a "mosaic of interest" or selected subset of a larger mosaic.

We discuss this distinction from two perspectives — physical and conceptual. Physically, a given collection of trees in close proximity that we call a forest, or a given collection of interacting people that we call a family, or a given collection of bolted together pieces of metal, glass, and rubber that we call a car — all exist separately from an observer. They are all legitimate mosaics, or part of an overall mosaic. Any observer can choose (or not) to designate the collection of trees as a system; the interacting people as a system; or the metal, glass, and rubber as a system. We see in the following discussion that the concepts we call "forest," "family," or "car" are also part of the mosaic.

Now consider the conceptual case. There are many entities that we think of as existing even though physically they do not exist. Examples are markets and organizations. Organizations such as a country, a state, a school district, a professional society, a government agency, or a commercial company exist in a conceptual sense but not in a physical sense. IBM does not exist physically even though it has physical assets such as buildings, equipment, and people; and it can buy and sell other organizations and it potentially can be bought and sold. We think of organizations as having their own separate existence, so we include them as legitimate, although conceptual, mosaics. We agree that IBM exists; but whether an observer chooses to designate IBM as an organizational system is a separate decision.

The "we" in the foregoing statements indicates a community. In the case of IBM, the United States economy, or the International Council on Systems Engineering (INCOSE), the community that agrees that they exist probably includes most humans on Earth, or at least those who have heard of these entities. In other cases, the community that agrees that a conceptual entity exists is contained within a more narrowly defined world. For the purposes of this book, those defined worlds are typically disciplines or domains.

There is another type of entity that can be part of a mosaic: category or general concept. We agree not only that the individual company IBM exists, but also that the category company exists; and not only that Bob's individual car exists, but also that the category car exists. Table 3.1 shows some examples of mosaic entities and the distinctions between physical and conceptual, and between individual and category.

All the entities listed in Table 3.1 exist in the mosaic of various defined worlds, or potentially a single defined world. The table entries, and the following statements that discuss the table, all reflect the view or ontology of MOSES. First, the lower-left cell has no entries, indicating the view that all categories are conceptual. Every tree on Earth is a physical individual, but the category "tree" that we have defined for all of those individuals has a conceptual existence, not a physical one. One might argue that some physical templates, forms, or molds (such as cookie cutters, industrial molds, or machine tools for producing tools and parts) are physical categories and belong in the lower left cell. In addition, some philosophical stances may designate categories in general as existing physically. We view the decision regarding categories as a modeling choice. Our choice is to model categories as conceptual, and physical templates as physical individuals. Regardless of where one allocates elements to this table, the real point is that conceptual elements can exist in a mosaic.

Now bring the systems concept into the picture, and apply it to Table 3.1. Consider two situations: (1) What entities in the table *could* be designated as systems — that is, what entities satisfy the definition of system? (2) Of those entities that we could designate as systems, which would we *choose* to designate as systems? The designation of system is not applied to all possible systems; we only do it if there is a purpose for it, if it is useful to us.

Table 3.1 Types of Mosaic Entities

	Physical	*Conceptual*
Individual	The person whose name is Bob; Bob's car with serial no. 1234; Bob's house with address XYZ; the U.S. power grid; the General Sherman tree in Yosemite National Park; the George Washington forest in Virginia; planet Earth	Bob's job; the company Bob works for (e.g., IBM); the New York Stock Exchange (NYSE); the names Bob, IBM, and NYSE; the serial number 1234; the address XYZ; the names General Sherman, George Washington, and Earth; the set S = {1, 2, 3, 4}
Category	— (Physical categories do not exist. See discussion in text)	The categories person, car, house, power grid, tree, forest, planet. The categories job, company, stock exchange, name, serial number, address, set.

1. What entities could be designated as systems? At first glance, we would probably all agree that all the physical individuals — the person, the car, the house, the power grid, the tree, the forest, the planet — qualify as systems and could be so designated. To cut to the chase, the MOSES position is that every entity in the table could be designated as a system. We explain this position with a couple of examples on the conceptual side.

   The name Earth: Designate the name as the whole system and E, a, r, t, h as its parts. Then this satisfies the definition of system: it defines a clear boundary that includes those five elements and excludes all other elements, it commits to the five elements as the whole system, and it has at least two related parts. The relations are that in this system, E precedes a; a precedes r; r precedes t; and t precedes h. If those relations do not hold, it is not the same name. Hence, the name Earth can be designated as a system.

   The category set: In MOSES, all categories are defined or described as models (discussed in this chapter in the section entitled "Model"). Our model of the category set would be something like this: A set has two entity types: collection and element; and one relation: member of, as in element is member of collection. Each individual set has one collection and zero or more elements. Now let us designate the whole system as the model of the category set, and designate collection and element as the parts of that system. Since element is related to collection, this designation satisfies the definition of system.*
2. What entities that could be systems would we choose to designate as systems? That, of course, depends on the modeling or engineering situation. But in the expanded SE, probable candidates for system designation would include all the listed physical individuals and the conceptual individuals (i.e., Bob's job, IBM, and the New York Stock Exchange). In certain circumstances, SE could also designate the individual names, numbers, and addresses, as well as models of all the various categories, as systems.

The systems approach exists in relation to, but separately from, the mosaic. The mosaic is unaware of the systems approach and systems concepts, but is indirectly affected by it. Systems produced by SE or other engineers using the systems approach become part of the mosaic — but as products or other entities, not as systems. An organization, a car, or a power grid becomes part of the mosaic on those terms; their designation as a system during their development or after they become part of the mosaic, remains part of the systems overlay — the systems view of the mosaic. More generally, the expansion of SE means that SE is considering a broader mosaic. SE is choosing to designate more of the total mosaic as systems, and especially as systems that we think we can engineer.

One other relation between system and mosaic should be noted. In many cases, a collection of entities exists in the mosaic before it is designated as a system. But in other cases, a collection of things may be designated as a system before they exist in the mosaic. Engineers do this routinely. We designate a non-existing entity as a system, give it a name, write requirements for it, and then some time later — after it is produced — it becomes part of the mosaic when it is deployed in the problem space.

System Boundaries

Part of designating a system is deciding where to draw the boundary. That decision can depend on several factors, including the purpose of the system and the interconnecting patterns of the mosaic

* This holds even if the element part is empty, that is, an empty set. The model has both collection and element parts, regardless of their content. An individual set S can be designated a system for a different reason: The members of a set have the relation co-member with each other.

on which the system is drawn. Consider these examples: human being, car, taxicab. Suppose Bob owns and drives a Toyota with serial number 1234. We start with the notion that we can classify or categorize Bob in many different ways. Three of those ways are: car driver, human being, and system. Likewise, we can categorize Bob's car in many ways, four of which are: Toyota, car, vehicle, and system. Both sets of categories are given in increasing levels of generality. The reason we present these categories is to point out that the designation of entities as a system is exactly the same activity that we perform every day when we categorize entities; for example, when we think of Bob as a driver or as a human being, or when we think of Bob's car as a Toyota or as a car. In fact, categorizing in general involves bounding and classifying entities.

Now consider the Toyota-related entities: windows, metal frame, engine, tires, controls, driver, passengers, fuel that enters, air that enters, air that leaves (exhaust), road, force of gravity. The boundary of the car system seems straightforward. Inside the boundary are all the metal, glass, engine, tires, parts, and control elements (steering wheel, brake, accelerator, etc.) that make up the Toyota. The environment includes the road, the force of gravity, outside air, and outside fuel. How about air and fuel that enter the car system — at what point do they become part of the system? Is fuel in the gas tank part of the system? For some purposes, it might be modeled as part of the environment — for other purposes, as part of the car. Examples of the latter include measuring the weight of the car for riding on a ferry or crossing a bridge, or determining acceleration or braking limits of the car.

What about the driver and passengers? A reasonable stance is that passengers are the users or the service consumers of the car, and therefore are outside the car system. Similarly, the driver is the operator of the car and therefore outside the car system. The passengers and driver are part of the environment. But now consider a situation in which Bob's Toyota is a taxicab. What is the boundary of the taxicab system? A reasonable modeling choice here is again to regard the passengers as outside the system but to include the driver (Bob) within the system. The Toyota system is one part of the taxicab system. Bob is a human system outside the Toyota system but inside the taxicab system. Why the difference in the conceptualization of the driver in the two cases? Because when we buy and use a personal car, we do not buy or use a driver with it. When we buy the services of a taxicab, we expect both car and driver as a system, as a service entity. If we saw a taxicab car without a driver in it, we would not regard it as a complete taxicab system; it has a missing component.

Other boundary issues are discussed later in the book — permeability in Chapter 6, and fuzziness and uncertainty in Chapter 11.

Model

Model Definition

The MOSES definition of model is derived from (IEEE 1989, 12). This standard defines model as the "approximation, representation, or idealization of selected aspects of the structure, behavior, operation, or other characteristics of a real-world process, concept, or system." Our definition is:

> Model: Explicit approximation, representation, or idealization of selected aspects of the structure, behavior, operation, properties, or other characteristics that can be associated with one or more systems.*

* In a broader sense, one could model mosaics and worlds as well as systems. In this book the focus is on modeling systems. Because in many cases mosaics and worlds can be designated as systems, this focus is not overly restrictive.

Why change the front end and back end of the IEEE definition? *The front end:* "Explicit" is added to distinguish explicit representations from implicit concepts. This distinction in MOSES is between model and specification, and is discussed in more detail in the section entitled "Specification and Its Relation to Model." *The back end:* First, "properties" is added because properties are commonly used characteristics in modeling. Second, the IEEE definition couples a model with a system ("representation … of a real-world … system."). We changed "characteristics of" to "characteristics that can be associated with" to decouple model from system. That is, we want a model to exist on its own as a distinct entity. Its relation with a system should be a separate activity or decision, not hardwired into the definition. Third, we changed "a real-world process, concept, or system" to "one or more systems." Reason: We include conceptual systems as well as physical systems, and our definition of system includes processes and concepts. In addition, we do not want to couple a model to one system. A model, in addition to being an entity that stands on its own, can target or apply to one or more systems, as we discuss next.

Model as System

A common relation among systems is that one system describes — is an abstraction of — another system or systems. This distinction leads to an important taxonomy view in the ontology of science and engineering: *model system,* which describes, and object or *target system*, which is described. That is, a model system describes a target system. Describe is used here as a broad term, to include represent, approximate, idealize, specify, or abstract, and the description may range from very general to very specific and detailed, and from formal to informal, and from mathematical to graphical to textual. It can also represent any modality, as discussed in Chapter 4.

The same system can be both a model system and a target system, such as when a target system is, in turn, a model system of another target system. As an example, suppose we are building an information system, and we define an architecture of the desired system and we define a data view of the architecture; that is, we describe the data elements that are visible and exchanged among the parts of the system. Then, the data view can be seen as a model system whose target system is the architecture, and the architecture is a model system whose target is the information system. The architecture in this case is both a target system and a model system. We define an end target system to be a target system that is not a model system. In this example, the information system is an end target system. Note that any model used in the engineering of a system can have the engineered (end target) system as one of its target systems. Thus, in this case, the data view can also have as its target system the information system; that is, the data view can be regarded as a model of the information system.

In this book, unless otherwise noted, we use the shortcut terms "model" and "system" to mean, respectively, "model system" and "target system."

Characteristic

Systems have numerous characteristics. In this book the term "characteristic" is used as a broad concept, to include not only system properties such as performance, but also behavior, state, existence, and category or type of system. It is in the spirit of the usage of the term in the definition of model taken from (IEEE 1989, 12) in the previous subsection.

In both science and engineering, we want to be able to identify, describe, focus on, examine, measure, evaluate, and compare characteristics that are of interest and to ignore or hide those that are not.

What we focus on (and what we ignore) varies according to our purpose or perspective. In addition, in engineering we want to specify and bring about characteristics of interest and avoid characteristics that are detrimental. The use of models is the dominant way we achieve this focus and separation.

Correspondence and Region

Correspondence*

The dominant relation between a model and a system is *correspondence*.† It exists in the context of a defined world. A system and a model correspond (or have correspondence) if the system resides in the defined world and has characteristics that match or are consistent with the model.

One of the conditions of correspondence is the law of requisite variety introduced in Chapter 2. A model defines certain characteristics of interest, and ignores other characteristics. Within the set of characteristics of interest, the model must represent a variety that is at least as great as the variety of the system being modeled. The law of requisite variety does not apply to the characteristics being ignored by the model. For example, suppose we define a model of the acceleration performance of a car. Acceleration is of interest to the model but other characteristics of the car, such as color, are not. The model must be able to represent the range of potential acceleration of the car to be an adequate model but it does not have to represent the range of possible colors of the car.

Correspondence is not directional; it is a symmetric relation. If a system S corresponds to a model M, then M corresponds to S, and more generally both are in each other's region: S is one of the systems that correspond to M, and M is one of the models that correspond to S. This is a practical issue, not just an academic or mathematical property issue. The symmetric nature of correspondence allows models and systems to be built or defined in any sequence, as we discuss in Chapter 4. We can define a model to correspond to an existing system, or build a system to correspond to an existing model, or find an existing system and model and determine if they correspond.

Correspondence is fully described in terms of a correspondence set, which is introduced and discussed in detail in the context of decoupling model, system, and modality in Chapter 4.

Region

We discuss region from two perspectives: (1) a given model, and (2) a given system.

> *Perspective (1):* In general, multiple systems correspond (at least potentially) to a given model. The exceptions are cases where a model is specifically designed to identify individuals, such as a fingerprint, a DNA model of a human, or a model of a car that includes the vehicle identification number (VIN). In general, a *region* is the set of all systems that correspond to a given model, and we refer to it as the region of the model, or more explicitly, as the region of the model in the defined world. Examples in traditional engineering include alternative systems that satisfy a given set of requirements, or alternative systems that adhere to a given reference architecture.‡

---

* Correspondence in this book is distinct from the correspondence theory of truth (see overview of the latter in David, 2005). As discussed in Chapter 4, we are not interested in whether a model is true; what we are interested in is whether it relates (corresponds) to a system or category of systems in a useful way.

† In this section, including the examples, the correspondence mode is assumed to be descriptive (as opposed to prescriptive, for example). Chapter 4 extends this to a full range of modes or modalities.

‡ Architecture and requirements are defined in a MOSES context in Chapter 12.

Perspective (2): In this perspective, a region is the set of all models that correspond to a given system, and we refer to it as the region of the system. Examples in traditional engineering include a set of documents (requirements, design, etc.) associated with a given system, or a set of views of the architecture of a given system.

Examples of Correspondence and Region

In the first example, we designate a car as a system and explore model-world-region relations. Figure 3.1 illustrates Example 1. In the figure, the systems are represented by dots, the regions by the various boxes, the models by braces {…}, and the world (W1) by the outside box. Correspondence between a system and a model is implied by the appearance of the system in the region where the model is defined. The blue dot (circle with B) labeled S1 is the system, the blue Honda, in Example 1. All blue dots (blue cars) are in the region of M2 and all red dots (red cars) are in the region of M3. All black, blue, and red dots (all cars) are in the region of M1. All white dots are systems in W1 that are not cars. Examples of white dots: Any tree, person, pencil, house, city, etc., that is in W1. Any system that is outside of W1, whether car or non-car, is not a dot in this figure; it is outside the box. All dots of any color (all systems in W1) are in the M0 region.

Example 1 — Model M2:

- System S1 = the blue Honda I had in the United States in 2006
- Model M0 = {*} (M0 is an unconstrained model; every system in the defined world corresponds to it and is in its region)
- Model M1 = {isCar}
- Model M2 = {isCar, isBlue}
- Defined world W1 = The United States from 1/1/2006 through 12/31/2006

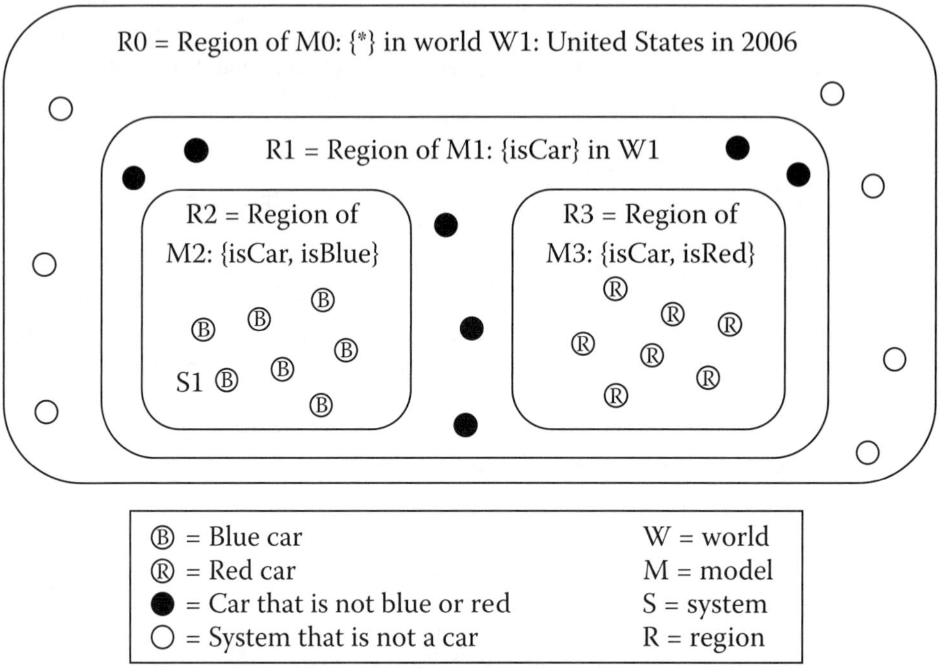

Figure 3.1 Models, systems, correspondence, and regions.

- Resulting correspondences:
  - M0 and S1 correspond in W1 because S1 is a system in W1
  - M1 and S1 correspond in W1 because S1 is a car in W1
  - M2 and S1 correspond in W1 because S1 is a blue car in W1
  - Region R1 of M1 in W1 = All cars in the U.S. in 2006 = {..., S1, ...}
  - Region R2 of M2 in W1 = All blue cars in the U.S. in 2006 = {..., S1, ...}
  - The region of S1 in W1 = all corresponding models in W1 = {M0, M1, M2}

Models M0, M1, and M2 illustrate that, in general, the larger a model in terms of the number of constraints, the smaller and more homogeneous the corresponding region. The most general model is the one with no constraints, M0; the corresponding region is unconstrained, that is, it is the space or region of all systems. Models M1 and M2 are successively larger and their regions are successively smaller than M0.

Example 2 — Classical engineering model regions:

- Region R1 of a given requirements model includes all implemented systems that satisfy the model.
- Region R2 of a given implemented system includes all architectures that are consistent with the system.
- Region R3 of a given requirements model includes all architectures whose corresponding systems satisfy the model.

These classical engineering examples illustrate that determining correspondence is not simple. Correspondence, like many elements of SE, is often fuzzy or uncertain. We often have to make a judgment, such as "We believe that this system satisfies the requirements model" or "We believe that this model represents the class of human genes."

Specification and Its Relation to Model

In MOSES, specification is defined as the shared understanding of an approximation, representation, or idealization of selected aspects of the structure, behavior, operation, properties, or other characteristics that can be associated with one or more systems.

This definition looks similar to the earlier definition of model. The basic relation between model and specification is that a model is the explicit part of a specification. A specification does not have to be explicit; it can be partly or even mostly implicit. But a specification has an added requirement: the idealization must be shared between at least two agents (which can be people or machines). A model can exist independently of context; it is a set of symbols. A specification, however, can only exist in the context of at least two agents that both have the same understanding of its meaning.

The goal of SE is to produce useful systems that satisfy desired purposes. To achieve this goal, engineers produce artifacts that characterize the systems that they engineer. The goal of the artifacts is to designate information about characteristics of the systems and to communicate this information to all parties — specifically among all the stakeholders of each artifact. Stakeholders include people in various roles, such as users, buyers, suppliers, development managers, requirements engineers, architects, design and development engineers, fabrication engineers, testers, integrators, operators, and auditors. Stakeholders also include tools and machines that produce or analyze artifacts or systems, or in general must "understand" certain aspects of the artifacts or systems, such as software editors and compilers, hardware simulators, analyzers, probes, sensors, and measurement mechanisms.

The MOSES concept of modeling and model orientation is meant to increase the shared understanding of system characteristics. The approach is to make artifacts explicit and precise. Note that precise does not mean specific or detailed. In MOSES, precision is orthogonal to specificity. This orthogonality can be illustrated with the models in Figure 3.1. Models M0 = {*}, M1 = {isCar}, and M2 = {isCar, isBlue} represent differing levels of generality: M0 is the most general and M2 is the most specific. However, they are all precise: M0 represents all systems, M1 represents all cars, and M2 represents all blue cars.

The concept of specification is meant first to provide the possibility of an implicit element of shared understanding, and then to support the natural tendency of maturing disciplines to make the implicit part of such understanding as large as possible without sacrificing the common understanding. Note that shared understanding between parties is not restricted to person-to-person sharing. It also includes person-to-machine (including software), machine-to-person, and machine-to-machine sharing.

It would be fair to say that a goal of the MOSES concept of model orientation is to make characterizations more explicit, while a goal of the MOSES concept of specification is to make characterizations more implicit. Is this inconsistent? Are models and specifications working at cross-purposes?

Figure 3.2 shows that they are not. The concepts of model and specification work in concert. Making information explicit in models works to drive out misunderstandings, both in characterizations that are not articulated (2) and in those that are (4). Modeling therefore aims to move characteristics from a lack of shared understanding to mutual comprehension. A predominant means of doing this is to make explicit certain things that are implicit (move from (2) to (3)). After achieving shared understanding, information that is explicit in models is ready to move into the implicit category (1). Thus, a model is always explicit; it may (3) or may not (4) be shared. A specification is always shared; it may (3) or may not (1) be explicit.

Shared understanding is the primary goal — the left column in Figure 3.2, and situations (1) and (3). If it takes a 1000-page document or model to achieve shared understanding of selected characteristics of a system, then that is what we should produce.

A secondary goal is to achieve this understanding as parsimoniously as possible, that is, situation (1) in Figure 3.2. If we can achieve shared understanding of 1000 pages of information with a

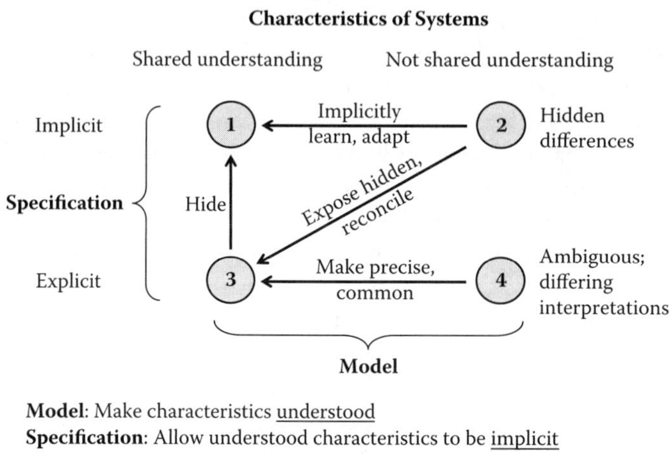

Figure 3.2 Role of model and specification concepts in MOSES.

10-page document or model, then that is what we should do.* In this case, the shared understanding (i.e., the specification) would be 1% explicit and 99% implicit. The 10-page model is not the specification; it is the explicit part of the specification.

Hidden differences in many cases can be resolved implicitly, as when two parties learn and adapt to each other's understanding without explicitly defining what it is, that is, without defining a model. This process is represented by the arrow directly from (2) to (1) in Figure 3.2. This process limits how much shared knowledge of a domain can be captured in explicit form (3) and therefore captured in the model space. Nevertheless, it is a significant part of any discipline or domain, including SE, and therefore we discuss it in a bit more detail.

Most of the knowledge that becomes implicitly shared, from (2) directly to (1), is probably tacit knowledge, which is a subset of implicit knowledge. Tacit knowledge can be used but not readily articulated and thus not readily made explicit. (1) is shared tacit knowledge, while (2) is personal tacit knowledge.

Our definition of tacit knowledge is based in part on the concept of tacit knowing in Polanyi (1958). Polanyi emphasized personal knowledge, and therefore the tacit knowledge in (2) only. We include tacit knowledge that is shared among two or more parties, none of whom can readily articulate it.† This is the type of knowledge represented in cultural, social, and organizational groups that persist over any significant period of time, and for which storytelling and narratives may be a primary means of communicating (Denning 2000). Storytelling is also gaining traction as a means of communication in organizational contexts and as an engineering method. For these reasons, tacit knowledge, and the flow from (2) directly to (1), is an important part of the repertoire of the expanded SE and MOSES, although most of it will not be captured in models. It is a part of SE that occurs outside the model space but it is reflected and supported in the MOSES concept of specification.

A community BoK consists of (3) and (1). In addition, each person or engineer in such a community has his own BoK, which consists of the sum of knowledge in (1), (2), (3), and (4), although (2) and (4) elements are outside the realm of specification. This depiction can apply to a BoK in some domain, such as SE. It can also apply to a set of information about a specific system or set of systems, such as a set requirements or an architecture. In the latter case, we would replace general or domain knowledge with individual system knowledge. In MOSES, explicit information in both cases resides in a model space: the BoK resides in general models in the knowledge space, and system information resides in specific models associated with the system, in a workspace.

Let's look back at a couple of the approaches to learning driving behavior in the first section entitled "MOSES in Action: A Story about Building an Automated Car" from this "model versus specification" viewpoint. In fact, for context, let's add in another approach, the manual case where a human driver is learning to drive. Now characterize these approaches as "human learns how to drive"; Approach 1: human defines rules for robot; Approach 2: robot learns via simulation; where Approach 1 and 2 refer to the approaches to learning in the first section. How do these map to Figure 3.2?

* Note that we are simplifying the situation here. Later we will discuss the issue of different stakeholders having different concerns, which brings in the concept of view (Chapter 10), and needing different representations and concepts, which brings in the dimension of conceptualization (Chapter 8). But the roles and goals of models and specifications described here apply in all cases.

† Gostoli (2008) proposes a model of the process by which a set of symbols, initially without meaning, acquire shared meaning among interacting agents.

- *Human learns:* Suppose Bob is teaching his son how to drive and his son has read a driving manual. At the beginning, Bob's son has very little knowledge of how to drive, while Bob has extensive knowledge but most of it is implicit. They share a small amount of explicit knowledge. Therefore, the starting state resides mostly at (2) in Figure 3.2 — implicit and not shared understanding. As the training proceeds, two things happen. First, Bob explains certain basic rules and things to watch for while driving. This aspect of driving rules is explicit and thus represents the path from (2) to (3) in Figure 3.2. Second, Bob's son implicitly learns a significant amount about driving simply by doing it, beyond what Bob says explicitly. This represents the path from (2) to (1) in the figure. At the point where Bob's son has assimilated all the necessary knowledge (from Bob's explicit teaching and from implicit experience), we are at (1) in Figure 3.2 because now Bob and his son both have implicit shared knowledge and there is no need to explicitly discuss how to drive.
- *Approach 1/human defines rules for robot:* Clearly, at the start of the process the robot has no understanding of driving and Bob's knowledge is implicit, so again the starting point is (2) in Figure 3.2. In this case, the robot cannot learn but must be explicitly told everything about driving, so there is no path from (2) to (1) in the figure. Bob must work to make all his knowledge of driving explicit in various models in the form of rules and patterns, which means everything is on the path from (2) to (3). Furthermore, the robot continues to rely on the explicit rules of driving, so nothing ends up at (1); the final state of all communication and understanding between Bob and the robot is at (3).
- *Approach 2/Robot learns via simulation*: The starting point is again (2) for the same reasons given in Approach 1. There are two sources of information for the robot. First, Bob provides a little starting information in the minimal initial model; that represents the path from (2) to (3) in Figure 3.2. But beyond that, all knowledge about driving is obtained directly by the robot. What path in the figure represents the robot learning?

To answer that question, let's assume the following scenario. Bob never looks at the learned rule base that emerged and was captured in some form from the robot's learning experience — in fact, it may not be in human readable form. But Bob puts it into the model space where other car-driving robots can access it and use it, although it is not accessible to human drivers. At this point, the BoK for experienced drivers is partly in (3) (e.g., driver manuals) and partly in (1) (i.e., learned through experience). The BoK for the car-driving robot community is all in (3) — the rule set that was learned and shared. The model space contains only the driver manual part of the human driver BoK but it contains the entire robot driver BoK.

Core Concepts II: Connection Ontology

The concepts of system, model, and correspondence defined in the previous section are building blocks for the core concepts of connection among systems* (including services) that are the subject of this section. Connection here means relation or interaction or both. The basic terms and definitions are given first, followed by a discussion of roles.

---

* System is used here in the sense of holon. That is, the relations and interactions discussed in this section can occur between systems, between a system and entities in its environment, or between parts of a system.

Terms and Concepts

For each named concept, a definition is given; and in many cases, explanatory notes and examples are also given.

- **Interaction:** Action among two or more parties.
  Note 1: The number and types of parties involved in any interaction are two or more end-point parties, one or more connector parties, and zero or more intermediary parties. (These types of parties are defined below.)
  Note 2: An interaction can occur between more than two parties at the same time. For example, one party can deliver an item for use by all parties registered for a particular service. That is why the definition above says "among" parties instead of "between" parties.
- **Relation:** An interpersonal, mathematical, logical, or causal connection among two or more parties.
  Note: Interactions often occur or are allowed to occur due to some relation among two or more parties. For example, two parties might not interact until they are within a certain distance from each other. Mathematical relations include things such as equality, order, topology, and functionality. Causal relations include such things as cause-effect, forces, and fields.
- **Party:** Entity involved in an interaction.
  Note: Usually a party plays the role of system, external entity, component within a system, or connector. It can be any type of system, such as a person, an organization, a machine, a nano-system, a software unit or program, a hybrid system, a "system of systems," or any connector type.
- **End point party:** Role played by a party that is not a connector and not an intermediary.
  Note: In a multiparty end-to-end linear process interaction, the end-point parties are the parties that anchor the two ends of the interaction, and the intermediary parties are in between. All interacting end-point and intermediary parties are connected via connector parties. (More details of topologies and protocols of multiparty interactions are given in Chapter 6.)
- **Intermediary party:** Role played by a party through which a second party and third party interact.
- **Connector party:** Role played by a party that transports stuff between other parties in an interaction.
- **Action:** Act; smallest unit of an activity or process.
- **Activity or Process:** Action or sequence of actions taken in the course of an interaction.
- **Stuff:** That which is exchanged among parties in interactions. The most common examples (these are not mutually exclusive) include goods, information, currency, symbols, people, effect, control, matter, energy, force (such as gravitational, nuclear, magnetic).
  Note: This list contains the preponderance of types of exchanged stuff. All interactions involve an exchange of stuff; if no stuff is exchanged, no interaction occurs. The latter constraint derives from the broad range defined for stuff, which includes both direct and indirect, tangible and intangible, real and virtual, and concrete and abstract types. Two notable and deliberate omissions from this sample list are service and resource. These omissions are explained in the definitions that follow.
- **Service:** (1) Activity performed by one party intended or interpreted to benefit another party; (2) Activity performed by one party intended or interpreted to benefit another party and the result or effect of the activity on the other party.

Note 1: The service activity may or may not be an interaction. In many cases it is performed separately by the service provider, not in interactive mode. The effect (as defined below) of a system or a service is in all cases involved in an interaction.
Note 2: Service is a type of process; it is not a type of stuff, and in many cases it is not part of the interaction between provider and consumer; that is why it was omitted from the examples of "stuff."
Note 3: There is a distinction between activity and result. Result is defined as the effect (see below).

- **Service provider:** Party that performs a service for another party and provides the effect to that party, directly or indirectly (e.g., through an intermediary party).
  Note: The phrase "provides a service" is often used as shorthand for the more precise but cumbersome "performs a service and provides the effect of the service."
- **Service consumer:** Party for whose benefit a service is performed and that receives the effect of that service.
  Note: The phrase "consumes a service" or "receives a service" is often used as shorthand for the more precise but cumbersome "receives the effect of a service."
- **Service initiator:** Party that initiates the request for a service to be performed.
  Note: The service initiator may also be the service consumer.
- **Disservice:** Activity taken by one party intended or interpreted to be for the detriment of another party, directly or indirectly.
  Note: An activity is decoupled from its designation as a service or a disservice. That is, service and disservice are not inherent characteristics of an activity, but rather roles or interpretations of an activity. (The examples in the next section illustrate this decoupling.)
- **Disservice provider:** Party that performs a disservice for another party and provides the effect to that party.
- **Disservice consumer:** Party for whose detriment a disservice is performed and that receives the effect of that disservice.
- **Effect:** Result, actual or interpreted, of system behavior on an environment (including service or disservice being performed). It can be:
  - An exchange of stuff between system and environment elements (e.g., tax preparer gives client/consumer filled-in tax forms)
  - A state or condition of environment (including consumer) elements, which may represent change or stasis (e.g., for a machine: proper functioning; for a human: improved, changed, degraded, or maintained physical, mental, or financial condition)
  - Access to a resource
  - Achievement of a goal of the environment element. Note: A consumer of a safety deposit box service could have either of two potential views—(1) transaction view: consumer gives provider valuables at time T1; provider gives back valuables to consumer at a later time T2; or (2) service view: consumer wants to keep valuables safe from T1 to T2; he needs someone to provide this "keeping safe" service; the provider provides this service. In this case the service view is the dominant view. It represents the fulfillment of the consumer's goal.
  - Different from the intended effect. E.g., the intended effect may be positive but the actual effect may be negative – to the detriment of the consumer rather than to his benefit.

  Note: Effect, in many cases, is an interpretation that differs by observer.

- **Intended effect:** Effect of a system (including a service or disservice activity) that is consistent with the intent.
- **Unintended effect:** Effect of a system (including a service or disservice activity) that is inconsistent with the intent.
  Note: These effects include side effects and unanticipated effects, often contrary to the intended effect. They often occur with complex systems. Typically they arise because the design did not take into account that a system changes its environment (e.g., mutual adaptation). Unintended effects can be interpreted as positive or negative.
- **Service interaction:** Providing and consuming a service is, in general, a composite activity or process. Assume that P is the provider and C is the consumer. Then two activities occur in the service interaction:
  1. A service activity performed by P for the benefit of C.
  2. An interaction between P and C in which P provides the effect of the service to C.

  Note: If the service effect is indirect, one or more intermediary parties may also be involved in the interaction.
- **System behavior:** In the context of an interaction between systems, the actual observed behavior of each system (e.g., pulling stuff in and pushing stuff out).
  Note: Behavior of a system in an interaction is visible at its ports.
- **Direct interaction:** Interaction in which the stuff exchanged is real or concrete or directly connected to system behavior.
- **Indirect interaction:** Interaction in which the stuff exchanged is virtual or abstract or indirectly connected to system behavior.
- **Cooperative interaction:** Interaction in which all parties have a common goal.
  Note: In the case of machines, which do not have goals, *per se*, they are designed to have, or behave as though they have, a common goal.
  Examples: People on an SE team with the common goal of engineering a system. The parts of virtually all Traditional Systems Engineering (TSE) systems (machines) are designed to work together, to have a common goal.
- **Competitive interaction:** Interaction in which different parties have conflicting goals.
  Examples: Players in a game in which one wins and the other loses; armies in a war.
  Note: Many interactions represent a mix of cooperation and competition. This is a hallmark of complex systems. For example, people in an organization often work cooperatively toward common organizational goals, while at the same time compete with each for recognition, promotions, resources, etc.
- **Asymmetric interaction:** Interaction in which different parties play different, typically conjugate, roles.
  Examples: client-server; publisher-subscriber; sender-receiver; service provider-consumer; leader-follower.
- **Symmetric interaction:** Interaction in which all parties play the same role. A prominent example is peer-to-peer interactions.
  Note 1: Both cooperative and competitive interactions can be symmetric.
  Note 2: Even in symmetric interactions, at a finer grained scale the interactions are often asymmetric. Example: conversation. A talks while B listens; then B talks while A listens; etc.
- **Agent:** Entity that has the capacity to continually sense its environment, make decisions based on that sensory input, and to act out those decisions in its environment without requiring control by or permission from entities with which it is associated. The hallmarks

of an agent are that it is reactive (timely response to changes in the environment), goal-oriented (not simply responding to the environment according to a predetermined script), autonomous (having its own agenda), interactive (with its environment and other entities), flexible, and adaptable. (From Krupansky, 2008.)

- **Transaction:** Interaction in which each party both gives and receives stuff.
  Examples: Party A gives currency to Party B, and receives goods that B gives, or receives information that B gives, or consumes a service that B provides.
  Note: Transaction is often a sequence of actions (i.e., a process).
- **Resource:** Party that serves as the source of stuff for another party in an interaction.
  Note 1: A service provider can be a resource. The stuff for which it is the source is not the service, but rather the effect of the service.
  Note 2: Resource is a source of stuff — it is not stuff — in an interaction. That is why it is omitted from the examples of stuff.
  Note 3: Resource is a role. A party can play the role of resource in one interaction and the role of stuff in another interaction. Example: Interaction 1: I go to the bookstore and buy a dictionary. Interaction 2: I look up the definition of resource in the dictionary.
  In Interaction 1, the bookstore and I are the interacting parties. I give the bookstore currency and the bookstore gives me goods — the dictionary. The bookstore is a resource for me as the source of the dictionary.
  In interaction 2, the dictionary and I are the interacting parties. The dictionary gives me information — the definition of resource. The dictionary is a resource for me as the source of the definition.
  The dictionary thus plays the role of stuff in the first interaction, and the role of resource in the second interaction. This does not violate the statement above that a resource is not stuff. In interaction 1, the dictionary is stuff and is not a resource. In interaction 2, the dictionary is a resource and is not stuff.

Roles

Each party plays a role in the interaction. First, two roles are intermediary and end-point parties, as defined above. A given party may be an intermediary in one interaction and an end point in another interaction. Roles of end-point or intermediary parties can include peer; client, server; publisher, subscriber; sender/giver, receiver; provider, and consumer. A sample list of roles of specifically intermediary parties includes adaptor, repeater, façade, transformer, interpreter, gateway, mediator, and proxy.

The connector party plays one role — to serve as the transport vehicle for moving stuff between end parties. Connector is discussed in more detail in Chapter 6. The connector party in this section corresponds to the assembly connector in Chapter 6.

Stuff plays different roles. The varieties of stuff are, in general, roles, not inherent types. Example: I trade you a bushel of wheat for a bushel of apples. I would probably regard the wheat as currency and the apples as goods. You would have the conjugate view: You would regard the apples as currency and the wheat as goods. In most cases, however, the parties share the same view of the stuff exchanged — money as currency, merchandise as goods, information as information.

An activity or process can also play different roles or have different interpretations. For example, a single activity can be interpreted by two different parties, or by the same party at different times, as a service and as a disservice. The following example illustrates both of these situations in an organizational setting.

At time T1, manager Mitch is prepared to fire employee Ed but employee Ellen intervenes with Mitch to save Ed's job. However, unknown to Ellen (and Mitch), Ed wanted Mitch to fire him. Later at time T2, Ed finds out about Ellen's intervention and tells Ellen that he wanted to be fired. Later at time T3, Ed is promoted and has the job he really wants. What are the interpretations of Mitch's readiness to fire Ed and Ellen's intervention at T1?

- At T1: Ed anticipates the (desired but aborted) firing action of Mitch to be a service provided by Mitch to Ed. He is disappointed when it does not happen.
- At T1: Ellen regards her intervention as a service she provides to Ed.
- At T2: When Ed finds out about Ellen's T1 intervention, Ed regards the intervention as a disservice that Ellen provided to Ed.
- At T2: When Ed tells Ellen that he wanted to be fired, Ellen then regards her T1 intervention as a disservice that she provided to Ed (agreeing with Ed on this point at T2).
- At T3: After Ed gets his promotion, both Ellen and Ed regard Ellen's T1 intervention as a service that Ellen provided to Ed.

The bottom line: the effect of an activity or interaction taken at a certain time can be interpreted differently at that time by the parties involved, and both interpretations can change over time. The same activity, which is of course historically fixed and does not change, can take on different roles in different views and at different times.

Summary

The example and core concepts presented in this chapter give an introduction to MOSES and its basic perspectives. The MOSES features described in Chapter 4 give a more complete basis for the model space and other details in the subsequent chapters.

References

David, M. 2005. The Correspondence Theory of Truth. *Stanford Encyclopedia of Philosophy*. http://plato.stanford.edu/entries/truth-correspondence/. (Accessed January 11, 2009.)

Denning, S. 2000. The Springboard: How Storytelling Ignites Action in Knowledge-Era Organizations. Boston, MA: Butterworth-Heinemann.

IEEE. 1989. IEEE Standard Glossary of Modeling and Simulation Terminology. IEEE Std 610.3-1989, IEEE, May 1989.

Krupansky, J. 2008. Definition: Agency. http://www.agtivity.com/def/agency.htm. (Accessed January 11, 2009.)

Martin, J. and T. Ferris. 2008. On the various conceptualizations of systems and their impact on the practice of systems engineering. *Proc. 18th Annu. Int. Symp. INCOSE*, The Netherlands, 15–19 June.

Polanyi, M. 1958. *Personal Knowledge: Towards a Post-Critical Philosophy*. New York: Harper & Row.

RM ODP. 1996. *Open Distributed Processing — Reference Model Part 2: Foundations*. International Standard 107462, ITUT Recommendation X.902.

Selic, B., G. Gullekson, and P. Ward. 1994. *Real-Time Object-Oriented Modeling*. New York: John Wiley & Sons.

Zeller, M. and R. Mock-Hecker. 1993. Modeling autonomous agents in a knowledge based simulation environment. *Proc. 1993 IEEE Int. Conf. on Tools with AI*, Boston, MA, November, p. 412–415.

Chapter 4

Features of MOSES

"Remember that all models are wrong; the practical question is how wrong do they have to be to not be useful."

—George E.P. Box and Norman R. Draper,
Empirical Model-Building and Response Surfaces

This chapter builds on the core concepts of Chapter 3 by describing model orientation and the modularity of MOSES. The latter discusses concepts that are decoupled in MOSES.

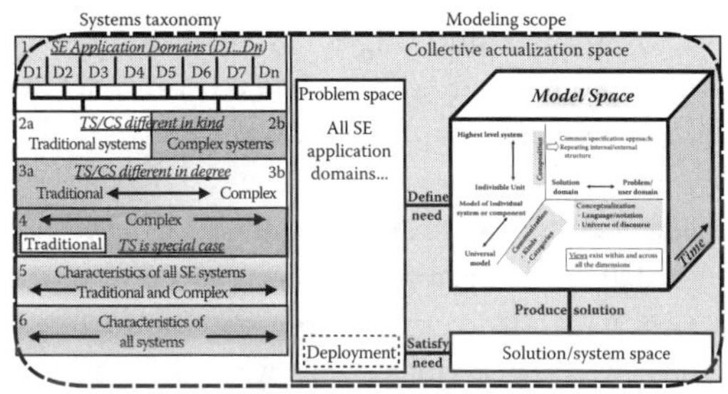

Model Orientation

We are now in a position to describe the concept of model orientation that is an important element of this book. First we review types of models now available or being used in SE, and then we discuss how model orientation in MOSES extends this to all engineering artifacts, and unifies the existing and extended set.

Basic Concepts of Model Orientation

Model orientation is the most important factor in unifying SE, and is pervasive throughout the taxonomy introduced in Chapter 2. The basic concept of model orientation is that one regards all artifacts across the engineering spectrum, as well as the scientific and mathematical foundation, as models. Models simplify and clarify our work. That has been true of traditional systems engineering and traditional science; it is even more critical for complex systems. Models not only help us focus on characteristics of interest, they also help us focus on what is common or what is different, and on what is permanent or what changes, depending on the purpose of the model. Because of the power of abstraction, the modeling approach scales up and down in concert with the vast range of system sizes.

The model-oriented method brings the traditional and complex approaches into a common framework as part of unification. The traditional CONOPS of a system and the influence of environment on a complex system are both treated in MOSES as models of the interaction of a system with its environment. The difference in the models is that a traditional model assumes the system does not adapt to the environment and does not change the environment except in a predetermined manner, while the complex model assumes that when a system enters an environment, both the system and the environment adapt to each other. MOSES assumes the latter as the general case.

A traditional requirements specification and the focus on a complex system as a whole can both be mapped to models of the externally visible behavior and properties of a system.* A traditional architecture or design and the focus on a complex system as a set of interacting parts can both be treated as models of the internal system components and their interaction. Component relations and interactions are more dominant in complex internal system models than in traditional internal system models. All these models can be defined in terms of rules, constraints, and other relations.

Clearly, simulation models fit the definition of model that we have given. What about the other artifacts listed — CONOPS, requirements, architecture, design? What is the benefit in thinking of these as models?

This is an important question. We claim that calling a document a model, while it does not change the document, does tend to change the perspective of the observer. We may have been content with the material as a narrative or graphic in a document, but when we think of it as a model, we start thinking about what variables, properties, behaviors, interactions, and constraints are involved, and which are the most important. In a word, we start thinking more precisely.

For example, suppose we have a set of "shall" statements in a requirements document. "The system shall do function *A*," "the system shall do function *B*," and the like. If this is simply a document, we tend to think in terms of whether the functions are properly stated, whether we have included all the necessary functions, etc. But if we now think of this as a model of the desired system, it naturally causes us to think more comprehensively about the system and its environment. Under what conditions do we want the system to do *A* or *B*? Do *A* and *B* interact or interfere with each other? If there is a conflict, which is to take precedence? Can *A* and *B* occur together? If so, under what conditions? Or if not, how do we make that clear? Other questions involve dependencies, ambiguity, what levels of quality and performance are needed, and whether the models are too general, too specific, or over-constraining. The number of questions and the quality of system thinking increase just because we think of the artifacts as models. We know how to create and use models, and what to expect of models.

Our view is that model orientation is the culmination of a trend that has gained momentum over the past decade. The practice of modeling and simulation has long been a significant element of TSE — for example, Sage and Armstrong (2000) and Maier and Rechtin (2000). In recent years the flavor and scope of modeling has evolved and extended beyond simulations. A number of modeling languages, standards, and methods have emerged in SE and software engineering. In software engineering, two early discussions of model orientation were Turski (1986) and Blum (1992). Wayne Wymore's (1993) book, *Model-Based Systems Engineering*, describes an approach to SE that uses discrete mathematical models for all artifacts. Wymore's approach, which has recently

* This mapping of requirements to an external model holds in some cases but not in the general case. Chapter 12 describes in some detail the more general mapping of requirements to models.

been receiving more visibility at INCOSE, could be a possible mechanism for formalizing MOSES and for exploiting MOSES concepts as a model-oriented SE (MOSE) method.

More recently, OMG defined an approach called Model Driven Architecture (MDA) (OMG 2003 and OMG 2007a) that has been instrumental in increasing the emphasis on model orientation in software and systems engineering. INCOSE's *Technical Vision 2020* (INCOSE 2006) discusses a more general approach to model-based SE. Friedenthal (2006a) discusses the benefits and requirements of model-driven SE, and Friedman (2006a) discusses the opportunities and challenges of unifying SE through modeling.

Over the past decade, the Unified Modeling Language (UML) — the latest version is specified in (OMG 2007b, c) — has brought substantial elements of model orientation to software engineering. Specifically, UML explicitly defines software design, and to some degree software requirements, in the form of models. It includes static models such as class structure diagrams, component structure diagrams, and composite structure diagrams. It also includes dynamic models such as activity diagrams, interaction sequence diagrams, state machines, and use case diagrams.

Over the past several years, OMG, in collaboration with INCOSE and ISO AP 233, has developed the Systems Modeling Language (SysML), which is an extension and customization of UML for systems engineering (OMG 2007d). Extended features for SE include blocks, with associated block definition diagrams; and requirements, with associated requirement diagrams. SysML is partially model oriented due to its derivation from UML but it has not taken model orientation appreciably further. For example, the essence of a requirement diagram is still a text requirement statement, presumably expressed in natural language.

Closely related to this is the collaboration between OMG and INCOSE on defining model-based systems engineering (MBSE). A survey of MBSE-related methodologies has been produced (Estefan 2007) and augmented (Estefan 2008). Among the methodologies in the 2008 survey are the Object-Process Methodology (OPM) (Dori 2002), Embedded Computer System Analysis and Modeling (ECSAM) (Lavi and Kudish 2005), and Model-Based (Systems) Architecting and Software Engineering (MBASE) (Boehm and Port 1999; Koolmanojwong and Boehm 2007).

Some SE management literature has defined a more engineering and modeling approach to management. For example, Garvey (2000) describes probabilistic cost models for systems engineering. More generally, the closely related field of Operations Research (OR) has been strongly model based for decades. OR can be expected to contribute to model orientation in SE.

A trend in federal acquisition practice is to emphasize model-based acquisition processes (CIO 2008). Models are encouraged as a way to define more precise requirements and architectures and to facilitate the use of patterns, test harnesses, and automation in general.

Taken together, the foregoing examples show that model orientation is being extended from traditional simulation models to a more pervasive concept of models. In MOSES, the full expression of model orientation considers all the key artifacts of engineering and the theories, laws, and observations of science to be models of systems. This perspective aligns systems engineering with related disciplines, based on our belief that models are the naturally dominant perspective, repository of knowledge, and vehicle of communication in all engineering and science disciplines.

In systems engineering, many qualitatively different things are modeled:

- Modeling systems:
  - Whole systems, and interacting parts of systems
  - The system's environment and the interaction of the system with that environment
  - Structure, behavior, entities, interactions, and relations

- Properties, including traditional ones such as performance or quality, as well as fuzzy, and stochastic ones
- Attributes, qualities, categories
- Operations or processes
- System change and evolution
- Constraints, including what is allowed, what is mandated, what is prohibited
- What is common and what is unique
- Views
- Information about the models themselves

- Modeling the engineering or actualization process:
  - Life span phases (traditionally called life ‘cycle’ phases)
  - Process types (e.g., waterfall, spiral, evolutionary, etc.)
  - Environment change to influence an adaptive system.
- Modeling management of SE projects, and of enterprises:
  - Models of staffing, cost, resources, schedules, etc.
  - Decision models

These areas are addressed in the chapters that follow.

Types of Models in the Extended SE

Generic Types

Table 4.1 identifies and briefly defines various types of models that have relevance to SE. Many of the definitions are modifications of those given in Kulakowski et al. (2007).

Models in engineering and science range from analytic models that represent exact closed form solutions, to simulation and other informal models that represent multivariable but still organized solutions, to probabilistic and other uncertainty models that represent explicitly non-deterministic solutions. This view of Table 4.1 corresponds to Weaver’s (1948) one-dimensional range of complexity from organized simplicity to organized complexity to disorganized complexity.

Trends toward Complex System (CS) Models

There is debate in the SE community about whether traditional SE addresses complex systems. In one camp are those who say that SE does address CS. It has always been multidisciplinary, has always dealt with the most complex systems that humans have built, and has always addressed human factors in man-machine interfaces, as well as political and economic issues in the funding and management of SE projects. In the other camp are those who say that traditional SE has treated systems as linear, deterministic, reductionist, analytic, and mechanistic, and that such treatment deals with complicated systems but not CS, where complex means non-linear, non-deterministic, autonomous, emergent, adaptive, self-organizing, and continually changing.

MOSES is based on a view that accepts the basic claims of the first camp but also accepts a fundamental CS extension of SE beyond TSE. We characterize and summarize the difference between TSE and the extension as mechanistic models versus organic or autonomous models. In this book we use the terms “mechanistic” and “TSE” somewhat interchangeably, and the terms “organic” and “CS” or “CS engineering” (CSE) somewhat interchangeably.

Table 4.1 Types of Models

Variable	*Type*	*Brief Definition, Example*
Linearity	Linear	Involves a single dimension, or all variables are of first order, or output is proportional to input; or elements are additive. Ex: F(x) = ax + b; linear differential equation
	Non-linear	Not linear; at least one variable is second or higher order; output is not proportional to input; or elements are not additive. Ex: F(x) = ax^2 + b; nonlinear differential equation.
Distribution	Lumped or centralized	Dependent variables are independent of spatial coordinates Ex: ordinary differential equations
	Distributed	Dependent variables are functions of spatial coordinates. Ex: partial differential equations; network model
Variability	Static or time invariant	Model parameters are constant across time. Ex: differential equations with constant coefficients
	Dynamic or time varying	Model parameters vary across time. Ex: differential equations with time-varying coefficients; simulations
Continuity	Discrete	Dependent variables defined only for distinct values of independent variables. Ex: difference equations
	Continuous	Dependent variables defined over continuous range of independent variables. Ex: differential equations
Determinism	Deterministic	Outputs are known, certain, predictable given inputs. Ex: equation with fixed variables; deterministic cellular automaton model
	Stochastic or probabilistic	Outputs have a degree of uncertainty given inputs. Ex: equation with random variables; probabilistic cellular automaton model
Analysis method	Quantitative	Uses and produces numbers; based on interval and ratio measurement scales. Ex: differential equation
	Qualitative	Uses and produces descriptions, categories, comparisons; based on nominal and ordinal measurement scales Ex: taxonomy
Executability	Non-executable	Model does not imply a process; or no agent exists that can execute it or translate it into executable model Ex: requirements document; design document
	Executable	Model implies a process and an agent exists that can execute it or translate it into executable model. Ex: simulation; executable architecture; UML design model for which code generator exists; requirements written in Prolog

(*Source:* Kulakowski, B.T., J.F. Gardner, and J.L. Shearer. *Dynamic Modeling and Control of Engineering Systems, 3rd ed.* New York: Cambridge University Press. 2007. Modified with permission.)

We presented the contrasting system characteristics of TSE and CS in Chapter 1. CS models have been around as long as TSE and software engineering but they have remained on substantially separate paths and in different communities — especially separate from the TSE community. The distinction between the perspectives of TS models and CS models is analogous to the distinction between the perspectives of Newtonian physics and quantum physics. CS models also represent an important step toward unification of multiple disciplines.

The study of CS has been explicit in its use of models, including mathematical models such as differential equations, difference equations, cellular automata, game theory, Markov processes, genetic algorithms, graph theory, and time series analysis (CSCS 2005). Work in dynamical systems (e.g., Hasselblatt and Katok 2003; Kulakowski et al. 2007; Martelli 1999) represents a bridge between TS and CS, in part because it includes models of chaos, and chaos is usually considered part of complex systems theory although it is deterministic.

The foregoing models have been used to represent system behavior and change in multiple disciplines, including biology, psychology, sociology, organizational studies, economics, and meteorology. Doyle (2006) identified the dominance of network models across multiple complex system domains, including networks in *technology* (Internet, wireless, energy, transportation, nanotechnology); *biology* (genomics, bioinformatics, medicine, epidemiology, ecology); *physics* (multiscale, turbulence, chaos, fractals, criticality, self-similarity), *social sciences* (economics, sociology, politics, management, law, small worlds); and *mathematics* (graph, information, control, dynamical systems, formal methods).

We noted in Chapter 1 that TSE regards the human as being outside the system. The IEEE 1220 standard (IEEE 2005) illustrates this view: in the data package section on humans, it "documents cognitive, physical, and sensory characteristics of humans who operate, maintain, and support the system…" The CS extension of system boundaries to include humans inside the system is a significant step forward in SE because it requires a more comprehensive understanding and capture of human behavior in models. Some work in dynamical systems has arguably included humans but they have been rather abstract with regard to human behavior. The arena of CS can bring in social and behavioral models to improve the fidelity of dynamical models of human interaction.

Other examples are the inclusion of evolution as a model of continuing change and the use of probabilistic models, as described in Chapters 9 and 11, respectively. A given class of CS models corresponds to systems in many domains or disciplines. Models of natural systems, such as biological evolution of populations, are being applied to artificial systems such as artificial life (Langton, 1995; especially Ray 1995). Models of artificial systems, such as information systems, are being applied to natural systems, such as living systems from cells to humans to societies (Miller 1995). Another example of the latter is the Physiome project, to which the April 2006 issue of the *Proceedings of the IEEE* was devoted (Physiome 2006). This project is an example of using a modeling approach to integrate different areas of knowledge at an extended range of space and time scales. Physics, chemistry, mathematics, and engineering knowledge are helping design biological and physiological models. Modeling is explicitly seen as a major integrating tool.

Similarly, Dayan and Williams (2006) also emphasized the use of — and relationship between — computational modeling and mathematical modeling in neuroscience.

We would make a key distinction regarding complex systems modeling. Modeling has been ubiquitous in the *science* of complex systems over the past several decades. Many of these models, for example, are described in Boccara (2004). The *engineering* of complex systems, however, is a relatively new area, and it is here that some further modeling research could be useful. One early collection of results in engineering complex systems is Braha et al. (2006).

Illustrations of Modeling Choice

In our definition of system, we noted that systems do not exist as systems *per se* in the world. A mosaic exists, and a system is all or part of the mosaic that an observer chooses to designate as a system.

In model orientation, choice in modeling is related to choice in systems but choice plays an even more prominent role in modeling. The concept of choice in the definition and use of models is sufficiently important to discuss it in detail. In any given situation, we typically have many choices in how we model. There is not a "right" or "wrong" way to model, or a "right" or "wrong" model. The question to ask is not "What is the model of this system?" The question to ask is "What do we want to achieve, and what is a reasonable model (or set of models) to help achieve it?" There are alternative ways to model, and some may have greater benefit than others for a given purpose, and some may cost more than others. So, cost/benefit ratio is one factor in selecting among the modeling choices.

Another factor in modeling choice is the prior knowledge, personal experience, views, perceptions, and biases that a modeler brings to a modeling situation. In the general case, Warfield (2006) calls this the cognitive burden that each person brings to problem situations.

We illustrate modeling choice with the following examples.

In physics, Newton's concept of gravitation is that it exerts a force that causes a body to change direction in linear space, while Einstein's concept of gravitation is that matter curves spacetime and a body's momentum simply takes it along the contours of the curves. The distinction may be summarized this way (Pogge 2006):

- Newtonian gravity:
  - Matter tells gravitation how to exert a force.
  - A force tells matter how to accelerate.
- General Relativity:
  - Matter tells spacetime how to curve.
  - Curved spacetime tells matter how to move.

These are two very different, even radically different, conceptualizations. It is tempting to say, based on experimental evidence, that Einstein's theories of special and general relativity are "true" and Newton's corresponding theories are "false." Yet we are still teaching Newtonian physics along with special and general relativity — and quantum physics, which in some respects replaced them both — in college and university physics courses. Why? Because the depiction of relativity as true and Newtonian theory as false is incomplete, misleading, and unhelpful. A more practical characterization is that in many circumstances, the relativity and Newtonian models yield basically the same results, while in some circumstances, the relativity model yields better results, that is, has noticeably greater fidelity to the physical universe (the system) than does the Newtonian model. More specifically:

- For common circumstances where weak gravitational fields and speeds much slower than the speed of light obtain, the two models give results that are indistinguishable.
- For circumstances in which gravitational fields are strong or speeds approach the speed of light, the relativity model gives significantly more accurate results.
- The mathematics of the Newtonian model are notably simpler than the mathematics of relativity.

These characteristics of the two models suggest a choice. Two generally desirable goals of modeling are fidelity and simplicity. For the large number of applications in which the two models give indistinguishable results, we would choose the Newtonian model because it is simpler. In cases where the relativity model gives greater accuracy, and we need greater accuracy, we would select it; in this case, our "choice" would be highly restricted. An example of the latter case is the Global Positioning System (GPS) navigation system. To determine the desired accuracy of location on the Earth's surface, extreme accuracy is required in clock synchronization between satellite and ground sites. Clock adjustments are needed from both relativity models — the special theory and general theory — to achieve this. GPS precision would be difficult to achieve without both theories.

Part of the choice in modeling is deciding how to conceptualize a situation. This is illustrated in the radical conceptual differences between the Newtonian, relativistic, and quantum models of gravitation, although in many cases they give similar results. It is also illustrated in the differences between biological, psychological, and social models of humans. Another example is the duality of light as wave and as particle. They represent different ways of modeling light.

The two examples of humans and light also represent another aspect of choice — choosing multiple models. Sometimes we choose one model instead of another, but often we choose multiple models because no single model does everything we need. Different models are useful for certain (different) purposes. In the light example, the wave and particle models are complementary models, both of which are necessary for a complete understanding of light (and actually all objects, as quantum mechanics tells us). In the human example, the biological, psychological, and social models are complementary models, all of which are necessary to fully account for human characteristics. Not every human modeling situation requires all three model types; in many cases we may only need one or two of them.

In MOSES, such complementary models or dualities are called *views*. There is a wave view and a particle view of light. There is a biological view, a psychological view, and a social view of humans. Views including dualities are an important part of MOSES and are discussed in more detail in Chapter 10.

The following example illustrates how different models result from varying purpose and focus. Suppose water flows from a spring to lake L1, then through a channel to lake L2. How would we model this? Here are some possible purposes and corresponding models:

1. We are interested in comparing the two lakes and how the water flows between them. Our model is something like this: we define our system of interest (SOI) to consist of the two lakes and the channel. More specifically, L1 and L2 are components, they interact via a connector, which is the channel; the stuff that is exchanged is water; and the flow of stuff is always one-directional, from L1 to L2. This is an internal or "system as parts" model.
2. We are interested in building, or rebuilding, the channel. The channel is our system/SOI; the environment includes water that will be transported by the system; the land/terrain across which we will build the system; L1, which is the supplier to the system (connected to its input port); and L2, which is the consumer of the system (connected to its output port).
3. Suppose the channel is a natural river and we are interested in the properties and dynamics of water flow through the river. The system/SOI is the water, and the environment is the river bed, especially its shape and contours.
4. Suppose we find possible pollutants in the water coming out of L1, and these pollutants are not in the water coming into L1 from the spring. Our system/SOI is L1. We want to investigate two things. First, what exactly is the difference between the (clean) input and the (polluted) output, that is, which pollutants, how much, etc.? Second, what are the characteristics and flows internal to L1? For example, is there a relatively smooth gradient from "clean" on the input/spring side to "polluted" on the output/channel side? Or is there a localized spike in pollution at one or more points in the lake? Our first question leads to

an external input-output model of L1, and our second question leads to an internal model of L1 defined to analyze the gradient.

5. Suppose that the answers to the modeling and analysis in (4) above are as follows: First, we identify the pollutants coming out of the lake, and they include a highly toxic substance, which increases the importance and urgency of finding an answer and solving this problem. Second, results of the internal analysis of L1 are inconclusive. We have not isolated where or how the pollutants are appearing. These results indicate that we need to define a different, more detailed model, whose purpose is twofold:
    a. Safety issues: What is the current or potential damage from the polluted water coming out of L1, and who or what are the vulnerable parties?
    b. Determining the cause: What is the source or cause of the pollution? This will point us toward a remedy.

   To investigate in more detail, we define a model that includes elements that we ignored in our previous model because we considered them to be negligible. Mostly these additional elements relate to expanding the environment as it is connected to input and output ports of L1. Addressing purpose 1, safety, is associated more with output ports. These include: (a) Water evaporates into the air above L1; (b) Water may seep into the ground and water table below L1; (c) People swimming in the L1 contact the pollutant on their skin and may get the pollutant in their eyes, or drink the polluted water; (d) People or predators eating fish from the lake (assuming there are any) may ingest the pollutant from the fish. We need to examine all these possibilities, and perhaps first take preventive measures, such as temporarily banning swimming and fishing in L1.

   Addressing purpose 2, cause, is associated more with input ports. These include: (a) Water precipitates from the air above L1 – perhaps acid rain is a source of the pollutant; (b) Pollutants may be seeping out of the ground under L1 into the lake water; (c) People may be dumping pollutants into L1; (d) Maybe there is another unknown source, like a nearby factory, that is producing runoff that flows into L1; (e) Maybe the pollutant is emerging from a reaction between a non-polluting substance that is entering the lake and some other non-polluting substance in the lake.

An example of a rather stark modeling choice is the contrast between the complex systems models of self-organized criticality (SOC) and highly optimized tolerance (HOT). Carlson and Doyle (2002) present a comparison of the nearly opposite assumptions and features of these models. More discussion of the SOC and HOT models from a MOSES perspective is presented in the context of views in Chapter 10.

All these cases illustrate how modeling involves choices and decisions; there is not an inherent way to model. Our choices are influenced by our purposes for modeling and by constraints existing in the situation, and sometimes by our interpretation of a situation.

Model Space

The model space is a structure for containing and organizing all models — not only systems engineering and enterprise engineering artifacts but the models of all supporting sciences and mathematics. Earlier forms of this model space applied to software engineering were presented in Hybertson (2001 and 2005) and a summary of the extended form for systems engineering in Hybertson (2006). Precursors of the model space for traditional engineering disciplines exist in engineering handbooks, such as *Perry's Chemical Engineers' Handbook* (Green and Perry 2007). Other forms of partial model spaces include published guides to bodies of knowledge (BoKs), patterns books, reference architectures, and standards. In the future, we expect online hyperlinked virtual repositories of models to be the norm, and they are the pattern for the envisioned SE model space. In the case of SE, a considerable amount of content exists, although not necessarily in a form that is ready for the model space.

Our intent in this book is to provide a structure for the repository or model space, and to point to some of the content for these traditional practices, as well as new ones. The model space is structured by five primary elements that separate SE concerns. The elements are

1. A *composition* spectrum that represents a whole–part hierarchy ranging from the largest system to the smallest unit. It exhibits a repeating self-similar pattern in that a given whole is part of a larger whole.
2. A *conceptualization* spectrum that ranges from problem domain languages to engineering domain languages, including mathematics. This highlights the complicating fact that SE, and therefore the model space, is fundamentally multilingual.
3. A *commonization* spectrum that represents "kind-of" and "is-a" hierarchies ranging from universal models to instance models. It exhibits a repeating self-similar pattern in that a model that is a generalization can, in turn, be generalized further.
4. A *time* spectrum that enables the modeling of system change and the comparison of models at different points in time.
5. A concept of *view* to define certain basic perspectives on models and systems; a view can be a model, any subset of a model, or information that cuts across multiple models or model space spectrums. The model space defines a starter set of generally applicable views and viewpoints.

The model space, and the motivation for the elements, are described in more detail in Chapter 5. Much of the material in the remaining chapters is an extended description of the elements of the model space.

Limits of Modeling

It is important to stress that we do not consider models or model orientation a panacea. Models can be misused like every other tool in the arsenal of SE. A few examples serve to illustrate the limits and misuse of models.

The Soviets used models extensively for their 5-year plans but the results did not seem to match their predictions. We can reasonably assume that the models were biased by political considerations, and that the models did not have good fidelity to the economic and social characteristics of the situation they purportedly modeled. More specifically, Holloway (1976) described the dominance of the cybernetic model of government in the Soviet Union in the 1960s and 1970s. The application of cybernetics by the Soviets, however, emphasized hierarchical control and a clear separation of two subsystems: the controlling and the controlled. The Soviet concepts of scientific management of society and optimized planning were explicitly tied to the earlier Taylor school of scientific management. Their emphasis on external control and society as mechanistic, rather than self-control and society as organic, can be seen as a significant contributor to the weakness of their models.

The book *Useless Arithmetic* (Pilkey and Pilkey-Jarvis 2007) gives many examples of the misuse of models, such as models of animal populations to justify continued over-fishing and other activities. They point to too much use of quantitative models of poor fidelity and too little use of qualitative models. For example, they discuss (p. 22–24) the U.S. Government's misuse of quantitative modeling in the Viet Nam War. One of the most prominent models was the use of an enemy body count as a measure of success. This had several deficiencies, to include:

- *Political objectives biased the numbers.* It was important to give the impression that the war was being won.
- *Career objectives biased the numbers.* Higher numbers meant commendations and promotions for individual combatants.
- *Determining which bodies were the enemy was difficult.* That factor, combined with the pressure to increase the count, resulted in a trend toward counting all dead Vietnamese bodies as the enemy.
- *No one verified the accuracy of the numbers or the model.*
- *The model used the wrong variable.* Body count in that situation was not a good measure of success.

The last item trumps all the others. If you select variables for a model that do not fit its purpose, then whether the variables are measured correctly does not matter much. On the other hand, if the purpose of this model was to increase domestic support for the war, then it was perhaps suitable for its purpose, at least for a while.

In a discussion of complexity science, Phelan (2001) cautions against the use of "resemblance thinking" and the use of analogies, calling this a practice of pseudo-science rather than science. For our purposes, this can apply as a limit or caution on applying models that were derived from one domain to a completely different domain — which, of course, is an important aspect of CS models. The caution here is that we need to be careful to have sufficient verification of the correspondence of a model derived from one domain with systems in another domain. MOSES provides a structure in which to calibrate and position the correspondence: the commonization dimension of the model space (see Chapter 7). That is, this structure allows us to separate what is common from what is unique. Applying a model from one domain to another may have some things in common and not others — it is not all or nothing. An example is evolution; not everything about biological evolution applies to other domains in which evolution may be applied. We have to abstract or generalize the biological evolutionary model (variation generator, selector) to reach a point on the commonization hierarchy that is appropriate for other domains.

The article "War Gaming in the Information Age:…" (Bracken and Shubik 2001) discusses trends in war gaming since the advent of computer modeling. The gamers and the computer analysts have pushed for more and more details and "realism" because computers gave them the capability to do so. One could argue that the models have increased fidelity to the war situation. However, fidelity is not the only criterion. Fitness for purpose is also critical. In this case, the older-style war games, such as board games, had to be simpler and therefore forced attention to the key variables. The result was that they supported intuitive decision making. But when many more variables are added, the key ones become obscured and the purpose of serving military decision making has been lost. The authors argue that military leaders need to apply to the war gaming field the same pattern that business leaders have applied to corporate IT systems. That is, they need to take control of the war gaming technology and make it serve the decision-making process.

Sometimes we actually need models of *less* fidelity to a system to be more useful. Consider the historical use of birds as a system we attempted to model in building flying machines. It was reasonable — even obvious — to use models of birds to design flying machines. But we modeled it too closely, with the wing providing two functions, floating (maintaining altitude), and power (by flapping), in accordance with the bird. We were not entirely successful in building systems to this model.

To achieve success, we had to define a model with less fidelity to the bird. It took a significant insight to back off the bird model and realize that we could separate the surface (floating) function from the power function, and could retain the floating function for the wing, but could allocate the power function to a different part of the system. The lesson is that a good model is one that recognizes what characteristics to retain and what characteristics to ignore (leave unspecified) for the purpose it has. Sometimes, over-specifying is more detrimental to envisioning a solution than is under-specifying. The model of the flapping-wing flying machine was over-specified; it was not sufficiently minimal.

As a last example of limits, models can be fuzzy, approximate, and ambiguous, but still be useful. Two well-known illustrations of this type of limits in science are the periodic table of chemical elements and the biological taxonomy of living things. Both are useful but both have significant exceptions and ambiguities that engender debates about where a specific entity should reside. The more closely one examines them, the messier they become.

The key point of this discussion is that models need to correspond to the systems they are intended to model, in a way that serves the purpose of the model. How do we validate this correspondence? Moss (2008) discusses differing validation approaches, ranging from formal approaches to narrative scenarios, for models across the spectrum from theory-driven models to evidence-driven models.

Modeling in MOSES is intended to emphasize abstraction, focus, and purpose. Fidelity is important but there is a trade-off — a point of diminishing returns beyond which increasing details and number of variables begin to mask the important variables. This is a delicate balance for CS because small changes can lead to huge differences. Second, models can be wrong in the sense that they do not correspond to the intended systems; it is not easy to define a model of high fidelity. A model is good for its purpose if it ascertains and distinguishes between what is most important for that purpose and what is less important. An investment is required to mature models to where they can be useful for certain purposes.

The point of this section is that models, like all engineering tools, require care and healthy skepticism in their use. Models should be defined transparently, with explicit assumptions, so that others can assess their fidelity and fitness for purpose. The maturity and degree of validation of models should also be made clear because less trust should be placed in immature or un-validated models.

Modularity of MOSES: Decoupled Concepts

Modularity or separation of concerns is a traditional and well-known principle. MOSES applies it to systems in an unremarkable way that, from a traditional viewpoint, does not need discussion. However, MOSES also applies the modularity principle to itself, and that creates the need to discuss the issue of decoupling concepts that are often highly coupled in SE — differently in the traditional camp and in the complex camp.

One of the key drivers of MOSES is the need to capture important concepts and principles that apply to the broad new spectrum of SE. Many of these core concepts have built up an attachment with other concepts until the multiple concepts, in the minds of many in the SE or CS community, are fused into a single concept. In some cases, this fusion has tempted the CS community to throw out these multiple concepts or assign them only to TSE, not CSE, which in our view is inappropriate and is the result of this unfortunate fusion. To achieve the goal of capturing and using applicable principles, we find it necessary to decouple or "unfuse" the concepts in question, to recover the simpler core concepts.

We are not trying to remove the coupled concepts; rather, we are trying to remove the coupling and allow the individual concepts to exist separately on their own merits. In many situations the concepts do belong together — even reconciled in a common framework. But connecting them should be a separate decision, not an *a priori* fixed condition of their existence.

We make a point of spelling out the modularity of MOSES and some of its manifestations because it forms an important part of the core world view of MOSES. Some might call it extreme modularity but we view it as natural and necessary for the new expanded SE.

The following characteristics are decoupled in MOSES:

1. Modularity and reductionism
2. Control and hierarchy
3. Model and system and modality
4. Model derivation and application
5. Specified behavior and valuation
6. Usefulness and "truth"

We will describe each of them in turn.

Modularity and Reductionism

We need to tackle this issue first because we claim that MOSES is modular. Some in the CS community argue that modularity and separation of concerns reflect a discredited, reductionist way of thinking, and that we need to focus on holistic thinking to address CS.* If they are right, there is no need for MOSES to exist. Russell Ackoff (Ackoff 1979) was one of the first to make this argument in the 1970s when he disparaged the direction of OR. His argument, which has carried forward to some degree to the present time, can be summarized this way: The traditional way of thinking is machine age and has this orientation: analysis, parts, bottom up, only causal relations, determinism, reductionism, knowledge, and machines without purpose. The new way of thinking is systems age and has this orientation: synthesis, wholes, top down, emergence, multiple relations, nondeterminism, understanding, and purposeful systems using machines. Synthesis should be used instead of analysis, wholes instead of parts, top down instead of bottom up.

The MOSES response to this is that we need both perspectives. A system is of interest both as a whole and as parts. The best process is an opportunistic combination of top down and bottom up. Modularity is still an important principle in the context of systems thinking and holistic thinking, and is separated from reductionism. Modular design of a system does not imply a reductionist disposition; modularity is orthogonal to the issue of reductionism versus emergence. In fact, MOSES embraces both modularity and emergence.

* For the purposes of this section, a reductionist view focuses on understanding a system by analyzing its parts, while holism focuses on viewing the system as a single whole entity. Emergence is the view that the whole cannot be adequately understood by analyzing its parts. Emergence is discussed further in Chapter 6.

Control and Hierarchy

"Hierarchy" is often coupled with top-down control, which reflects the origins of the term in religious and military structures. In MOSES, we separate control and start with a minimal concept of hierarchy that consists of a general mathematical relation. Following is a description of several types of hierarchies that differ by the relations involved:

1. Basic minimal definition of *hierarchy*: a structure generated by a *repeating one-to-many relation*—a tree structure. This basic definition implies that the relation is asymmetric and non-reflexive, but it is not restricted to directional versus nondirectional, transitive versus intransitive, or binary versus N-ary. Those constraints may appear in specializations of the relation, some of which are given below.
2. If the relation is that many are *part-of* one, then it is a whole–part or composition (or aggregation) hierarchy. This relation generates the well-known and ubiquitous whole–part repeating structure. The study or theory of wholes and parts is mereology. The part-of relation is transitive: If A is part-of B, and B is part-of C, then A is part-of C. Direction and navigation may be added to a composition hierarchy: top-down decomposition implies proceeding from top (or root node) to bottom (or leaf nodes). Bottom-up composition implies proceeding from bottom (or leaf nodes) to top (or root node). This directional or sequence constraint may be imposed on the basic composition hierarchy, which itself is not directional. As we discussed in Chapter 2, Koestler (1967) proposed the term "holon," which is both a whole and a part (of a larger whole). In this sense, the holon concept generates a composition hierarchy or holarchy. We agree with the concept of holon in the sense of allowing us to consider the same entity as a whole and as a part. We interpret whole and part as separable views or roles. We should be able in one context or in one view to consider an entity as a whole, ignoring it as a part; and in another context or view to consider the same entity as a part, ignoring it as a whole. That approach supports problem solving and engineering better than does conflating the two views into one inseparable concept.
3. If it is composition and the many parts at each level are *interrelated*, then it is a *system* composition hierarchy. This is a dominant concept in systems engineering — most famously in traditional SE but also in complex SE. An important discussion of this is Herb Simon's 1962 "Architecture of Complexity" article, published in Simon (1962) and Simon (1996).
4. If the relation is *control* of many by one, then it is a control hierarchy. This relation is exemplified by the relations implied in a traditional organization chart, and most strongly by a military organization/command structure.
5. If the relation is that many are subtypes of one *type*, then it is a specialization (or generalization) hierarchy. This relation is transitive: If A is a specialization of B, and B is a specialization of C, then A is a specialization of C. The relation is exemplified by a class-subclass hierarchy in object-oriented software engineering, by a type-subtype hierarchy, and by the specialization hierarchy in the biology taxonomy.
6. If the relation is that many are instances of one *category*, then it is a categorization hierarchy. This relation is not transitive: if A is an instance of category B, and B is an instance of category C, it does not follow that A is an instance of category C. This relation is exemplified by a class-object or class-instance relation in object-oriented software engineering, by a model-metamodel relation, and by the categorization hierarchy in the biology taxonomy.

These various types of hierarchies all exist in the MOSES model space and are discussed in later chapters. But the main point of this decoupling is that when we say hierarchy without further qualification in MOSES, we mean only a repeating one-to-many relation, and we do not otherwise constrain what that relation might be. In more specific contexts, we add appropriate constraints to the relation.

Model and System and Modality

Introduction

This section considers the decoupling and integration of the concepts of model, system, modality, and world. Discussed first is the separation of model and system, followed by the separate concept of modality. The integration of the concepts is then discussed in terms of a correspondence set. Finally, complex systems variations of modality and correspondence sets are briefly discussed.

Model and System Separation

Models and systems have separate existences. Each model exists in its own right, and each system exists in its own right.* Models have relations with other models, models have relations with systems, and systems have relations with other systems, but we can always view a model or a system in its own separate existence. For any given model or system, this gives us the freedom to establish, discover, define, consider, or ignore relations with any other model or system.

This decoupling removes built-in dependencies between models and systems. Models can be defined for which no corresponding systems exist, or for which many systems exist. We place no constraint on which came first — the model or the system. Or the sequence may be systems exist, models are created, and then more systems are created that correspond to the model.

In mathematics, mathematical models have been defined before or after applications for the models are found and linked. That is, the model may exist first, or the corresponding system may exist first. An example of a model existing first is Riemannian geometry. It was defined in the nineteenth century. In the early twentieth century, Einstein discovered that Riemannian geometry provided a mathematical model he needed for spacetime curvature and the general theory of relativity (Schubin 1997).

Chapter 5 introduces the basic model types, external and internal, which represent external and internal perspectives of a system, and which are referred to throughout the book. However, a more detailed view of an external model is that it includes a set of interfaces or ports that can be defined separately and then used in one or more external models. Therefore, in this detailed view, interfaces and external models are also decoupled. This is mentioned here because it is needed for the detailed treatment of specifications in Chapter 11.

Modality and Its Separation

Modality is concerned with the mode of propositions such as existence, contingency, necessity, possibility, impossibility, or temporal mode such as formerly or eventually. It is reflected

* Note that existence here refers to the definitions of system and model in Chapter 3: it is not *a priori* existence, but a choice or definition made by an observer, that these entities constitute a system, and those entities constitute a model.

in engineering specifications or models in the distinction between terms such as shall, should, must, is, will, may, is permitted, is prohibited, etc. MOSES decouples modality from model and system, to allow models and systems to be associated in different contexts with different modes. We discuss the separation here in terms of descriptive versus prescriptive modes, but it applies to any mode. Modes are an open-ended set. Many modes come from modal logic and its variations, including deontic logic and temporal logic. We discuss some possible added modes for complex systems later in this section.

An example that illustrates why MOSES decouples modality from models is from Gigerenzer and Selten (2001, 1):

> "... Over the last centuries, models of rationality have changed when they conflicted with actual behavior, yet, at the same time, they provided prescriptions for behavior. This double role — to describe and to prescribe — does not map easily onto a sharp divide between descriptive and normative models..."

The problem is the sharp divide, the view that a model is inherently prescriptive or inherently descriptive. In MOSES, a model at a given point in time may simultaneously be considered descriptive of one system ("This model describes certain characteristics that system A has") and prescriptive of the same system or of another system ("This model prescribes certain characteristics that System B shall have"). The attributions of modes such as "prescriptive" or "descriptive" are not only relative to what system is being related to the model, but also to some extent a matter of choice. The model itself does not have an inherent property of being prescriptive or descriptive; modes are not in the model's field of view. The model is simply a set of defined rules, such as: characteristic C1 = X; C2 < Y; C3 > Z. Bringing together this model with a system and certain mode is a separate activity that is discussed below under the topic of correspondence set.

In general, modality types include fuzzy and probabilistic modes. In MOSES, the generally preferred way to address uncertainty and fuzziness is in the specification (addressed in Chapter 11). However, there are exceptions. One exception could be if the systems engineer wanted to use the same model to associate with different systems, and wanted to allow different probabilities for the different systems. Instead of defining a different model for each case, varying only the probabilities, the engineer might define a single deterministic model M1 (i.e., probability of behavior = 1.0), and then put the varying probabilities in the modalities (system S1 shall with $P \geq .999$ satisfy model M1; S2 shall with $P \geq .99$ satisfy M1...).

The separation of modality means that the most salient aspect of a requirements document or model is not the "shall" form of the statements; it is rather the actual rules. The whole set of rules in a model M can be made prescriptive by saying "System S shall satisfy M," or made descriptive by saying "System S does satisfy M." In fact, in the MOSES approach the term "shall" (or any other mode) should be removed from requirements documents/models and applied separately, using the correspondence set approach described next.

Integration via Correspondence Set

The vehicle for bringing together models, systems, and modes is called a correspondence set. Its form is {M, MP, W, S}, where

- M = a specific model or set of models or * model, where * means all or left open
- MP = modal phrase or modality = a specific mode or set/concatenation of modes

- W = a specific world or a set of worlds or * world
- S = a specific system or set of systems or * system (or service*)

A correspondence set represents a commitment to a model-mode-system association in a defined world. The model and the system do not need to exist to be included in a correspondence set. The */undefined construct makes an open commitment. If system is left open (S = *), then the correspondence set depicts all systems in the defined world that correspond to the defined model with the defined modality. Likewise, if model is left open (M = *), then the correspondence set depicts all models in the defined world that correspond to the defined system with the defined modality.

When we want to establish a correspondence relation between a model and a system, we can do so — in any of the following ways: (1) by creating a model and creating a system that corresponds to it, (2) by creating a model to correspond to an existing system, (3) by creating a system to correspond to a preexisting model, or (4) by finding a correspondence between an existing model and an existing system.

The evolution of a correspondence set can be illustrated with a very common SE example. Suppose an airline determines at a point in time that they need to buy about 20 new commercial aircraft, and after writing requirements and doing trade studies, they select the Boeing 777. Boeing builds them the planes and the airline puts them into operation. We can define correspondence sets for four points in the course of this process.

- Time T1:
  - *Situation:* The airline decides they need to buy some commercial aircraft, which they designate as system S1, but they do not know what specific aircraft S1 will be. They plan to write a requirements document called R1.
  - *Correspondence set 1:* {M = R1 (identified but not defined); MP = shall when operational; W = W1 = world in which S1 is acquired and operated (sketched below); S = S1 (identified; constrained only to be in commercial aircraft category; approximately 20 in number; they do not exist)}.

The mosaic of world W1 at T1 includes the airline, competing airlines, companies that build commercial aircraft (including Boeing), existing commercial aircraft including Boeing 777s, availability of new aircraft of existing types including Boeing 777, models of existing aircraft including performance parameters, air transportation infrastructure, air traffic regulations, etc.

- Time T2:
  - *Situation:* the airline has written their requirements document R1 and is ready to start a trade study of existing aircraft types.
  - *Correspondence set 2:* {M = R1 (identified and defined); MP = shall when operational; W = W2 = W1 + model R1; S = S1 (identified; constrained by this correspondence to be in system region of model R1; approximately 20 in number; they do not exist)}.

---

* The distinction in a correspondence set between system and service is this: if an external model corresponds to a system, its scope is the full set of system ports that provide and require services and input and output stuff, and S in the correspondence set refers to the name of the system. If the external model corresponds to a service, its scope is only the port that provides that service, and S in the correspondence set refers to the name of the service.

- Time T3:
  - *Situation:* The airline finishes its trade study, selects the Boeing 777 as the S1 aircraft type and places an order for 20 777s with Boeing.
  - *Correspondence set 3:* {M = R1 (identified and defined); MP = shall when operational; W = W2; S = S1 = Boeing 777 type (20 in number; they do not exist)}.
  - *Correspondence set 4:* same as correspondence set 3, except MP = will when operational (this is a prediction based on existing 777s, not a prescription).
    - How can 777s exist (in the W part of the correspondence) and not exist (in the S part)? Answer: The 20 S1 777 aircraft the airline needs do not yet exist, but other 777 aircraft do exist, and those are the basis for the correspondence set 4 prediction that the 20 S1 777s will correspond to the requirements model R1.
- Time T4:
  - *Situation:* the airline has received the 20 777 aircraft from Boeing and has put them into operation.
  - *Correspondence set 5:* {M = R1 (identified and defined); MP = shall when operational; W = W3 = W2 + the 20 new 777 aircraft; S = S1 = Boeing 777 type (20 in number; they exist)}.
  - *Correspondence set 6:* same as correspondence set 5, except MP = *does* (this is a description of each of the 20 existing S1 777s now in the possession of the airline).

Over the time period from T1 to T4, the state or value of all correspondence set elements changed — model, mode, world, and system. The mode "shall when operational" remained across the full time period but additional mode "will" appeared temporarily at T3, and additional mode "does" appeared at T4. The airline hopes that "does" applies for the full operational lifetime of the 20 S1 777s. The same model and system were associated via two different modalities at both T3 and T4. The usefulness of these simultaneous modalities is one reason for the decoupling of model, system, and modality, which gives the necessary flexibility. Note also that the mode "shall when operational" can be viewed as a combination of prescriptive (shall) and temporal (when operational — i.e., during the period of its operational lifetime). This made explicit something that is usually left implicit because it is obvious: The system cannot satisfy the "shall" prescription for correspondence prior to time T4 because the system does not exist until T4. The temporal mode "when operational" included in all the correspondence sets states the conditions in which the prescription holds — namely, starting at the point when S1 becomes operational and ending at the point when S1 is retired from operation.

The requirements model R1 would probably change to some degree over this period and across the operational life of the 20 aircraft but we did not show that because it was not the purpose of this example.

Complex Systems Extensions of Modality and Correspondence

The foregoing discussion is based on a binary notion of prescriptive: in some context, a model is considered either prescriptive of a system, or not prescriptive of a system (in which case it is descriptive or some other mode). Prescriptive is mapped in traditional models to control: We prescribe system S to satisfy model M, and we control the building of S to ensure that it does satisfy M. But for CS such as people and other autonomous agents, prescriptive is mapped to a spectrum of influence that ranges from partial control to suggestion or desire. We want S to satisfy M, and we will arrange the environment and/or other factors to influence S to satisfy M. Over the course of time, we may also change M to increase the odds of S satisfying M, for example, in cases where the system (a person) convinces the requirements engineer or the user that the requirements M should change.

Maybe CS will add to the standard modes. Where in the past we might have stated "Organization P1 shall correspond to model M1," we might now say "Organization P1 is enabled to correspond to model M1" or "Organization P1's goal is to correspond to model M1." Or instead of "You *shall* correspond to model M1," we could say "I would *really appreciate it* if you would correspond to model M1." Or modalities could include incentives to encourage agents or organizations to satisfy a requirement model: "I will *pay you $X* if you correspond to model M1."

Model Derivation and Application

An important distinction in the support of SS for SE is that between the derivation and the application of a model. The derivation of a model is the set, class, or category of the systems from which it is derived or abstracted.* The application of a model is the set of systems that correspond to it. If we assume correspondence of a model to both its derivation and its application, then clearly the application set is at least as great as, and includes, the derivation set; that is, the derivation set is a subset of the application set. In terms of model regions, the derivation is a "starter set" of the model region, and the application set is the full model region.

In the larger scheme of things, the derivation of a model is of little importance, while the application of a model is of critical importance. We do not care where a model came from; we do care what it applies to.

What if we try to apply the biological evolution model beyond its derivation set, that is, beyond populations of organisms or living things as defined in traditional biology? For example, do economies, organizations, societies, aircraft, software systems, and bridges follow the model of biological evolution? That is, do they exist in the region of this model? If it applies at all, it is helpful to determine what level of the model applies. Suppose we want to apply the model to organizations. The goal is to apply the model at the level that satisfies two competing constraints: (1) it has as much detail as possible but (2) it only includes characteristics that are common to all systems in the applied set — in this case, to all organizations. Why this goal? First, we want a model to apply to organizations, which means it must include characteristics that are common to all organizations. Second, we want a model that includes as many characteristics of organizations as possible because that gives us more information about organizations. Further discussion of the application of evolution to SE is in Chapter 9.

When a model is applied to a domain that was not part of its derivation, the generalization and specializations that occur should be made clear — in terms of the level and characteristics of the model that is actually applied and how this is specialized for the application domain. MOSES gives a framework in the model space, including the commonization dimension, to facilitate this clarity and precision.

There is another important way in which the distinction between derivation and application is important to SE and unification. The notion that only traditional models are appropriate for traditional systems does not hold in many cases. SE has made certain assumptions in TS models and imposed them on traditional systems; however, the systems have varying degrees of fidelity to those models and assumptions. CS models often have a higher fidelity to traditional

* The derivation systems can be an empty set. That is, a model does not have to be derived from systems; it could be created or defined independently.

systems than do TS models. One interesting example of this, as argued in Erdi (2008), is that clocks, as well as other machines, exhibit chaos in their normal operation. Clocks are often cited as an example of quintessential traditional systems in discussions of contrasts between TS and CS.

The point is that even if we decide we are engineering a traditional machine, that does not necessarily imply that we should use traditional models, or only traditional models. We need to consider the full range of models as candidates for any system that SE engineers or for which actualization is applied. This is another reason why the unification of TSE and CSE is needed.

Specified Behavior and Valuation

In traditional specifications of system behavior, we define what is expected or normal, and distinguish it from abnormal behavior. The assumption is that all parties agree on what behavior of the system is normal or good, and what behavior is abnormal or bad. However, in the comprehensive approach that includes CS, such as autonomous agents, not only are there more categories of possible behavior (including a "better than expected" category), but also the assignment of a behavior to a category (including the traditional normal and abnormal) in many cases depends on context, role, viewpoint, or interpretation. As an example, suppose a manager in an organization behaves in a way that entails a significant amount of risk to the organization with corresponding potential for considerable benefit to the organization. That behavior, regardless of the specific outcome, may be interpreted as good by a risk-taking organization and as bad by a risk-averse organization. Therefore, MOSES allows the valuation or interpretation of a behavior, such as good or bad, to be decoupled from the specification of the behavior itself, so the context can make its own interpretation.

In competitive games and conflict models such as war, two or more parties have opposing views of "good" and "bad" behavior or results of behavior. This situation requires decoupling of valuation from behavior.

Another reason for decoupling is that the valuation of many behaviors is not known for some period of time after it occurs, or it may never be known. If a company makes a significant change in direction, the results of that change may not be known for some time. And if the goal is to determine if it was better to make the change than to maintain the status quo, that may never be known.

Although behavior and valuation are decoupled, MOSES does not require them to be unconnected in all cases. For a large number of systems it is likely that all observers would agree on certain behaviors being good and certain other behaviors being bad, and for those cases it is reasonable to *commit to* a specific valuation. The point is that this decoupling is in the MOSES repertoire, and therefore commitment decisions can be left open or deemed relative to the observer in appropriate circumstances. The behavior and valuation relation is applied in Chapter 11.

Valuation of a given behavior by a given observer can also change, as was illustrated in the Chapter 3 section on "Roles," where an initial valuation indicated a service but later a disservice.

Usefulness and "Truth"

In engineering in general, and MOSES in particular, we do not tackle the issue of whether something is true.* We are interested in whether something is useful. For example:

* Truth here means contingent truth; we do allow and make use of logical truth, necessity, tautology, proof, and results implied by axioms and premises.

- Not "Is this really a system?" or "Is this element inside or outside the system?" but "Is it more useful to define this system to include this element or to exclude it?" In our earlier definition of system, we addressed this issue at some length.
- Not "Is this model true?" but "Is this model useful for some purpose?"
- Not "Is this the right view?" but "Is this view useful for some purpose?"
- Not "Is this system autonomous?" but "Is it useful to model this system as having autonomy?" or "What degree of autonomy should our model attribute to this system?"
- Not "Is this system unchanging?" but "Is it useful to model this system as unchanging in this space-time region?"

Summary

The decoupling described in this subsection provides flexibility in MOSES. This, together with the model orientation described earlier, enables MOSES to support the broad range of extended SE. The chapters that follow spell this out in more detail.

References

Ackoff, R. 1979. The Future of Operational Research is Past, *J. Opl. Res. Soc.*, 30(2): 93–104, Pergamon Press.

Blum, B. 1992. *Software Engineering: A Holistic View*. New York: Oxford University Press.

Boccara, N. 2004. *Modeling Complex Systems*. New York: Springer-Verlag.

Boehm, B. and D. Port. 1999. When models collide: lessons from software systems analysis. *IT Professional*, 1(1):49–56.See also http://sunset.usc.edu/research/MBASE/. (Accessed January 11, 2009.)

Box, G. and N. Draper. 1987. *Empirical Model-Building and Response Surfaces*. New York: Wiley.

Bracken, P., and M. Shubik. 2001. War gaming in the information age: theory and purpose. *Naval War College Rev.*, Spring 2001, p. 47–60. http://findarticles.com/p/articles/mi_m0JIW/is_2_54/ai_77034455/print. (Accessed January 11, 2009.)

Braha, D., A.A. Minai, and Y. Bar-Yam (Eds.). 2006. *Complex Engineered Systems: Science Meets Technology*. New York: Springer.

Carlson, J.M. and J. Doyle. 2002. Complexity and robustness, *Proc. Natl. Acad. Sci. U.S.A.*, 99(1):2538–2545. http://www.pnas.org/content/99/suppl.1/2538.full.pdf. (Accessed January 11, 2009.)

CIO. 2008. *Enabling the Mission: A Practical Guide to Federal Service Oriented Architecture*. Draft Version 1.0. Feb. 15. Federal CIO Council.

CSCS. 2005. University of Michigan Center for the Study of Complex Systems. About the Science of Complexity. http://www.cscs.umich.edu/about/complexity.html. (Accessed January 11, 2009.)

Dayan, P. and J. Williams. 2006. Putting the computation back into computational modeling. *Pharmacopsychiatry*, 39(Suppl. 1):S50–S51.

Dori, D. 2002. *Object-Process Methodology*. Berlin: Springer-Verlag. See also also http://www.objectprocess.org and http://www.er.byu.edu/er2003/slides/ER2003PT1Dori.pdf (Accessed January 11, 2009.)

Doyle, J. 2006. BAST1 [Network Science 2006: Board on Army Science and Technology] Slide 7, 11.

Erdi, P. 2008. *Complexity Explained*. Berlin: Springer-Verlag.

Estefan, J. 2007. Survey of Model-Based Systems Engineering (MBSE) Methodologies, Jet Propulsion Laboratory, California Institute of Technology, Pasadena, California. http://www.omgsysml.org/MBSE_Methodology_Survey_RevA.pdf. (Accessed January 11, 2009.)

Estefan, J. 2008. Survey of Model-Based Systems Engineering (MBSE) Methodologies, Jet Propulsion Laboratory, California Institute of Technology, Pasadena, CA.

Friedenthal, S. 2006a. Reaping the benefits of model driven systems engineering (MDSE). *INCOSE INSIGHT*, 8(2):14–15.

Friedman, G. 2006a. On the unification of systems engineering. *INCOSE INSIGHT*, 8(2): 16–17.

Garvey, P.R. 2000. *Probability Methods for Cost Uncertainty Analysis: A Systems Engineering Perspective.* New York: Marcel Dekker.

Gigerenzer, G. and R. Selten. 2001. *Bounded Rationality.* Cambridge, MA: MIT Press.

Green, D. and R. Perry. 2007. *Perry's Chemical Engineers' Handbook, 8th edition.* New York: McGraw-Hill.

Hasselblatt, B. and A. Katok. 2003. *A First Course in Dynamics: With a Panorama of Recent Developments.* New York: Cambridge University Press.

Holloway, D. 1976. The political uses of scientific models: the cybernetic model of government in Soviet social science. In *The Use of Models in the Social Sciences,* L. Collins, Ed. Boulder, CO: Westview Press, p. 110–129.

Hybertson, D. 2001. A uniform component modeling space. *Informatica,* 25(4):475–482. http://www.mitre.org/work/best_papers/01/hybertson_space/hybertson_space.pdf. (Accessed January 11, 2009.)

Hybertson, D. 2005. Using a specification approach to facilitate component testing. In *Testing Commercial-off-the-Shelf Components and Systems,* S. Beydeda and V. Gruhn, Eds. New York: Springer-Verlag, p. 213–237.

Hybertson, D. 2006. Using models and abstraction to extend and unify systems engineering. *Proc. IEEE Int. Conf. System of Systems Eng.* (SoSE 2006), Los Angeles, CA, 24–26.

IEEE. 2005. IEEE Standard for Application and Management of the Systems Engineering Process, IEEE Std 1220-2005, New York: IEEE Computer Society.

INCOSE 2006. Systems Engineering Vision 2020, Version 2.0. Seattle: INCOSE.

Koestler, A. 1967. *The Ghost in the Machine.* New York: Macmillan.

Koolmanojwong, S. and B. Boehm. 2007. An Empirical Study on MBASE and LeanMBASE. *First Int. Symp. Empirical Software Engineering and Measurement, 2007.* ESEM 2007. 20–21 September 2007, p. 496–496.

Kulakowski, B.T., J.F. Gardner, and J.L. Shearer. 2007. *Dynamic Modeling and Control of Engineering Systems, 3rd ed.* New York: Cambridge University Press.

Langton, C.G. (Ed.) 1995. *Artificial Life: An Overview.* Cambridge, MA: MIT Press.

Lavi, J. and J. Kudish. 2005. *Systems Modeling & Requirements Specification Using ECSAM: An Analysis Method for Embedded and Computer-Based Systems.* New York: Dorset House.

Maier, M. and E. Rechtin. 2000. *The Art of Systems Architecting,* 2nd ed. Boca Raton, FL: CRC Press.

Martelli, M. 1999. *Introduction to Discrete Dynamical Systems and Chaos.* New York: Wiley.

Miller, J. 1995. *Living Systems.* Liwot, CO: University Press of Colorado. (Originally published 1978.)

Moss, S. 2008. Alternative Approaches to the Empirical Validation of Agent-Based Models. *J. Artificial Societies and Social Simulation,* 11:1(5). http://jasss.soc.surrey.ac.uk/11/1/5.html. (Accessed January 11, 2009.)

OMG (Object Management Group). 2003. MDA Guide Version 1.0.1. OMG/2003-06-01, 12 June 2003. http://www.omg.org/docs/omg/03-06-01.pdf (Accessed January 11, 2009.)

OMG (Object Management Group). 2007a. Model Driven Architecture. http://www.omg.org/mda/. (Accessed January 11, 2009.)

OMG (Object Management Group). 2007b. Unified Modeling Language (OMG UML), Superstructure, V2.1.2. OMG Document Number: formal/2007-11-02 Standard document URL: http://www.omg.org/spec/UML/2.1.2/Superstructure/PDF. (Accessed January 11, 2009.)

OMG (Object Management Group). 2007c. Unified Modeling Language (OMG UML), Infrastructure, V2.1.2. OMG Document Number: formal/2007-11-04 Standard document URL: http://www.omg.org/spec/UML/2.1.2/Infrastructure/PDF. (Accessed January 11, 2009.)

OMG (Object Management Group). 2007d. Systems Modeling Language (OMG SysML™), V1.0. OMG Document Number: formal/2007-09-01 Standard document URL: http://www.omg.org/spec/SysML/1.0/PDF (Accessed January 11, 2009.)

Phelan, S. 2001. What is complexity science, really? Emergence, 3(1):120–136.

Physiome, 2006. Special Issue on the Physiome and Beyond. Proc. IEEE, 94(4):671.

Pilkey, O.H. and L. Pilkey-Jarvis. 2007. Useless Arithmetic: Why Environmental Scientists Can't Predict the Future. New York: Columbia University Press.

Pogge, R. 2006. Lecture 32: Space, Time & Gravity: General Relativity. *Astronomy 162: Introduction to Stars, Galaxies, & the Universe. Ohio State University, 2006.* At: to http://www.astronomy.ohio-state.edu/~pogge/Ast162/Unit5/gr.html. (Accessed January 11, 2009.)

Ray, T.S. 1995. An Evolutionary Approach to Synthetic Biology: Zen and the Art of Creating Life. p. 179–209 in (Langton 1995).

Sage, A.P. and J.E. Armstrong, Jr. 2000. *Introduction to Systems Engineering.* New York: Wiley.

Schubin, M. 1997. Riemannian Geometry and General Relativity. Course syllabus. http://www.math.neu.edu/~bratus/diffgeom/intro.htm. (Accessed January 11, 2009.)

Simon, H.A. 1962. The Architecture of Complexity. *Proc. Am. Philosophical Soc.,* 106(6):467–482. http://www.jstor.org/sici?sici=0003-049X%2819621212%29106%3A6%3C467%3ATAOC%3E2.0.CO%3B2-1 (Accessed January 11, 2009.)

Simon, H. 1996. *The Sciences of the Artificial, 3rd ed.* Cambridge, MA: MIT Press. (Originally published 1969.)

Turski, W. 1986. And no philosophers' stone, either. *Information Processing '86, Proc. IFIP Congress,* H.J. Kugler (Ed.). Dublin, North-Holland, Amsterdam, p. 1077–1080.

Warfield, J. 2006. *An Introduction to Systems Science.* Danvers, CO: World Scientific.

Weaver, W. 1948. Science and complexity. *Am. Scientist,* 36:536. http://www.ceptualinstitute.com/genre/weaver/weaver-1947b.htm. (Accessed January 11, 2009.)

Whitehead, A. 1911. *An Introduction to Mathematics,* New York: Henry Holt & Co. (Republished 1958. New York: Oxford University Press.)

Wymore, W. 1993. *Model-Based Systems Engineering.* Boca Raton, FL: CRC Press.

Chapter 5

SE Model Space

"I am myself a great lover of these processes of division and generalization; they help me to speak and to think."

—Plato, in Phaedrus

The purpose of this chapter is to introduce the model space, which is where most of the action takes place in MOSES, and which defines the variables and dimensions that are addressed in succeeding chapters of this book. The chapter is laid out in the form of a traditional engineering sequence: scope, context, requirements, design, and implementation.

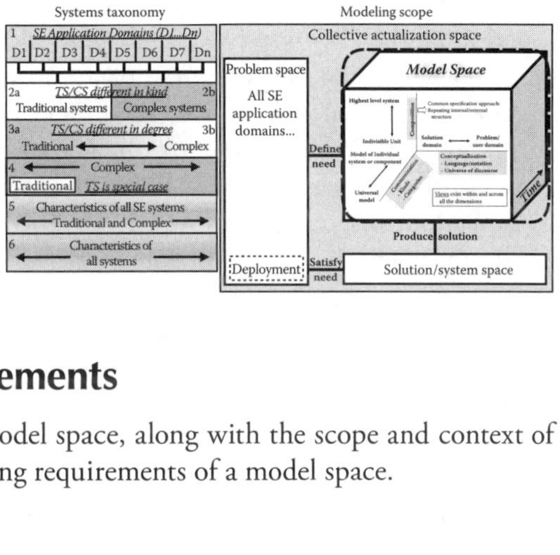

Model Space Context and Requirements

This chapter introduces a general concept of a model space, along with the scope and context of the SE model space and a statement of the resulting requirements of a model space.

What Is a Model Space?

A preliminary answer to this question is that a model space is a futuristic model-oriented combination of two traditional concepts: (1) a domain body of knowledge (BoK) and (2) a collection of project libraries or project databases. We call these, respectively, a knowledge space and workspaces.* The concept of BoK is discussed in some detail in Chapter 7 but for now we will

* The term "project" here is meant to be general, to include SE development project, research project, and education. Any effort that either uses the knowledge space or contributes to it is in the scope of project.

explain both the knowledge space and the workspace by means of an example. Traditional engineering handbooks provide examples of guides to mature BoKs. Two examples are *Perry's Chemical Engineers' Handbook* (Green and Perry 2007) and *Marks' Standard Handbook for Mechanical Engineers* (Avallone, Baumeister, and Sadegh 2007). Engineering handbooks contain a very large number of models. Some of the models are based on theory, others on experimentally observed data (captured in tables or formulas), and others on rules of thumb or heuristics. Many are expressed as mathematical formulas, others as graphics, including block diagrams.

For the example here, we select the domain of chemical engineering, and a company of chemical engineers that has the task of designing and developing a new chemical product. Perry's handbook first covers the necessary supporting domains — especially mathematics, and physics such as thermodynamics, and then explains the many unit operations (discussed in Chapter 2) and related activities: heat transfer, fluid dynamics, reaction kinetics, process control, distillation, separation, etc. It therefore is self-contained in the sense that it includes chemical engineering knowledge plus needed material from supporting and interacting disciplines.

The workspace portion is where the chemical engineering team develops and keeps their artifacts while doing the necessary research, experimentation, development, and management to engineer a new chemical product or modify an existing one. These artifacts could include descriptions of the desired properties of the chemical product, design documentation, records of experimental trials, and equipment used in the various activities, as well as management artifacts such as staffing plans, budgets, schedules, and status reporting. The process would include drawing on the knowledge contained in Perry's handbook and tailoring it to their specific product.

That is a description of today's incarnation of a model space. What is the futuristic view? Continuing the *Chemical Engineers' Handbook* example, suppose we transformed and parsed the material in the handbook from book form into model form in a database or other machine-manipulated repository that has a set of modeling and other repository management tools. Much of the material in the handbook already consists of models. The work involved would be to organize the models; represent them in appropriate languages such as mathematical models, System Modeling Language (SysML), Unified Modeling Language (UML), description logics, constraint languages, scripting languages, and software programming languages; and associate the appropriate metadata to enable machine manipulation and enable search by the engineers. The variables or dimensions introduced later in this chapter provide a basic organizing structure. All the foregoing models would be entered in the community knowledge space portion of the model space.

And suppose we did something similar with the company's project and management artifacts, and the project documents, descriptions, plans, schedules, budget, staffing, and reports. Although in principle, the model space and model orientation allow physical models, as well as conceptual models, this book primarily focuses on conceptual models such as mathematical models, designs, requirements, constraints, and rules. If we stayed within the conceptual model space, we would exclude the physical equipment the engineers used but we would put all other artifacts in the company project workspace portion of the model space.

The model space would now contain the BoK in the form of general models in the knowledge space that are accessible to the whole chemical engineering community, and would contain the workspace of the company project in the form of specific models for this product, which perhaps would be kept private to that company only. Each chemical engineering organization would then have a different view of the chemical engineering model space: the public common knowledge space and their own private workspace.

The foregoing futuristic version of the chemical engineering model space represents the concept of a model space as described in this chapter and this book. It is a general concept that can be applied to any field of inquiry, such as a domain, discipline, or multidiscipline, that satisfies two rather loose constraints: First, that it is a reasonably coherent field with a community that generates and evolves a BoK, and second, that the BoK can be represented as models. Because we view most, if not all, human knowledge as being representable as models, the second constraint is more or less redundant — its satisfaction being implied by the first constraint. However, we include it because some may not share our view of knowledge as models.

There are other possible variants of the workspace portion of the model space. One is an organization or enterprise workspace. Because the model space is designed to include models of enterprises in the form of projects, it can also support models of organizations as a whole, as a context in which projects exist or come and go. Organizational planning, budgeting, staffing, and all forms of management (or enterprise engineering/management as discussed in Chapter 1), including enterprise architectures, could be contained and used in an enterprise workspace. And because the knowledge space of the model space includes the knowledge and patterns of management in general from the management science community, the enterprise workspace could draw from these general models and tailor them to manage a specific enterprise. This type of workspace is within the scope of the expanded SE of the future.

Other workspace variants are team and personal workspaces. A project might have multiple teams, each of which could have their workspace, and each member could have a personal workspace. There might be coherence within the team workspace but not yet in the project workspace. Model coherence is not necessary at all levels. It should be decided what portions of the workspace at each level need to have consistency, coherence, congruence, etc.

Consider the possibility that all information available on the Internet could be a knowledge space. In one sense, we might argue that this body of information maps into a knowledge space as we define it in this chapter, although the Internet information exploits only a fraction of the structure available in a knowledge space. Ultimately, however, we would say no, this body of information does not constitute a knowledge space because it does not separate the wheat from the chaff.* A knowledge space is not an organized pile of material; it is an organized set of models that represents knowledge as determined by an associated community. Specific workspace models either evolve into the knowledge space, leave the model space when no longer needed, or are archived in the model space for historical purposes. This is part of the overall collective actualization process that we describe in Chapter 13.

How does a community determine what is accepted into its BoK? The traditional process for this acceptance or vetting has been primarily journal or conference refereeing and usage of textbooks and other material at institutions of learning. This process is evolving to reflect more online publications and more mass collaboration and mass review. MOSES does not define what the vetting process must be. Each community defines its own process and standards. The SE of the future is a composite of multiple disciplines, as indicated in Chapter 2, and therefore the SE model space will draw from multiple communities.

---

* To the extent that Internet mass collaboration communities practice self-monitoring, review, and correction of material, it could be considered a community BoK. But what keeps the Internet as a whole from being a knowledge space is that material that is not accepted is still available, and there is no demarcation between accepted material and not accepted material.

Scope of the SE Model Space

In the chemical engineering example, the handbook was a single source representing a partially self-contained BoK that included the supporting disciplines. In other disciplines, and particularly SE, no such single source exists. The body of knowledge of SE is spread across multiple supporting disciplines and associated domains, most of which have their own defined body of knowledge. Because of this, the SE model space must be modular and composable from the individual model spaces of its supporting disciplines. The SE model space should draw from the constituent ones rather than duplicating them. However, the composite SE model space must integrate its constituent model spaces, and thus it has its own models as well — for example, models that are common or generalized across multiple constituent disciplines.

Given that approach, the scope of the model space we describe in this book is the expanded multidiscipline of SE and its supporting disciplines, as sketched in Chapters 1 and 2. In particular, this scope includes both system models and management models as reflected in Table 1.3. Throughout the book, when we refer to the model space, we mean this SE model space unless specifically designated otherwise.

Content type provides another aspect of the scope. The model space contains only models; things that are not models are outside the model space. For example, model processors are outside the model space, and systems — that is, target systems that are not models — are outside the model space. Implicit parts of specifications are not in the model space. We should note, however, that entities that are sometimes called metamodels or meta-metamodels are in the scope of the model space. In MOSES, these entities are models.*

Context of the Model Space

The basic context of the model space, shown in Figure 5.1, is a problem space and a corresponding solution space, which covers the full range of problems for which SE provides solutions. The relations among the elements are that the model space defines models of problems or needs in the problem space, and defines models of solutions. Solutions satisfy needs in the problem space. The figure does not constrain the sequence of SE activities that take place in the collective actualization space. One sequence is the traditional forward engineering or need-driven process: first a need in the problem space is identified, then models of the problem and solution (e.g., requirements, design) are defined that exist in the model space, a solution is produced from the solution model and appears in the solution space, and finally the solution is deployed in the problem space.

For example, a need is identified to transport people and goods between locations on the land surface of the Earth. An aircraft engineering organization performs SE in the actualization space, and produces all the specifications, designs, drawings, simulations, and other models in the model space for building a type of aircraft. A manufacturing plant builds multiple aircraft based on these models, and the aircraft exist in the solution space. When an airline buys aircraft and puts them into operation, the aircraft satisfy the need. The foregoing assumes the necessary supporting environment, such as airports and air traffic control. Another sequence is the traditional technology-driven process. A technology appears before a need is identified. The appearance of the technology causes recognition of a need that the technology satisfies.

---

* A metamodel is a description or category of a model. If a model is a category of individual systems, then a metamodel is a category of individual models. Chapter 7 discusses related concepts of commonality, category, class, and instance. Chapter 8 discusses related concepts of taxonomy, conceptualization, and ontology.

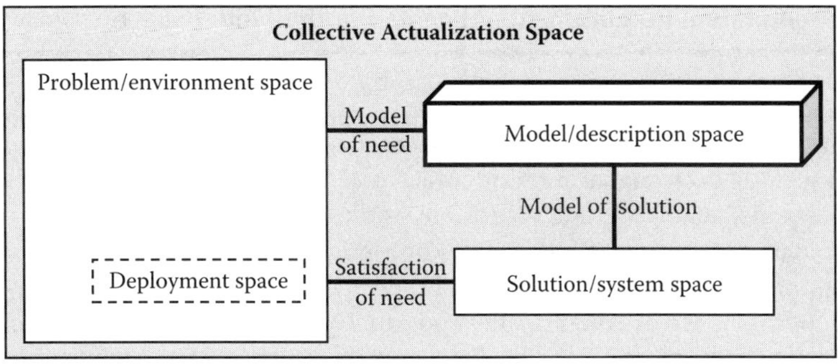

Problem space: Problem domain environments in which engineered systems are needed – and eventually used in the deployment space
Model space: Models of problems and solutions, ranging from general models representing knowledge of multiple systems to specific models of a problem or solution
Solution space: Solution engineered with the help of models and that satisfies one or more needs in problem environment; in the preponderance of cases, the solution is a system
Collective actualization space: Engineering agents, self-organizing and managing agents, interaction, modeling/engineering/management processes and tools

Figure 5.1 SE model space interaction context.

To get a more complete picture of the context and a better understanding of the model space, we need to consider the multi-domain nature of the SE problem and solution spaces. Systems engineering has always supported multiple application domains, such as aerospace, transportation, and construction, and it is expanding to more domains as we have discussed in previous chapters. Each solution, by definition, needs to provide help in the application domain that corresponds to the problem, so the solution needs to appear in understandable form to that domain. Therefore, the domain context is that multiple application domains feed into the model space, and the model space feeds solutions, or feeds models to solution builders who build solutions, that go back into the same application domains. This context is illustrated in Figure 5.2, showing both problem and solution spaces divided into application domains.

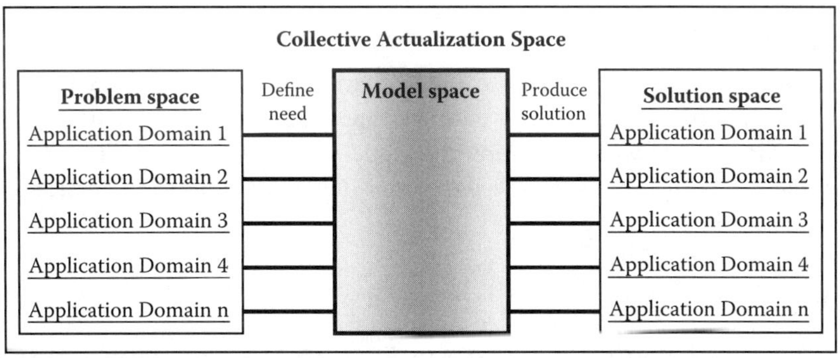

Problem space: SE problem space includes multiple application domains
Model space: Models must address problems and produce corresponding solutions in each application domain
Solution space: Each solution exists in the corresponding application domain

Figure 5.2 SE model space domain context.

Table 5.1 SE Foundation Requirements Allocated to the Model Space

No.	*Requirement*
5	Support scale-up and scale-down of systems, and multiple scales in a single system, to include both spatial and temporal scales.
10	Be flexible, adaptable, and extensible, and evolve as needed.
11	Facilitate incorporation and use of knowledge and experience.
13	Support multiple views and perspectives of a given system or class of systems.
16	Support balance of commonality and variability: find everything in common; but do not force one size fits all; recognize imperative and benefits of diversity.
17	Support balance of stability and change: find everything that is stable but do not force false stability; recognize imperative and benefits of change.
18	Support clear communication among all SE stakeholders, but recognize that not all clear communication is explicit; support both explicit and implicit communication.
20	Support analysis of alternatives.
24	Provide a unified approach to organizing and harnessing the knowledge from the array of disciplines supporting SE.

Requirements of the SE Model Space

Table 5.1 lists the Chapter 1 requirements of an SE foundation that were allocated to the model space.

Based on the context discussion thus far, the derived requirements for the SE model space are to receive and understand problems in a wide variety of application domains, and produce and provide solutions for that same variety of domains — in each case, matching the problem domain and the solution domain, and matching the specific solution to the specific problem. In many cases, the solution will be an engineered system. Furthermore, it must do this in a reasonably efficient manner. This implies that the model space must support the capture and exploitation of characteristics and models that are common across a domain, or across multiple domains, to minimize duplication of work.

Model Space Design

How should the SE model space be designed and structured to satisfy the requirements, in the context of multiple application domains and multiple supporting disciplines? This section describes first a simplistic design that would partition the model space into each application domain, and then a more realistic design that supports the greater interrelation between application domains and engineering disciplines.

Approaches

Approach 1: Partition into Separate Application Domain Model Spaces

We could have a partitioned model space, or equivalently multiple separate model spaces, one for each application domain, such as aerospace, transportation, construction, medicine, or organizational management. We would have domain-specific problems, which we describe in the model space in domain-specific models, from which we produce domain-specific solutions, which satisfy the problem in that domain. That would satisfy the functional requirements, and seems at first glance to be a reasonable design.

But if we look at the world of SE, this solution is very inefficient and not realistic because of the relation between application domains and all the disciplines and specialties of SE that support solutions. An application domain sees only the external view of each solution or system. While that external view looks to that domain like something that fits nicely into the domain, the internal view is another world, the engineering world that goes well beyond the problem application domain. Flying on an aircraft from the user perspective involves getting a ticket, boarding, and sitting in a seat on the plane while it flies from one city to another. This external view does not need to see all the internal domains of power plant, avionics, etc. that make up the modeling and the overall aircraft solution to transportation. But the model space needs to address all these constituent modeling domains and disciplines, and they are incorporated or at least reflected in the solution — that is, the built aircraft.

Furthermore, in many cases the same general models apply to multiple disciplines and domains. That has been true of TSE, with mathematical, computational, and engineering models; it is even more pronounced with the incorporation of CS models, which have been explicitly investigated across a variety of domains. Many internal models apply across multiple external application domains, representing a many-to-many relation. Therefore, this approach of keeping problems, models, and solutions partitioned along application domain boundaries in a one-to-one relation does not work. We need an approach that supports the relationship between internal models and application domains.

Approach 2: One Common SE Model Space

The selected approach is one common model space for all domains and disciplines, as shown generically in Figure 5.3. In the context of this book and of SE, we include all domains and disciplines that are in the new SE: the classic domains of traditional SE that we have mentioned; the domains of extended SE, such as organizations and nanosystems; and the supporting disciplines such as SS, CSS, and mathematics. The model space has very general models that cut across all domains and has other general models that are oriented to a specific domain.

A reinterpretation of the aircraft example given earlier can be used to illustrate the many-to-many relation between application domains and internal models or domains. A simplified view of the situation from the perspective of Figure 5.3 is this: the application domain is commercial air transportation. An external view of the solution is an aircraft that transports goods and people through the air from one airport to another. An internal view of the solution is a set of subsystems representing multiple domains, such as airframe structures including fuselage and wings, a power plant including one or more engines, and avionics including communications, navigation, and flight control. The model space includes models from all engineering, supporting science, mathematics, and computational domains needed to engineer the aircraft.

This illustrates the many internal domains related to the one air transportation domain. The other side of the many-to-many relationship is that most of those internal and supporting domains clearly also support multiple other application domains. This relation is supported by having all domains and disciplines together in a single model space. Any given application problem can draw on the full spectrum and variety of models.

As in the previous figure the sequence is not constrained. In a traditional forward or top-down SE process, the lines and arrows in Figure 5.3 would all be arrows pointing to the right.

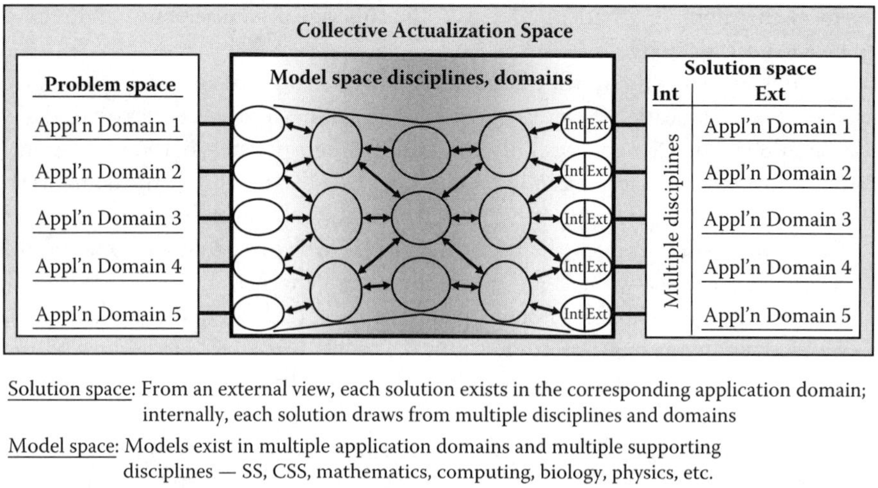

Figure 5.3 SE model space multi-domain texture.

The models on the left side of the model space would be the desired or "to be" models defined up-front for any given system in each domain, such as requirements documents. The models on the right side would be the actual "as built" models defined after development of those systems. If we assume that the models on the left are external descriptions such as requirements specs, then the "Ext" or external models on the right are the "as built" models that can be compared to the models on the left.

If no changes occurred in the requirements during the engineering of the system, then the external models on the right should be identical to the models on the left, for any given system in each domain. In this case there is one difference of interpretation between the left and right models, which are the same model. Let's call this model M, which characterizes functions {F1, F2, ...}. The left interpretation is prescriptive: the system shall do M = {F1, F2, ...}. The right interpretation is descriptive: the system does M = {F1, F2, ...}. These are two different pieces of information, represented as correspondences with different modalities, and both are important to know.

Many times, systems engineers receive descriptions of solutions, such as user-written requirements documents, that do not adequately solve the problem, are inconsistent, ambiguous, or incomplete. The MOSES stance is that it is the responsibility of the systems engineers to understand the problem, the gap, and the context so they can define an appropriate solution. In terms of Figure 5.3, systems engineers need to reach into the problem space to ensure that the models on the left side of the model space are appropriate characterizations of the application domain problem they need to solve.

Internal Model Space Structure

Based on the discussion above, we want to design a structure that organizes models in a way that satisfies all the external and derived requirements identified to this point. Recall that SE domains exist at the level of whole systems, and also at the level of components, so that a system in one domain includes components from multiple domains. Also recall that domain taxonomies can be sliced in different ways. One way of slicing it is between TS and CS, as we did in the Chapter 2 taxonomy. We want the model space to be capable of supporting that taxonomy.

How do we satisfy these requirements? Our approach is to use basic principles such as modularity, which allows us to define small pieces that we can combine in multiple ways, which gives us considerable flexibility. We want to identify the major variables that distinguish and structure models. The variables we selected, along with a trace to requirements and a general motivation for each one, are as follows:*

1. *Composition:* The whole–part concept, and the requirement to model multilevel systems, lead to the notion of composition of parts into wholes. The variable is composition level in a multilevel context. This element responds directly to requirement 5: Support scale-up and scale-down of systems, and multiple scales in a single system.

   General motivation for composition — wholes, parts, and environments have been a recognized part of SE and engineering and modeling in general for decades. Now with the expanded scope of SE, especially scale-up and -down, the extension to a larger number of levels requires even more focus on and support for whole–part structures.
2. *Commonization:* Another key need is distinguishing similarities from differences, what is unique from what is common, for which we invoke the concepts of categorization and generalization/specialization. Categorization tells us what is common among a set of individuals. Generalization tells what is common among a set of categories or kinds. We combine these into the commonization variable. At a basic level, this element responds directly to requirement 16: Support balance of commonality and variability: Find everything in common but do not force one-size-fits-all; recognize the imperative and benefits of diversity. At a higher level, it directly supports requirement 11: Facilitate incorporation and use of knowledge and experience.

   General motivation for commonization — general models that apply to multiple systems, classes, and categories have been a significant element in learning, accumulating knowledge, and organizing and recalling knowledge for many years, even centuries. Now with the expansion of SE, this dimension takes on added significance because CS brings more disciplines and more models to the table. We need a way to capture all this commonality, and at the same time a way to capture the differences that exist.
3. *Conceptualization:* Conceptualization of a domain has to do with what things exist in that domain, how they are related, what they mean, and how we communicate about them. By its nature, SE is multilingual and has multiple ontologies and universes of discourse (UDs). The variety of application domains reflects a variety of languages and ontologies, typically even within the engineering of one system. Therefore, we need a model space structuring variable that supports and helps model this type of heterogeneity. This element responds directly to requirement 18: Support clear communication among all SE stakeholders but recognize that not all clear communication is explicit; support both explicit and implicit communication.†

   General motivation for conceptualization — we have had the notion of different languages and UDs throughout the history of engineering and mathematics, including the

* Some of these dimensions and associated graphics were presented by D. Hybertson in *Informatica* 25(4), 475–482, 2001.

† Although the model space includes only models, and all models are explicit, conceptualization exists in context of the model/specification relation described in Chapter 3, and therefore indirectly supports implicit communication.

languages of the environment versus those of the internal system and those of mathematical models and engineering artifacts. This has been an important area that has increased in significance with the rise of ontologies in the engineering arena. Engineers who define models that apply to multiple domains must be able to translate those models into the languages of those specific domains. Now with the expanded scope of SE, the greater number of disciplines, applications, and supporting sciences make this dimension even more important in the SE model space of the future.

4. *Time:* Models and systems have a life span in which they are created, exist, change, and terminate. We need a time variable to accommodate change and to address past and future, evolution, differences over time, and rates of volatility and stability. This element responds directly to requirement 5: Support scale-up and scale-down of systems and multiple scales in a single system, to include both spatial and temporal scales; and to requirement 17: Support balance of stability and change. Find everything that is stable but do not force false stability; recognize the imperative and benefits of change.

   General motivation for time — concepts of time and change, and temporal and dynamic models, have always been a prominent aspect of engineering and systems engineering. The expanded SE and its increasingly dynamic environment only increase the need for this dimension in the model space.

5. *View:* We need to look at cross-cutting issues, and to have the flexibility to combine into one entity or view any collection of items from anywhere in the model space. This element responds directly to requirement 13: Support multiple views and perspectives of a given system or class of systems.

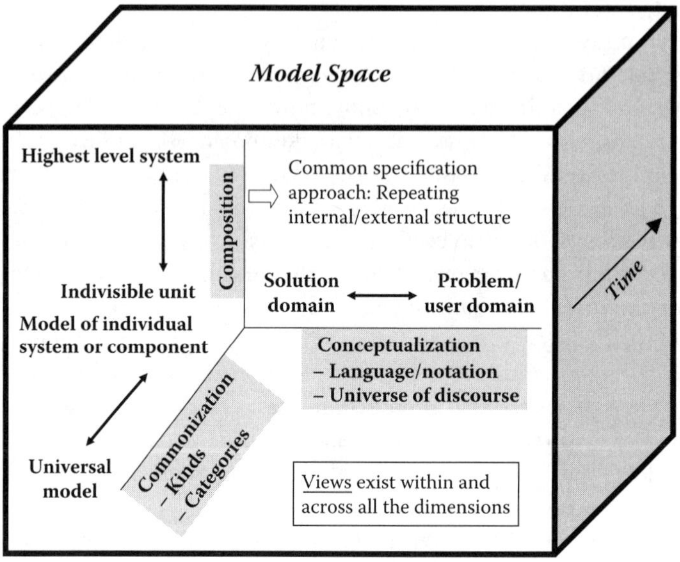

Figure 5.4 SE model space internal structure. (*Source*: Hybertson, D. A uniform component modeling space. *Informatica*, 25(4):475–482. Published by Slovenian Society *Informatika*. 2001. Used with permission.)

General motivation for view: One important motivation for views is that in general we cannot see or understand everything of interest about a system in one visualization or representation. Another motivation, which is consonant with the first, is that most observers or stakeholders are not interested in everything about a system, but rather the things or features related to their concern. A third motivation is the need for a free form construct that has the unconstrained flexibility to cut across all other dimensions and define any subspace of the model space for observation and engineering purposes.

The combination of these five variables responds to requirement 10: Be flexible, adaptable, and extensible, and evolve as needed. The variables lead to four dimensions that structure the model space, plus one free-form concept (view) that can cut across the structure, as shown in Figure 5.4. The figure also indicates a range or spectrum for each variable.

Conceptual and Physical Space

Models of physical systems often include parameters that designate the position or movement of those systems in classical three-dimensional (3D) space. In that context, discussions of space and time, or time as the fourth dimension, are also based on 3D classical physical space. In this book, however, the term "space," especially in the context of the model space, means conceptual space. The model space has three conceptual dimensions, with time as the fourth dimension. It is pure coincidence that the number of conceptual dimensions in the model space is equal to the number of dimensions in classical physical space, and that time is the fourth dimension for both.

Those models that parameterize physical space are of course included in the envisioned SE model space. Does that pose a contradiction? No, because they are conceptual models of physical systems. That is, the models position physical systems in physical space dimensions, while the model space positions the conceptual models in conceptual space dimensions. The physical space position of the physical system is important to the model but irrelevant to the model space. The model space cares about where the model is, not about where the modeled system is.

Therefore, unless otherwise noted, in this book all discussions of space, or space versus time, are referring to the model space conceptual dimensions.

Before defining these variables further, we need to acknowledge that there are a large number of potential variables that characterize models and that therefore could be used to structure the model space. We have attempted to define a "pantheon" of structuring variables that are dominant, universal, fundamental, and stable in the modeling and engineering of systems. Each of the variables or dimensions draws directly from a variety of disciplines over a long period of time; each has a very strong pedigree, and each is a significant factor in modeling systems. Still, which variables belong in the pantheon is certainly a matter of debate. Although we believe the variables we selected have proved their *gravitas*, that does not mean necessarily that there is consensus on this set, or that others will not be added over time.

The concept of the model space does not depend on this set of variables. The model space, and the MOSES approach, could survive changes in the set. But in the scope of this book, this is the set we use for discussion.

Composition

The definition of system contains within it the concept of composition. A system is something of interest as a whole and as a set of interrelated or interacting parts. In most systems, especially those of interest to SE, the parts are also systems that, in turn, have parts. Thus, systems typically exhibit a whole–part hierarchy, and this extended spectrum of wholes and parts forms the composition dimension of the model space. "System" and "component" can be used as relative terms. A system at one level may be a component of another system at the next higher level, and the same relations repeat at each level. The holons and holarchies discussed in Chapters 2 and 4 reside in this dimension. Thus, both structure (whole–part) and dynamic behavior (interacting parts) are important elements of the composition dimension.

The composition dimension is characterized by a general recursive pattern of composition levels that brings the important SE distinction between internal and external into play. At each level, an external view of a whole is decomposed into an internal view that includes the external views of its parts, as shown in Figure 5.5. In turn, each of these external views is decomposed into an internal view. The pattern is an intertwining of external and internal views. The external view

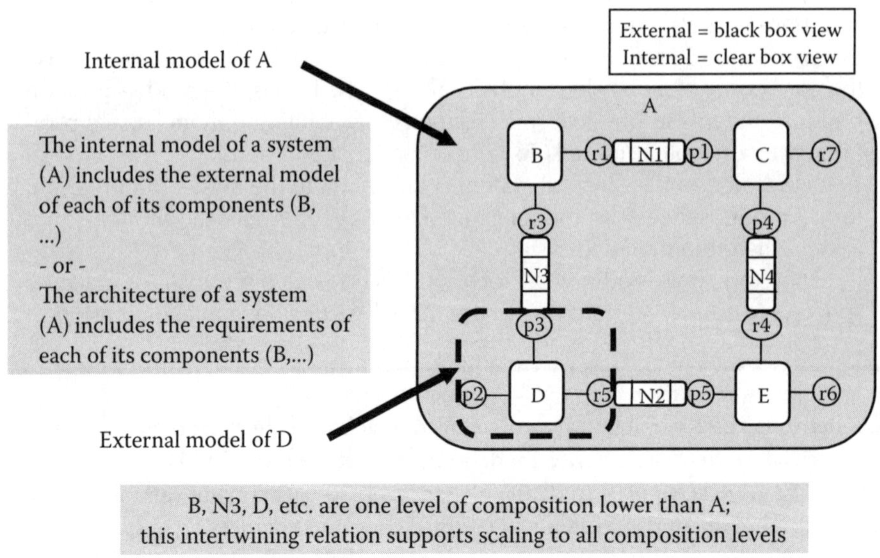

Figure 5.5 Intertwining composition structure. (*Source*: Hybertson, D. Using a specification approach to facilitate component testing. In *Testing Commercial-off-the-Shelf Components and Systems*, S. Beydeda and V. Gruhn, Eds. New York: Springer-Verlag, p. 213–237. 2005. Used with permission of Springer Science + Business Media.)

corresponds nominally to what we often call requirements, and the internal view corresponds to what we often call architecture or design.*

The intertwining nature of internal and external models in the composition dimension allows the same types of models and specification information to exist at every composition level, rather than defining a different kind at each level. This represents a significant simplification of specification types from multiple traditional types down to two types: external spec and internal spec.

Commonization

Whereas the primary distinction that generates composition structures is whole versus part, the primary distinction that generates commonization structures is similarity versus difference.

Commonization is a criterion for abstraction in which the relevant information is what is common among a set of entities. Common or general models are useful in multiple environments. One environment is the model space itself. This dimension enables the knowledge space, and distinguishes between the knowledge space (general models) and the workspace (specific models).

General models include conventional designs in engineering handbooks, complex system models such as evolution and predator-prey models that apply to multiple application domains, domain taxonomies, architecture frameworks, engineering patterns, standards, templates, architecture styles, reference architectures, product lines, abstract data types, and classes in class hierarchies.

The importance of general models is that they make engineering efficient and effective, and they make science possible. Without the commonality captured and exploited by general models, we would be forced to consider every natural and artificial system, every problem and solution, as unique. We would have no scientific laws, and no categories or taxonomies, because we would have to treat each individual, each atom, cell, human, tree, and planet as unique, with its own properties and behavior. Each engineering effort would start from a clean slate. We could not learn how to design and build a new aircraft from what we learned by building earlier aircraft because they have nothing in common.

Fortunately, a very large amount of commonality exists in both natural systems and engineered systems. General models capture that commonality. The generalization spectrum allows models of varying levels of generality to be defined.

Conceptualization

Each model in the model space is represented in some notation or language, or combination of languages, and represents a domain or universe of discourse. The conceptualization dimension ranges from models in the languages or notations of problem domains (such as banking or aircraft control) to models in the abstract notations of mathematics, and in the case of software engineering, languages of hardware processors. Corresponding to the representation differences are differences in concepts or ontologies. For example, in the banking domain, key concepts are account, withdraw, deposit, balance, and transfer. In the geospatial domain, key concepts are map, contour, elevation, feature, thematic layer, and projection. In the aircraft control domain, key concepts are frequency response, transfer function, feedback, airframe, equilibrium, and vehicle motion. In the computer processor domain, key concepts are load, store, add, branch, memory address, and register. The computer processor uses this basic set of concepts to solve problems in banking,

* A more complete mapping is discussed in Chapter 12.

geospatial information, aircraft control, and all other problem domains. Note that we understated the distance between the computer and problem domains because we described the concepts in all these domains using the same language — English; but even so, the conceptual distance between them is large. Thus, two models can represent two different universes of discourse and be represented in the same language. Conversely, two models can be represented in different languages and describe the same universe of discourse.

Two primary relations between models, which often occur in parallel, in the conceptualization dimension are (1) translation from one language to another and (2) transformation from one universe of discourse to another. A universe of discourse is the set of concepts to which the words of a language refer. A common example is translation and transformation from the language and universe of discourse of users to a design language such as SysML or Specification and Description Language (SDL), and then to a mathematical form such as differential equations or difference equations. An example of software system model translations: problem domain notation and concepts to formal specification to UML to Java to binary machine language.

In the multidisciplinary world of SE, even different components of the same system can come from different universes of discourse. For an example given earlier, a commercial aircraft, the application domain is transportation. The internal parts represent domains that include power systems, airframe design, communications, navigation, flight control, computer hardware and software systems, hydraulics, and many others. Each of these has its own universe of discourse.

A third relation is semantics, which is a mapping or interpretation of a model into another model — for example, interpreting a formal model into an application model. A given model can have multiple interpretations.

Figure 5.6 shows the semantic range of concepts and universes of discourse of systems and models in the model space context. The application domain has its universe of discourse in the problem space. The solution space and model space need to address that application universe, at least in terms of external views of the solution. The models and internal solutions in the model space and solution space typically add other universes of discourse that extend the semantic range far beyond the problem space, as described above.

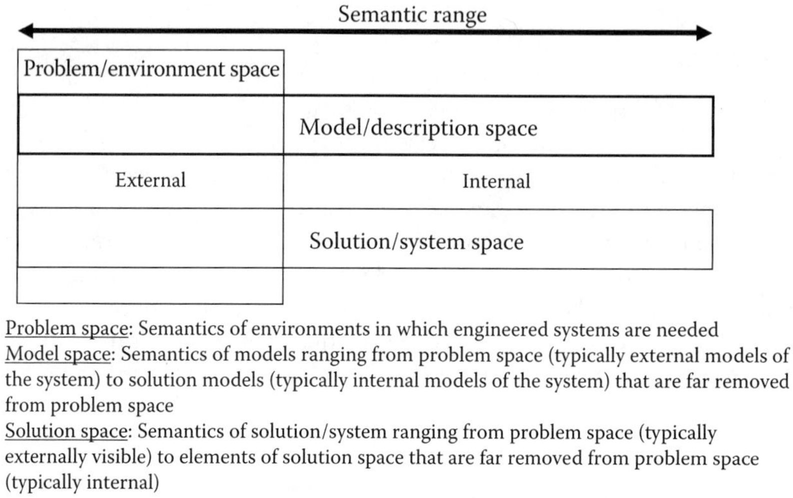

Figure 5.6 Semantic range view of model space and its context.

Time

While the other three dimensions of the model space are regarded as structural or static, time is the temporal or dynamic dimension. The aspect of time that motivates this dimension is *change*. We believe that traditional systems engineering has given insufficient attention to change in systems engineering processes and in the assumptions about engineered systems and their environments. MOSES assumes continual change as the norm, not the exception. This stance reflects the inclusion of complex systems in SE. An important feature of complex systems is change, reflected in growth, learning, adaptation, and at a population level, evolution.

One view of the time dimension is a range of absolute or elapsed time. The concept of life span exists in this view. For example, a system has a life span. A mechanical system in traditional terms has a period of development, followed by operation, retirement, or termination. An organic system has a life span that proceeds from birth to growth to maturity to decline to death. Chapter 9 shows a combined life-span view consisting of formation, maturation, and full functioning. An alternative to linear time is the concept of cycles, also discussed in Chapter 9.

Another view of this dimension is a range or degree of change. Elements of the model space experience a whole spectrum of change across multiple time scales, from very rapid continual change to very slow change or, in the extreme, no change at all.

A third view is concerned with dynamic processes. MOSES defines several categories of dynamic processes, including dynamic behavior exhibited by engineered systems, change and maturation of engineered systems, change of models, and models of the process of collective actualization. The latter is discussed in Chapter 13. The others are discussed as levels of change in Chapter 9, which discusses the time dimension in more depth.

View

The concept of view is an important and flexible tool in MOSES. Informally, a view is a portion of one or more models, or an aspect of one or more systems perceived as being useful for some purpose. A view does not exist in isolation. It exists with respect to a thing being viewed, which we call the target or object of the view. The target can be more or less any ontological element in SE, including another view, a region, or a collection of things; but most often in our discussions the target will be a model or a system. A view is an abstraction of, or a subset of information about, its target.

Also associated with a view is a viewpoint, which is a point in a physical or conceptual space-time region from which a view is observed. Viewpoint carries with it an implication of an observer or a stakeholder, an entity that observes the view. Typically, the information that constitutes the view is defined on the basis of the interests or concerns of an observer or stakeholder, or a class of observers.

Two important requirements for a view are (1) separation and (2) integration. These sound somewhat contradictory but both must be true. It is self-evident that a view must be separable from the rest of the information about its target. That is, after all, the primary motivation for defining a view. But a view must also be integrated with the rest of the information about its target. In a minimal sense, a given piece of information about a target must be consistent in all views that include it. In a broader sense, the information included in all views of the same target must form a consistent integrated coherent whole because the total set of information about the target must be consistent and coherent.

The way views treat separation and integration is one example of an important use of views in MOSES, namely that of supporting and reconciling contrasts, contradictions, and dualities — common features of complex systems. This aspect of views is discussed further in Chapter 10.

Examples of how views have been used in engineering and building architecture include drawings from different perspectives. In information systems, database views and architecture views are commonly used. In MOSES, the concept is given increased emphasis and regarded as a core element. The approach is to define a common starter set of views and define an extensible method for defining additional views as needed.

Views appear in the model space structure in several ways. First, each dimension is a different view. That is, the engineering of a system involves elements of (1) composition, (2) the use of general and specific models, (3) various languages and universes of discourse, and (4) models of time, change, and processes. Each of the four dimensions is a view that supports separate consideration of one of these areas of concern, while at the same time recognizing that the four areas form a unified whole.

Second, we can define views within each dimension. For example, in the composition dimension, we can focus on a selected set of levels, which is a view of the full set of composition levels. A key distinction in the composition dimension is that between external and internal views, whose intertwining structure generates multilevel composition hierarchies. In the commonization dimension, we can define a view as selected levels of a class or taxonomy hierarchy, or we can focus on a class-subclass view versus a category-individual view.

Third, and perhaps most importantly, we can define views that cut across the four model space dimensions. One common example is the set of models associated with a given system. These models cut across all four dimensions. Another example is aspects, which are cross-cutting concerns that are naturally defined in terms of views. Examples of aspects include error-handling, security, and safety.

Relations among Dimensions

We have separately introduced each dimension of the model space, which means we have discussed the separation aspect of each dimension as a view. We now discuss some of the integration and unification aspects of the dimensions.

Any given model occupies a point, a line, or a region (3D or 4D) in the model space, in that it is positioned somewhere along the spectrum of each dimension. A model represents one or more composition levels somewhere in the composition dimension, one or more generalizations or categorization levels somewhere in the commonization dimension; a language point on the conceptualization dimension, and one or more points in time on the time dimension. Furthermore, while we do not claim that the four dimensions are mutually orthogonal (i.e., independent of each other), we do claim that models can vary separately along the four dimensions. For example, two models can be at the same position on three of the dimensions and differ on the fourth one. Some relations among dimensions are commonly recognized. One relation is that a transformation of one model to another often involves a simultaneous translation to a different language and a decomposition into parts. Another is that a general model is often at a high composition level, and a specific model is often at a low composition level. In fact, we tend to think that descriptions of large systems are naturally general and descriptions of small components are naturally specific.

Other relations among dimensions are less commonly recognized. One is that a model can be general and at a low level on the composition spectrum. An example is the quantum model of atoms that describes rules concerning atomic orbitals. This model is general because the same rules apply to atoms of all elements, which means that the rules apply to all atoms on Earth, and presumably to all atoms in the universe. That is fairly general. Another less commonly recognized relation is that a model can be specific and at a high level on the composition spectrum. For example, we can specify in detail the interfaces or public access points of a large organization. We deliberately call out these two relations because in MOSES it is important to exploit the ability to have both general and specific models at all levels of the composition dimension.

Relationships among the dimensions can be discussed further if we look back at Figure 5.3. In the model space domains box, a set of five models on the left side maps into a set of three models. Then on the right side, a set of three models maps into a set of five models. The five models on both sides correspond to models of problems and (external) solutions, respectively, in five corresponding application domains. This mapping has meaning simultaneously in the commonization, conceptualization, and composition dimensions of the model space.

- In the commonization dimension, it represents the transformation of relatively narrowly defined models of each relatively homogeneous application domain, where each model has a region of modest size, to the more broadly defined multidisciplinary models of the SE model space, where at least some models have a region of greater size. Examples are mathematical equations, software programs, and complex system models (we are not assuming here that these are three disjoint sets, e.g., a given complex system model may be a mathematical equation represented in software), all of which can apply to multiple application domains.
- In the conceptualization dimension, the mapping represents translation from application domain languages to modeling domain languages, or transformation from a UD of the application domain to a UD of the modeling domain, or both. Example 1: From a desired behavior model of an aircraft to a set of differential equations that characterizes the movement and force of rigid bodies through a medium. Example 2: From a model of banking transactions to a formal and general transaction model in a logic language.
- In the composition dimension, we decompose the application domain problem into parts. More specifically, we define external models of the desired system, and then define internal models of the desired system, and external models of each part.

Model Space Implementation

The implementation discussion in this section consists of three parts: (1) content, (2) management, and (3) existing repositories.

Content: Models

An important partition of the model space was discussed earlier: the general models that represent the SE BoK versus the set of models of specific systems, where each project is characterized as a workspace. To help understand model space content, we first discuss the vision of the SE model space of the future and then assess where SE is relative to this vision.

The Vision

The model space consists of models across the full scope of SE. This includes all application domains and external solution domains; all internal solution domains; all supporting disciplines; at all levels of composition and commonization, in all conceptualizations required, in a variety of languages including visual, computational, and mathematical, at any point in time or any time range; all views; properties of models and systems; relations among models; and information about models. The knowledge space for any application domain includes the models from both natural and artificial supporting sciences for that domain.

Four states of affairs exist in the path from the most traditional form of knowledge to the goal state of the model space. First is knowledge in the form of paper, such as books and journals. Second is the loading of the knowledge into a digital repository, albeit still in the image of the original book or journal article. Third is the parsing of that knowledge into individual units and defining relations among the units. Fourth — the goal state — is expressing the knowledge in the form of models that can be executed or semantically analyzed. An example of the third state is the DOORS (Dynamic Object Oriented Requirements System) tool for managing requirements. DOORS separates and manages each individual requirement statement and the relations among the requirements, but the text of each requirement is treated as a character string with no semantics for DOORS. SysML is an example of a language that to a large degree represents the goal state, but not entirely. The treatment of requirements statements in SysML is the same as that in DOORS, in that a requirement statement is a character string instead of a model with semantics. An important corollary of the progression of states is the evolution of configuration management from controlling paper documents to controlling or tracking models in the model space repository. In the goal state documents will still exist, but they will be in the form of reports, projections, or views that come out of the model space.

The models in the goal state reflect the kinds of information expressed in MOSES specifications, and are machine processable, as previewed in the chemical engineering handbook example in the section entitled "Model Space Context and Requirements." Desirable properties of languages used to represent models in the model space (associated with the conceptualization dimension) are that they be machine processable, capture appropriate semantics, and (collectively) support textual, graphical, and mathematical forms. A candidate family of languages for representing all these models in the model space would be Extensible Markup Language (XML) and various languages based on XML, including graphical languages such as Scalable Vector Graphics (SVG), Mathematical Markup Language (MathML), and ontology languages such as OWL. Another family of languages is UML and its related languages SysML and OCL, and its meta-language MetaObject Facility (MOF). XML and UML are integrated through the MOF-compliant XML Metadata Interchange (XMI) standard. A more extensive discussion of these and other languages, is given in Chapter 8. The question of what languages to use to represent the models in the SE model space, and the interchanges among them, needs considerably more investigation.

When using the model space to support the engineering of a specific system, the entry point is a specific application domain portion of the model space. In the section entitled "Model Space Context and Requirements," we sketched a model space vision of chemical engineering. For SE, the same approach would be used. Suppose we are engineering a new commercial aircraft. We set up a project workspace for this effort and begin by reaching into the general public space in the domain of commercial aircraft — including models of previous similar aircraft. The general models here have links to all the necessary parts of the model space, including all the domains that participate in the engineering of an aircraft. The models reflect requirements, designs, simulations, system or component static and dynamic properties, etc. in all these domains. These general models are tailored or specialized as necessary and put into the project workspace for this specific aircraft, using processes that are described in Chapter 13.

Current State

For SE, there are two issues. The first is maturity and completeness. Is the SE BoK mature and complete in some important sense? The second issue is readiness for the knowledge space portion of the model space. Is the knowledge in a form that can be captured in models? In this section the second issue is addressed because the context is the model space. The first issue is discussed in Chapter 7 in the context of the SE BoK.

The form of knowledge that exists in a domain or discipline affects its state relative to the model space view of BoK. The SE knowledge that exists is in a preliminary form for the model space, with some material machine manipulable but much of it in informal narrative and requiring significant work to get it into more usable model form.

More specifically, because the SE BoK is multidisciplinary, it is not uniformly ready to be captured in models. We briefly discuss the readiness of some key supporting disciplines and then SE proper.

- *SS:* Systems science, while not entirely in model form, is partially in model form and lends itself to being easily turned into models. Therefore, SS would score fairly high on this criterion of readiness for an SE model space.
- *CS:* The knowledge of complex systems is already in model form to a significant degree. In fact, ironically, because this newest area of SE is more oriented toward system models than engineering process models, CS knowledge is more ready for the SE model space than much of the traditional SE literature. From the viewpoint of the SE BoK, the larger issue with CS knowledge is that it has not been incorporated into the overall SE BoK, as discussed in Chapter 7. When it is incorporated, it will already be in the form of models.
- *Mathematics:* The mathematics that supports SE ranges from differential equations to logic, and is in model form and ready for the SE model space.
- *Software:* The software community has four perspectives on the issue of knowledge that is ready to be captured in a model space. First, it has defined a SWEBoK guide, which is based on the engineering process and life cycle approach, and is in the spirit of TSE more than CS. This knowledge is targeted more to the actualization space than the model space of SE. Second, the software community in the last decade has been a leader in bringing model-based approaches to the forefront, as mentioned in Chapter 4. Third, software has codified to some extent certain conventional designs in the form of design patterns, architecture patterns, and reference architectures. Fourth, software has a strong tradition in simulation modeling. The combination of the model-based engineering approach, the capture of knowledge in patterns and reference models, and the simulation knowledge mean that software knowledge is to this extent ready for the model space.
- *Main SE:* SE proper (as distinguished from all the supporting disciplines) has parallels to the software situation. It has three perspectives. First, like software, SE has defined an SEBoK guide, which is based on the engineering process and life-cycle approach, in the spirit of TSE, and targeted more to the actualization space than the model space of SE. Second, SE has recently begun leveraging the work of the software community in moving toward the model-based approach to engineering, although it is not as far along as software in this regard. Third, SE has a strong tradition in simulation modeling. SE has done much less work than software engineering in the area of capturing design knowledge in the form of patterns and models. As a whole, SE proper has some degree of readiness but still has a considerable amount of work to do to be ready for the SE model space.

Management: Tools and Processors

We briefly discuss tools to manage the model space as a model repository or database, and processors that manipulate the models. Processors can be humans, software, computer hardware, or maybe physical engines or agents.

Processors perform certain actions on models, including store models, transform one model into another, execute a model, and analyze a model to determine its properties or whether it matches or satisfies another model. There are several classes of processors:

- *Model storage and management tools:* These processors store and keep track of models, in the manner of typical repository or database management tools.
- *Navigators, query tools:* These processors help a user navigate, browse, and search the model space.
- *Analyzers, comparators, checkers, verifiers:* These processors analyze and help determine properties and qualities of models, compare models for differences, and evaluate models to see if they satisfy other models.
- *Transformers, generators:* These processors transform a model into another model, or generate new models.
- *Visualizers, viewers, report generators, projectors:* These processors read model contents in the model space and present them to users.

Existing Repository Approaches

We are not aware of any repository that corresponds in all respects to the model space described in this chapter. However, there is considerable activity in implementing repositories that have some of the model space characteristics, including varying degrees of content and tools. We briefly identify a few of these efforts.

IEEE 1220

The first example is not a repository but rather the prescription for an integrated repository for an SE project, as defined in IEEE SE Standard 1220 (IEEE 2005). The content corresponds closely to the project workspace portion of the MOSES model space. The repository data includes plans, schedules, designs, drawings, and many other items that cover the software, hardware, human interfaces, management, and overall system. The scope of this is consistent with the project workspace in the model space. The difference is that in the IEEE description, these items are conceived as data, while in the model space they are conceived as models. More specifically, the model space would have an integrated set of underlying models that would produce many of these data items as reports. This is a level of integration and built-in consistency that is not explicitly called for in the IEEE repository or offered by most traditional repositories.

A more model-oriented example is a proposed content standard for computational models (Hill et al. 2001). That article identifies 165 elements grouped into the following categories: identification information, intended use, description, access or availability, system requirements, input data requirements, data processing, model output, calibration efforts and validation, and metadata source.

Digital Library Reference Model

The DELOS Digital Library Reference Model (DLRM) (Candela et al. 2007) has the ambitious goal of identifying the concepts and relationships that characterize the essence of the digital library universe. It depicts this universe as a three-tier framework of digital library, digital library system, and digital library management system. The all-encompassing tier is the digital library. The digital library system is application software that provides the library functionality, and the management system provides the underlying system software support. The core concepts of DLRM are five digital library domains — user, content, functionality, quality, and policy — and one digital library system domain — architecture.

In addition to the domains, DLRM concepts that relate to the model space include resources, views, and ontologies. The library contents are conceived as resources, and are represented in views that serve the individual users. Ontology is used as a basis for the library concepts and the content categories. Not all the structure and specification information of the model space is visible in the reference model but the reference model seems to be close enough that it could serve as the basis for tailoring to the model space and MOSES.

Sample Repositories

A prominent European example of a repository effort is the DRIVER project (DRIVER 2008), now in the DRIVER II phase. DRIVER stands for Digital Repository Infrastructure Vision for European Research. It is attempting to put together a confederation of digital repositories to serve as a knowledge infrastructure of European research results, to make the results accessible to a wide audience. In DRIVER II, they are adding research data as well as papers.

OpenDOAR is a directory of academic open access repositories (DOAR 2007). It contains a list of 1000 repositories worldwide.

The Registry of Standard Biological Parts, or "biobricks," at MIT that we introduced in Chapter 1 (Registry 2007) is at a smaller scale but does represent another aspect of the model space.

An index of common crystal lattice structures, with a graphical representation and information about each one, is available at Navy (2008).

The availability of *Perry's Handbook*, which was discussed earlier, on CD-ROM is a small next step toward a repository. A big step forward would be to capture all the models in this handbook as models, instead of as text.

Boeing and its partners used a common development environment called the Global Collaboration Environment for the development of the 787. This contains a shared repository of all documents and specifications, as well as a means of interacting (e.g., virtual conference meetings). It corresponds to the workspace area of the MOSES model space, containing the specific models for a given project/system. It is global in the sense that this project was a global collaboration effort.

This small sample set illustrates a wide range of repositories, all related in some way to the SE model space.

Some Model Space Literature Sources

This section need only be read if the reader is interested in examples from the literature on model space ideas and dimensions.

The idea of a model space is not new. We briefly review here a few sample discussions of the concept in the literature. Some of the literature is from software engineering, which has focused heavily and increasingly on modeling. The first part of this section discusses the idea of a model space, and the second part discusses possible variables or dimensions inside it.

The Model Space Idea

Several discussions of concepts that map well to the model space are reviewed here.

Christopher Alexander: Actual World, Pictures, and Patterns

The idea. In 1964, Christopher Alexander published a book entitled *Notes on the Synthesis of Form* (Alexander 1964). He introduced three levels of design: unselfconscious, self-conscious, and formal. He describes unselfconscious design as learning a mature but implicit architecture and applying and adapting the resulting system, such as a simple house, to maintain the fit between form and context (Alexander 1964, Chapter 4). There is a direct correspondence between actual world context and actual world form. Self-conscious design is where the human forms an informal mental picture of the context and attempts to design to that mental picture (Alexander 1964, Chapter 5), which Alexander says is generally wrong. Formal design is where the mental picture is formalized and therefore can be examined and analyzed and made correct, and design is to that formal picture or model. Alexander states that the evolution of design starts with unselfconscious design, and that is followed by self-conscious design, and then by formal design, which is his desired method. In each case, he describes how form fits context, or how design satisfies requirements.

In the 1970s, Alexander emphasized a different approach: patterns for use in the field of physical building architecture and community planning (Alexander et al. 1977, Alexander 1979). A pattern is a general model or design — for example, of a room, a house, or a neighborhood — that is based on accumulated knowledge and experience, and can be applied and specialized for a given environment. In the past decade, patterns have become popular in the fields of software and systems design and architecture. More recently, the pattern concept has been used to describe both desirable and undesirable ("anti-patterns") behaviors and structures in a wide variety of other areas, including management and business-oriented fields. Examples of management and business-oriented patterns include configuration management and organizational structure.

Correlation with the Model Space

- Unselfconscious design — takes place entirely in the problem space ("context") and solution space ("form"). No model space exists but it introduces some elements of MOSES: common architecture, adaptiveness, and balance of stability and change. We would call this pre-engineering.
- Self-conscious and formal design — is consistent with problem space, model space, solution space relationships. Context maps to problem space; mental pictures map to specifications, and, if made explicit, model space; and form maps to solution space.
  - In the overall model space, the Alexander distinction between self-conscious and formal design (Alexander calls it "the program") maps to the MOSES distinction between informal and formal models.

- In the conceptualization dimension of the model space, the distinction between self-conscious and formal design can be conceived as mapping to the MOSES distinction between informal user-defined requirements represented in application domain terms and a requirements spec represented in more formal engineering terms.
- Alexander patterns concept maps to the model space, and more specifically maps to general models in the commonization dimension that capture knowledge of a class of solutions to a class of problems in an application domain. The actual Alexander patterns map to MOSES models of organic systems and hybrid (mix of organic and mechanistic) systems. The pattern language (Alexander et al. 1977) maps to the MOSES concepts of language and UD in the conceptualization dimension of the model space.

Thus, Alexander first emphasized formality in design and later emphasized knowledge in the form of patterns. MOSES uses ideas from both. MOSES adopts the Hume distinction (Hume 1952, 24) between matters of fact (or contingencies) and relations of ideas (or necessities). The former category includes statements such as "the sun will rise tomorrow" and, more generally, scientific theories. The latter category includes mathematical relations and logical propositions. The difference between Alexander's 1970s emphasis on experience-based patterns and his 1964 emphasis on formal models aligns rather cleanly to Hume's distinction. In the model space context, if we assume that only conceptual models exist inside the model space, then the relations to ideas exist in the models inside the model space, and the contingent relation to matters of fact exists in the interactions of the model space with the problem space and the solution space.

Engineering in general, including SE, includes both contingent facts and necessary ideas. Contingent facts are present in the physical constraints on a solution, and in the question of whether the engineered solution solves the problem. Necessary ideas are present in some of the relations among various models, such as formal requirements specs, designs, and simulations.

Bertrand Meyer: What Does Software Model

The idea. Meyer (1997, 229–231) makes the point that a software system is not a model of reality (i.e., a model of the real world in the problem space), but a model of a domain model of the problem space. In terms of the model space context, there is work done in the application domain in defining models of the real world, however informal, and those models set the stage for requesting and incorporating an engineered solution.

Example: Suppose the real world at some location has a river with people living on both sides of the river but with no bridge across the river. Suppose the application domain is transportation. The local government probably has some models of this portion of the real world, including descriptions of boat traffic across the river, boat traffic passing through up and down the river, speed and volume of the river, flood levels, number of people on both sides of the river, etc. These are models in the application domain — say, AM1, AM2, AM3, AM4, AM5. Now the government decides that they need a bridge across the river. They hire an engineering firm to design it. In the discussions, the government offers the models they have, and the engineering firm says they would like to have models AM4 and AM5, which the government provides. The engineering firm then examines these models and decides that they can use AM5 in their engineering effort, but not AM4 because it is out of date or in some way not adequate for their engineering purposes. The engineering firm also develops models EM1, EM2, …, EM9 in the course of their engineering. These include projected bridge traffic, environmental models, requirements specifications, design, etc. Given this situation, we would say that models {AM1, …, AM5} are application domain/problem space models, and models {AM5, EM1, ..., EM9} are SE model space models. Obviously, model AM5 exists in both the problem space and the model space.

Correlation with the Model Space

Meyer's idea about software being a model twice removed, and his distinction between the real world and the models of the real world, generalizes to SE. The primary contribution is to help clarify the scope of the model space and the boundary between the problem space and the model space. The application domain includes the real world associated with that domain, as well as the domain models of the real world used in that domain. We do not focus on those models in this book. We focus on the models in the model space — which include models of the application domain and models that are used in both the application domain and in the model space.

Michael Jackson: The World and the Machine

The idea. Michael Jackson (1995a), distinguishes between the world and the machine. He describes multiple overlapping facets or views of the world and machine, including description, phenomena, engineering, and problem facets. DW = descriptions true of the world. DM = descriptions true of the machine. The overlap consists of descriptions that are true of both. Jackson (1995b, 9–11, 73–76) refers to the world, the application domain, and the machine. He defines the application domain in two slightly different ways: first, as the specific environment in which a specific system operates; and second, as the intangible concepts that reside in the machine.

Correlation with the Model Space

To clearly distinguish Jackson terms from MOSES terms, we show the correlation of concepts in Table 5.2. In particular, the MOSES definition of application domain maps to Jackson's world concept, not to Jackson's application domain concept.

Bruce Blum: Models and Domains

Bruce Blum (1992, 33–35) describes the essential software process as a series of three transformations: from an application domain need to conceptual models; from conceptual models to formal

Table 5.2 Correlation of Jackson and MOSES Model Space Concepts

Jackson Concept	*MOSES Concept*
World	General problem space of application domain
Machine	Solution, system
Application domain: first definition (environment)	Specific system environment in application domain
Application domain: second definition (concepts)	Application domain universe of discourse applied to specific system
DW	Model of application domain inclusive of specific system, (i.e., with the system operating in the application domain)
DM	External and internal models of the system
Overlap (DW $\cap$ DM)	External model of system

models; from formal models to an implementation domain software product. More specifically, Blum describes the process this way:

- From a need recognized in an application domain into a conceptual model that describes the solution in terms of the domain vocabulary.
- From the solution as expressed in the conceptual model into some formal model that defines what the software is to do. The model is considered formal in that it defines what the product is to do and can determine if an implementation satisfies its definition. Blum (1992, 83) says the formal solution is documented in a software requirements spec.
- From the formal model into a software product that is correct with respect to that model.

What Blum treats as the essential software process MOSES treats as (the software part of) the collective actualization space. In MOSES, the model space is a structure, an organized model repository in which models are stored and processes are performed.

We show the correlation of Blum concepts with MOSES concepts in Table 5.3.

Jim Neighbors: Hierarchy of Domains

Jim Neighbors describes a hierarchy of domains for software engineering that consists of application domains, modeling domains, and execution domains.

"We define the *level of abstraction* of other domains with respect to an executable domain to be the longest acyclic path from any of the refinements of the domain's components to the executable domain. The domains with the highest levels of abstraction are called application domains, while the domains in the abstraction levels between the application domains and execution domains are called modeling domains" (Neighbors 1989, 307).

Application domain experts construct application domains, and modeling domain experts construct modeling domains. The basic process is transformation.

Neighbors' three domain categories map respectively to MOSES application domains in the problem space, the model space, and the solution space.

Wladislaw Turski: Specifications and Domains

Turski and Maibaum (1987, 9–12) discuss three entities and their relations: (1) application domain, (2) specification, and (3) programming language domain. The relations are that a specification

Table 5.3 Correlation of Blum and MOSES Model Space Concepts

Blum Concept	*MOSES Concept*
Application domain	Application domain in problem space
Need	Application domain problem or gap model
Conceptual model	User requirements model in application domain terms
Formal model	Formal external system model
Software product	Internal system model in programming language or computer processor language; exists both in the model space and in the solution space
Implementation domain	Solution space

abstracts an application domain, and a program in the programming language domain satisfies the specification. These concepts map to MOSES: the application domain is the same as in MOSES. The specification maps to an external specification of a solution, and the computer program maps to an internal specification of a solution, where the solution is software. The specification and the computer program are both in two places: the model space and the solution space. This formulation is a clear way to make the important point that an application domain sees the external view of a solution, so in that sense the external view or the specification is the intermediary between the problem space context and the internal solution, to some degree hiding both from each other (see the encapsulation discussion in the section entitled "Composition"). This point is similar to Jackson's concept of overlap discussed above.

Turski (1986) also discusses the essence of software, saying that the essence of software lies in its being a description of two in general different structures: (1) application domain and (2) hardware. This leads to a specific software view of the model space context that is presented in more detail in Chapter 8. For the purposes of this chapter, Turski's model of the essence of software maps to MOSES in that the external view of software addresses the application domain, while the internal view of software is represented in a language the computer hardware understands.

Model Space Dimensions

A considerable literature exists that discusses each of the variables in the model space. A few examples are presented here.

Philosophy, Especially Ontology

A significant element of ontology in philosophy is the categorical structure of reality (Honderich 1995). Philosopher ontologists argue about what occupies the second spot of the ontology hierarchy or the hierarchy of being. The top is "being" or "entity." "Some schemes take the division between universals and particulars as the next step in the hierarchy, others the division between abstract and concrete entities." (Honderich 1995, 634).

Simons (1987) wrote a treatise on mereology or part–whole theory in the context of ontology. Philosophical debates about the concept of emergence have occurred over the past century. Emergence is directly related to part–whole structures.

These concepts from ontology illustrate several points concerning MOSES model space dimensions:

- The ontology treatment of the hierarchical structure of categories of being clearly illustrates, and occurs in, the *commonization* dimension of the model space.
- The basis of the debate over the second ontology level is that there is one correct second-level set, and we need to determine what that is. The MOSES position on this issue is that categories and generalizations are created by humans and therefore there is not one inherently correct second level. There may be more than one second level that is useful; each would represent a different view of the hierarchy of being, and therefore the model space concept of *view* can inform and potentially resolve this debate. The two sets mentioned represent two views.
- Mereology and emergence illustrate and occur in the *composition* dimension of the model space.

Whitehead's Discussion of Generalization and Stability/Change

Alfred North Whitehead (1911, 11) made this statement: "To see what is general in what is particular and what is permanent in what is transitory is the aim of scientific thought." The quote has two interpretations in this book. In the context of Chapter 3, it means that a key goal of MOSES is to capture SE concepts that have some generality across the broad scope of SE and some stability across time. In this chapter, it reinforces the importance of the commonization dimension and the time dimension in terms of a locus for capturing SE-related knowledge in the form of general models and stable models.

Information Systems Engineering Sources

Information systems engineering has made significant contributions to several dimensions in the model space, including:

- Structured analysis methods made a strong contribution to modeling multilevel composition, which supports composition hierarchies.
- Object-oriented analysis methods made a strong contribution to modeling multilevel commonization — both generalization in the form of class hierarchies and categorization in the form of class-object structures and metamodel levels.
- In the conceptualization dimension, numerous contributions have come in the form of formal languages, modeling languages, formal ontologies, and semantics.

Conceptual modeling over several decades has embraced a variety of areas of study, especially in information systems, that include artificial intelligence, semantic networks, knowledge representation, and database/knowledge base modeling. An example discussion of knowledge representation in Mylopoulos and Levesque (1984) identifies the organizational axes or dimensions offered by semantic networks for structuring a knowledge base. These include:

- Aggregation, which relates an object (John Smith) to its components or parts (can be different views; as physical object, parts are head, arms, etc.), can be applied recursively to define parts of each component, etc.
- Classification, which associates an object (John Smith) with a generic type (Person); sometimes applied recursively to define meta-types.
- Generalization, which relates a type (Student) to a more generic type (Person), recursively organizes types into a generalization hierarchy.

Clearly, the aggregation axis corresponds to the composition dimension in the model space. The classification and generalization axes correspond, respectively, to the categorization and generalization aspects of the commonization dimension in the model space. The recursive feature of all three axes also corresponds directly to the repeating multilevel hierarchical structures of the model space dimensions.

Summary

The design of the SE model space has been presented. Because the model space is a dominant feature of MOSES, each of the variables is further elaborated in Chapters 6 through 10: Composition, Commonization, Conceptualization, Time, and View.

References

Alexander, C. 1964. *Notes on the Synthesis of Form*. Cambridge, MA: Harvard University Press.

Alexander, C., S. Ishikawa, M. Silverstein, M. Jacobson, I. Fiksdahl-King, and S. Angel 1977. *A Pattern Language: Towns – Buildings – Construction*. New York: Oxford University Press.

Alexander, C. 1979. *The Timeless Way of Building*. New York: Oxford University Press.

Avallone, E., T. Baumeister III, and A. Sadegh, Eds. 2007. *Marks' Standard Handbook for Mechanical Engineers, 11th ed.* New York: McGraw-Hill.

Blum, B. 1992. *Software Engineering: A Holistic View*, New York: Oxford University Press.

Candela L., D. Castelli, N. Ferro, Y. Ioannidis, G. Koutrika, C. Meghini, et al. 2007. The DELOS Digital Library Reference Model: Foundations for Digital Libraries, v0.96. DELOS Network of Excellence on Digital Libraries. November. http://www.delos.info/files/pdf/ReferenceModel/DELOS_DL ReferenceModel_096.pdf. (Accessed January 11, 2009.)

DOAR. 2007. The Directory of Open Access Repositories — OpenDOAR. University of Nottingham. http://www.opendoar.org/. (Accessed January 11, 2009.)

DRIVER. 2008. Digital Repository Infrastructure Vision for European Research. http://www.driver-repository.eu/. (Accessed January 11, 2009.)

Edman, I. (Ed.). 1928. *The Works of Plato*. New York: Random House. (Original copyright: Simon and Schuster.)

Green, D. and R. Perry. 2007. *Perry's Chemical Engineers' Handbook, 8th ed.* New York: McGraw-Hill.

Hill, L., S. Crosier, T. Smith, and M. Goodchild. 2001. A content standard for computational models. *D-Lib Magazine*, June 2001, Vol. 7, Issue 6. http://www.dlib.org/dlib/june01/hill/06hill.html. (Accessed January 11, 2009.)

Honderich. T. 1995. *The Oxford Companion to Philosophy*. New York: Oxford University Press.

Hume, D. 1952. *An Enquiry Concerning Human Understanding: And Selections from a Treatise of Human Nature*. La Salle, IL: The Open Court Publishing Company.

Hybertson, D. 2001. A uniform component modeling space. *Informatica*, 25(4): 475–482. http://www.mitre.org/work/best_papers/best_papers_01/hybertson_space/. (Accessed January 11, 2009.)

Hybertson, D. 2005. *Testing Comercial-off-the-Shelf Components and Systems*, S. Beydeda and V. Gruhn, Eds. New York: Springer-Verlag. p. 213–237.

IEEE. 2005. *IEEE Standard for Application and Management of the Systems Engineering Process*, IEEE Std 1220-2005, New York: IEEE Computer Society.

Jackson, M. 1995a. The world and the machine, in *Proc. 17th Int. Conf. SWE*, ICSE 95, 23–30 April 1995. Seattle: ACM, p. 283–292.

Jackson, M. 1995b. *Software Requirements & Specifications: A Lexicon of Practice, Principles and Prejudices*. Wokingham, England: Addison-Wesley.

Meyer, B. 1997. *Object-Oriented Software Construction, 2nd ed.* Upper Saddle River, NJ: Prentice Hall.

Mylopoulos, J. and H. Levesque. 1984. An overview of knowledge representation. In *On Conceptual Modelling: Perspective from Artificial Intelligence, Databases, and Programming Languages*. Brodie, M., J. Mylopoulos, and J. Schmidt, Eds. New York: Springer-Verlag, p. 3–17.

Navy. 2008. Crystal Lattice Structures. Naval Research Lab. http://cst-www.nrl.navy.mil/lattice/index.html. (Accessed January 11, 2009.)

Neighbors, J. 1989. Draco: A method for engineering reusable software systems. In *Software Reusability, Concepts and Models*, Vol. 1. T.J. Biggerstaff and A. Perlis, Eds. Reading, MA: Addison-Wesley/ACM Press, p. 295–320.

Registry. 2007. Registry of Standard Biological Parts. MIT. http://parts.mit.edu/registry/index.php/Main_Page (Accessed January 11, 2009.)

Simons, P. 1987. *Parts: A Study in Ontology*. New York: Oxford University Press. (Paperback edition, 2000.)

Turski, W. 1986. And no philosophers' stone, either. In *Information Processing '86, Proc. IFIP Congr.*, H.J. Kugler, Ed., Dublin, 1–5 September. Amsterdam: Elsevier Science Publishers B.V. (North-Holland), p. 1077–1080.

Turski, W. and T. Maibaum, 1987. *The Specification of Computer Programs*. Wokingham, England: Addison-Wesley.

Whitehead, A. 1911. *An Introduction to Mathematics*. New York: Henry Holt & Co. (Republished 1958. New York: Oxford University Press.)

Chapter 6

Composition

Composition is the dimension of the MOSES model space that addresses wholes and parts, based on the notion that a system consists of two or more interacting parts. This chapter discusses common static and dynamic models of composition, including static structures, dynamic interactions, and boundaries.

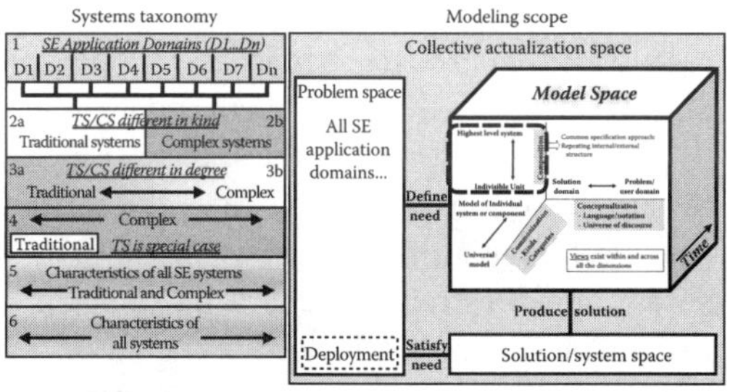

Familiar Composition Elements and Characteristics

Elements

We start with a common SE scenario, namely a situation in which we want to engineer a commercial aircraft. We will call this our system of interest (SOI). Associated with our SOI are five elements, or types of models, that are commonly defined in the course of engineering the SOI. Before we state what the five are, we discuss the example so that when we introduce the five elements, they will be better understood.

When we are ready to design and build the aircraft, we consider several different perspectives. One is the system as a whole. What characteristics should the aircraft have — what passenger capacity, what range, what required take-off and landing distances, what safety features, etc.?

Another perspective is the system as parts: what parts should the aircraft have to achieve the overall characteristics? We assume that we are designing and building a new aircraft early in the twenty-first century and we have several decades of experience available to us, starting with the well-understood operating principle of fixed-wing aircraft. From this we have evolved a common commercial aircraft architecture that reflects the internal domains and supporting domains mentioned previously. A power plant consists of a set of two or three jet engines; an airframe structure provides the necessary surface and a pressurized cabin for passengers; a computer subsystem consists of software, hardware, and network; and a navigation and flight control subsystem includes sensors and mechanical actuators and uses the computer subsystem. Each subsystem can be described separately and built separately. In fact, many of the subsystems are designed and built by engineers in different domains. Aeronautical engineers design the airframe structures and the jet engines, software and electrical engineers design the avionics subsystem, and mechanical engineers design the navigation and control actuators and hydraulics. Many of these, such as the jet engine, may be designed, built, and supplied by a company separate from the company building the aircraft. Systems engineers take all the subsystems and put them together into one coherent system.

In this discussion of system as parts, we have implicitly hinted at another architecture perspective: the relations and interactions among all the parts or subsystems. That is, how do all the parts or subsystems relate to each other, and how do they interact so that the overall functions of an aircraft are achieved?

Another perspective is the aircraft environment, which we define as the set of external elements that interact directly with the aircraft. One element of the environment is the physical atmosphere that surrounds the aircraft while it is flying — air pressure, clouds, temperature, wind, rain, sun, etc. Another element is senders and receivers of communications with the aircraft, such as air traffic controllers. Another element is the airport runways on which the aircraft lands or takes off, and the gates at which it parks for passengers and freight to enter or exit. Another element is the set of other aircraft in its vicinity, which it wants to avoid. Another element is the airline that owns the aircraft, and the mechanics and support staff who maintain and service it. Depending on the purpose of the model, passengers and crew may also be part of the environment.

Another perspective is the interacting elements of the larger encompassing system of which the aircraft is a part. This includes the relations and interactions among these elements: the aircraft itself; the airline industry, the laws and regulations (and lawmakers and regulators) that govern and constrain the characteristics and usage of the aircraft, and the entire infrastructure that enables air traffic, including the set of commercial airports, the air traffic control system, all the routes between the airports, the energy companies that supply fuel, the suppliers of aircraft and airline and airport equipment and services, the global positioning system, communications networks, and the like. It could also include national energy policy, business tax structure and rates, the overall national economic health, and international circumstances that can affect energy supplies and possible destinations that the airline can service.

Characteristics

The perspectives we have presented are all related in some way to the aircraft as the SOI. These perspectives are (1) the SOI as a whole, (2) the SOI as parts, (3) the SOI as a set of relationships and interactions among parts, (4) the direct environment of the SOI and its interaction with the SOI, and (5) the interactions among the elements of the larger encompassing system of the SOI.

In this section we discuss and extend these perspectives and their characteristics. A prominent characteristic implied by this set of five perspectives is the concept of multilevel composition hierarchy. An aircraft is part of a larger system of airlines, airports, etc. In turn, an engine is part of an aircraft. Although we did not mention it, each subsystem can be considered a whole system with its own parts. An engine is a system that has parts; this decomposition can be continued down to the nuts and bolts level — or even down to the level of atoms.* The computer subsystem is a system that has parts; it can be decomposed into computer hardware, software, and networking, and each

* TSE would view decomposition to the level of atom as extreme and out of scope; but because SE now includes nanotechnology, including for aircraft, this level would not be extreme.

of those can be further decomposed — for example, the software down to individual lines of code, or even further to individual symbol level.

This multilevel holarchy structure means that engineers or other stakeholders need to define their level and system of interest. That is what SOI is for — to designate clearly what system is being considered in a multilevel and multisystem context. We discuss this first in a traditional systems engineering (TSE) scope. To a policy maker, the entire airline industry might be the SOI. To a systems engineer, the aircraft might be the SOI. To a computer systems engineer, the system of computer hardware, software, and networks on the aircraft might be the SOI. To a software engineer, the software might be the SOI. And to an aeronautical or a jet propulsion engineer, the power plant or a single engine, might be the SOI.

In the scope of the extended SE discussed in Chapters 1 and 2, the systems engineer role can be expanded. For example, in addition to the aircraft, the systems engineer may bring an engineering approach to the larger encompassing system to address issues such as analysis of alternative airline regulatory approaches, energy policies, airport and route planning, and constraints based on projected increases in air traffic over the next decade; or maybe how all these factors affect each other. Another example is scale-down — maybe the systems engineer will be involved in examining how nano-materials can be used to achieve desired macro features at the aircraft level.

Common Structure Models

Each perspective on the aircraft SOI in the previous section is now generalized to a corresponding set of models or views associated with an SOI. These common views* are:

1. *System as whole:* What are the properties and behavior of the system as a whole?
2. *System as parts:* What are the properties and behavior of each part of the system?
3. *System as connection of parts:* How are the parts related? How do they interact?
4. *System environment connection:* What elements interact with our SOI? How does our system relate to those elements?
5. *Encompassing system connection*: What larger system encompasses our SOI? In addition to our SOI as a part, what other parts exist in the larger system, and how do all these parts relate and interact?

Our SOI is typically an end target system, as defined in Chapter 3. Traditional SOIs in SE include our example of a commercial aircraft, a major military weapons system, a power plant, etc. SOIs of the expanded SE include a military coalition, an organization structure, a transportation system, a health care system (including a response to potential epidemics), a power grid, or a societal system such as an overall retirement system.

Basic Model Types

Suppose we have an SOI S1 that has parts S11, S12, S13, and S14. S1 interacts with S2 and S3 in its environment, and S1 is part of system S, which has parts S1, S2, S3, and S4. The five views associated with S1, labeled V1 to V5, are depicted in Figure 6.1.

* Views are discussed in detail in Chapter 10. For the present discussion, simply assume that a view is a model.

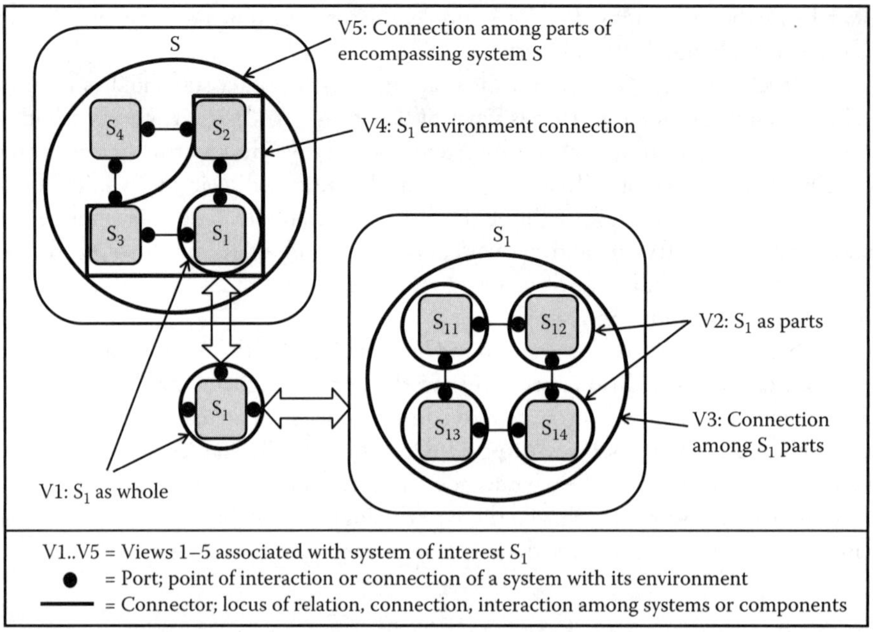

Figure 6.1 Common views associated with a system of interest.

In MOSES, when a model corresponds to or is linked to a target, it is important to make explicit what the target is, that is, what a model is modeling. This may sound like something that is trivial and obvious and that could be left implicit. But in our experience, the target of a model is, surprisingly, often not clear, or is assumed by different observers to be different targets.

We noted that the five views are all models or model types, and we can regard our SOI S1 as a target system. But there is a reason we said Figure 6.1 shows views *associated with* a system of interest, rather than views *of* a system of interest. An SOI is not the target system of all the models (views) associated with it. Views V1 and V3 are views of S1, so they are models whose target system is, in fact, the SOI S1.

V2 can be regarded as a view of the set of S1 parts, {S11, S12, S13, S14}, in which case S1 is the target system of the model V2. But V2 can also be regarded as a set of four separate models, an external model of each part (S11, etc.), in which case each V2 model has a different target system, and none of them are the SOI S1. Which is the preferable way to regard V2? We lean toward the second perspective; that is, V2 as a set of four models, a separate external model of each part. One reason is that the perspective of all parts considered together is addressed adequately in V3, the connection among all the parts. A second reason is that these four separate models are useful for specification matching in the composition of pre-built parts. This latter topic is briefly discussed in the section entitled "Composition-Related Processes" in this chapter, and addressed further in Chapter 11.

V5 is straightforward. It is a model of S; that is, it is a model whose target system is S. V4 is, like V2, interpretable in two ways. Because it is defined from the viewpoint of S1, and it is the environment of S1, one could argue that the target system of V4 is S1. But we argue for a second interpretation. V4 is not just a model of S1; it is a model of S1 interacting with its environment, so the target is the combination of S1 and its environment.

Variations

The basic model types in Figure 6.1 represent variations of both internal and external model categories.

Internal Model Variations

The first variation is the internal model we just discussed. A full internal model includes all the parts of a system and their connections, as in V5 (internal model of S) or V3 (internal model of S1). An environment connection model is defined with respect to, or from the viewpoint of, a given part: it includes only the *parts* that connect with the given part, and it includes only the *connections* with the given part, as in V4 where the given part is S1. Parts and connections not visible in the S1 environment, even if present in a full system such as S, are excluded from the environment model. Note that an environment model can be defined with respect to each part. Thus, in Figure 6.1 we could have defined separate environment models for S2, S3, and S4, respectively. We could also have defined environment models internal to S1 for S11, S12, S13, and S14, respectively.

What happens if a given part connects with every other part of a system? Does this mean the environment model of that given part is equivalent to the overall connection model for that system? Consider S in Figure 6.1. Suppose we added a connection between S1 and S4. Would that make the S1 environment model V4 equivalent to the S connection model V5? No. The S1 environment model V4 would still not include the connections between S3 and S4, and between S4 and S2, both of which are part of the connection model V5.

External Model Variations

The second variation of basic model types in Figure 6.1 is a variation of external models, specifically model V1 of S1. Notice that V1 points to two model instances of S1. These are the variations. The top one shows V1 as an external model of S1 in the context of S. The bottom one shows V1 as a stand-alone external model of S1. The top model defines a specific environment for S1 and is called a context-specific external model. The bottom model does not define any specific environment for S1 and is called a context-general external model. Although in some instances the two variations may be equivalent, in general this is an important distinction. We give three brief examples: a molecule, a software component, and a human.

1. Small molecules, called monomers, can compose themselves into large molecules called polymers. An example is shown in Figure 6.2 (Kirchner 2001; see also Miller 1995, 204–205). Two amino acid monomers, analogous to S1 and S2 in S in Figure 6.1, are shown in their stand-alone context-general form in the upper part of Figure 6.2, and in their composed context-specific form on the lower part. The two entities in the shaded box in the upper part (OH molecule and H atom) are removed, in the form of a water molecule, when the two monomers combine. This change corresponds to S1 in Figure 6.1, where the third dot on the boundary of S1 in its stand-alone form is removed when S1 is composed in S. It is also an example of mutable composition, which is discussed below in the section entitled "Composition Properties."
2. Suppose S1 in Figure 6.1 is a software component that offers three features or services, such as identification, authentication, and authorization, represented by the three dots on the boundary of context-generalized S1. And suppose that the S environment only needs

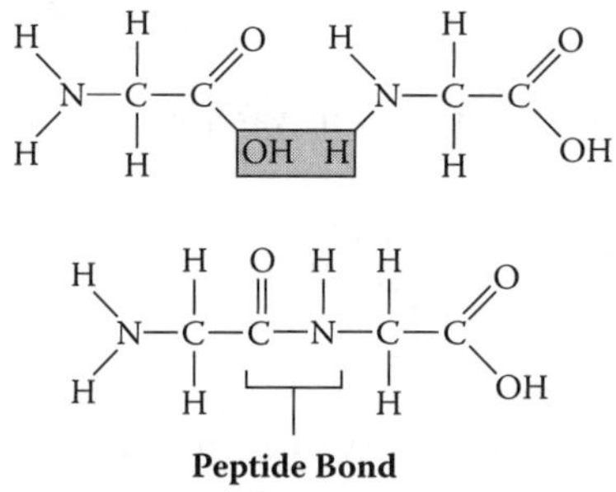

A molecule of water is removed from two glycine amino acids to form a peptide bond.

Figure 6.2 Example of context-specific and context-general external models. (*Source*: Kirchner, G. Self-Instructional on Biological Molecules: Organic Compounds. http://homepages.ius.edu/GKIRCHNE/biomolec.htm. 2001. Used with permission.)

two of these features, represented by the two dots in the S context. The general model of S1 specified by the component vendor includes all three features. When the component is brought into the S context, only two of the S1 features are used, so those are the only two included in the S1 external model in the S context.

3. Consider a human being as S1. What type of behavior will the person exhibit? We cannot say much knowing only this information. But if we put a person in a specific context or role, such as a systems engineer on a project team, a member of a poker club, a parent of a preschool child, or a coach of a baseball team, we can be much more definitive about what type of behavior the person will exhibit. The three dots on the stand-alone S1 boundary represent the many possible behavior types the person might exhibit, and the two dots on the S1 boundary in the S context represent the more specific subset defined in that context. That is, S might be any of the contexts listed above, such as a project team.

In general, the greater the variety of behaviors a system can exhibit, the bigger the difference between a context-specific model and a context-general model. Of the three examples given here, the human has, by far, the largest variety of potential behaviors. Greater variety implies ability to be useful in more environments.

The large difference in predictability between context-general and context-specific behavior of humans shows the importance of roles in narrowing their behavior and increasing their predictability. Atay and Jost (2004) discuss the tendency to consider a complex system as emerging from very simple parts, but they point out a fact that is important to SE and especially the new complex aspects of SE. Real-world complex systems typically have complex parts. Furthermore, and more importantly, the behavior of a complex part in an unconstrained environment is difficult to predict but the behavior of that part in an environment is highly constrained and much more predictable. This is especially true for complex systems that are considered autonomous. This phenomenon is what makes reasonable the proposition that we can specify the behavior of humans and other autonomous agents as components of a system (see Chapter 11).

Relation and Interaction Models

This section builds on the connection ontology defined in Chapter 3. In the discussion that follows, interaction is intended to include static relation and dynamic interaction. Traditionally, the emphasis in engineering documentation has arguably been more on parts and less on interactions among parts. This emphasis is reflected in the large majority of diagrams of boxes and lines, such as block diagrams, network diagrams, hierarchy charts, and process diagrams. These typically identify the boxes but not the lines, that is, identify the components but not the relations or interactions. However, the literature of complex systems science has made clear the importance of interactions of the parts of a system. The interactions are often at least as important as the parts in determining the characteristics of a system. Because of this, MOSES regards both systems and interactions between systems as first-class entities in the ontology of systems and systems engineering.

We first develop the basic concepts of interactions among parts or components using a binary interaction, that is, an interaction between two components. Then we describe additional concepts for N-ary interactions, that is, interactions among multiple components.

In our discussion of interactions in this chapter, we sometimes call the interacting elements components and other times systems. In general, we regard component and system as synonyms or roles, in the sense that both are holons: a system is a component of a larger system.

Binary Relation and Interaction Model

How do we model relations and interactions between two components in a way that provides some substance but at the same time supports a wide variety of situations? For example, we want our model to support relations and interactions such as:

- Two atoms combining into a molecule
- Two cells interacting
- A dam interacting with a river
- An aircraft interacting with an air traffic control system
- An aircraft owned by a company
- An aircraft operated by an airline
- Two people communicating, via face-to-face conversation, sign language, talking on the phone, exchanging e-mail, or via a Web site
- Two people having the same parents (i.e., they are siblings)
- Two people in a conflict, such as opposing soldiers in a battle
- Two people competing and cooperating, such as in a chess match or in organizational settings
- Two countries interacting as trade partners
- Two countries in armed conflict with each other
- Two countries having a treaty with each other
- A person interacting with a machine, such as a car
- A person interacting with a Web site via a Web browser
- A person operating a Web site
- A person interacting with an organization, such as applying for a job, or a customer with a company
- Software interacting with other software, such as sending a message or invoking a service
- Software contained as a module within other software

- An organization interacting with a natural system, such as the Army Corps of Engineers interacting with a flood plain
- Two organizations interacting, such as a company with a supplier organization, a competitor, or a government regulatory organization
- An organization having a long-term contract with a supplier

What is common to all these connections? In the discussion and model that follow, the focus is on interactions. We propose that the following set of elements, based on the connection ontology in Chapter 3, is common to all binary interactions: two entities that are the end parties of the interaction, an action involving the two parties, a medium called a connector party through which stuff is exchanged, roles that each party plays in the interaction, and a protocol or set of constraints on the interaction. We call the end parties the components, and we call the connector party the connector. In addition, in some cases, a service is provided by one component and consumed by the other component. In these cases, the components play the additional respective roles of service provider and service consumer. This general component-connector model was developed in the 1990s in the software architecture research community — see, for example, Shaw and Garlan (1996) and Allen and Garlan (1997).

The basic binary interaction model is shown in Figure 6.3, using a service provider and consumer pattern, which, as stated in Chapter 3, is intended to be the general interaction terminology. This model applies to any binary interaction between systems, including the interaction examples in the list given above.

In the binary interaction model, each of the two end parties S1 and S2 is modeled as an encapsulated system with defined access points in its boundary called ports. The interaction also includes the connector N, the stuff exchanged (s), the protocol or constraints on the interaction, and the system roles: S1 is service consumer and S2 is service provider. Note that the role played by a system in an interaction is defined by the connector, not by the system. This reflects the fact that,

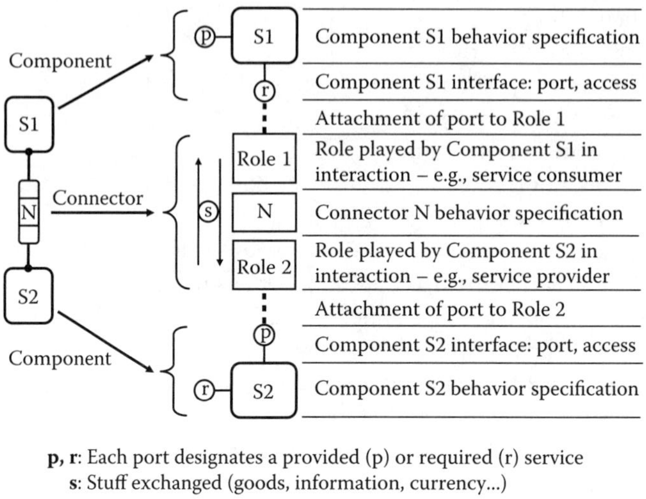

Figure 6.3 Basic binary interaction model. (*Source*: Hybertson, D. Using a specification approach to facilitate component testing. In *Testing Commercial-off-the-Shelf Components and Systems*, S. Beydeda and V. Gruhn, Eds. New York: Springer-Verlag, p. 213–237, 2005. Used with permission of Springer Science + Business Media.)

in general, an environment of a system defines the role played by the system in that environment, as we discuss later in the chapter.

Although Figure 6.3 shows a provided and required service, the concepts of competition and disservice from Chapter 3 are also implicit in this basic model.

The Unified Modeling Language (UML) defines two types of connectors: (1) a *delegation* connector and (2) an *assembly* connector (OMG 2007). The UML concept of assembly connector corresponds to the MOSES connector in the case where a service is involved in an interaction. The UML delegation connector corresponds partially to MOSES delegation and propagation, which are discussed further in the section entitled "Relations Between Wholes and Parts."

The foregoing discussion has emphasized features that are *common* to all interacting pairs of any type, along with some additional features of service interactions. The general interaction models given in this section, as well as the N-ary models of the next section, are intended to apply to both machines and humans. We discuss mechanisms for describing *differences* in later chapters, especially Chapters 7 and 11.

N-ary Relation/Interaction Models

Systems in general have more than two parts or components. The binary model gives us much information about interactions in general but we also need to look at relations or interactions in which one or more connectors link multiple components. A connection among multiple components can be modeled by a network. More generally, we can use ideas from the mathematical fields of graph theory and topology to model these relations. The graph concepts of node, point, and vertex correspond to components, and the graph concepts of edge and line correspond to connectors. In the general case, the lines or edges of a graph are topological elements, which simply means, among other things, that they are undirected and the concept of length does not apply. The lines or edges may be assigned a direction, in which case they are called arcs, arrows, or directed edges, and the graph is called a directed graph or a digraph. In addition, the edges may be assigned weights, which may be given interpretations such as length. A digraph with weighted edges is a network.

Networks extend the concepts introduced in the discussion of binary relations. A single connector or network now can connect multiple components. Protocols apply to the entire network. Stuff can be exchanged among multiple components. Most or all the examples listed under the binary interaction discussion can be extended to multiple components, such as sending an e-mail to multiple recipients.

Networks open up multiple issues and opportunities. Three general issues are (1) network structure or topology, (2) network performance or optimization, and (3) network security. With sufficient connectivity of nodes, multiple paths are available between any given pair of nodes, which can make a network more fault tolerant and robust. On the other hand, the inherent connectivity of a network increases security risks. Figure 6.4 illustrates some common network topologies (ATIS 2008).

In the bus topology, all nodes are connected by a single connector. The mesh topology is multiply connected; that is, at least two nodes have two or more connection paths between them. A network in which every node is directly connected to every other node is a fully connected mesh. Most networks of any size do not use a fully connected topology because it does not scale up well. However, a common connector like a bus can also provide a fully connected effect, and is much more scalable. A common connector also seems like a more natural model of N-ary or multi-party interactions than does a multiple pairwise topology, although in some cases the pairwise model gives more information. Many networks are hybrids of two or more of these topologies.

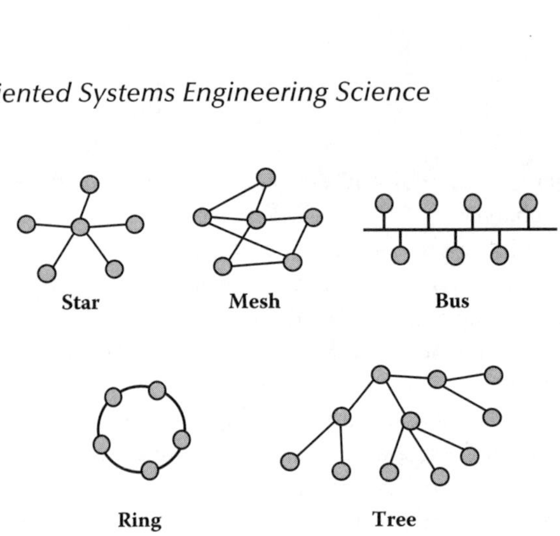

Figure 6.4 Common network topologies.

The recent increase in the study or science of networks* should benefit the development of complex system models, for two reasons. First, complex systems emphasize the relations among parts, which network analysis addresses. Second, the same network models seem to apply to multiple domains, such as transportation networks, social networks, and computer networks. More broadly, the study of networks and the resulting models should also encourage movement toward a model-oriented approach. Research in network models is finding common properties of networks such as "scale-free" networks (Barabási and Albert 1999) in which a minority of nodes, called hubs, are heavily connected, while the majority of nodes are lightly connected. Many real-world networks reflect this pattern, including the Internet and some airline networks (where a node connection represents a flight to/from a city). The "scale-free" term comes from the observation that as the network grows in size, it retains the same pattern, the proportion of hubs to total nodes remains stable, and apparently the performance is not significantly degraded. The performance, however, might depend on the situation, or even the viewpoint. The use of hubs for airline routes might provide the best performance from the viewpoint of the airline, but for a passenger, flying hundreds of miles out of the way through hub city C to fly from city A to B might not seem like good performance. Much of the research in this emerging science of networks area is captured in Newman et al. (2006).

The N-ary component-connector model is a very general one. It can be specialized in many ways. Some of the slightly more specialized types of interaction models are collaboration, coordination, orchestration, and choreography. Common protocol patterns used in both binary and N-ary interaction models include peer-to-peer, client-server, service provider and consumer, request-response, message oriented, event-driven interaction such as interrupts or publish and subscribe, and streaming. They can be synchronous or asynchronous. In some types of interaction, such as orchestration, one part directs the interaction. This also applies to many social situations where one person, in the role of manager, leader, coach, etc., is directing a group interaction. Other models, such as collaboration and choreography, distribute interaction responsibility among the parts, as in a heterarchy. An ant colony is an example of the latter type of model.

A topology can be modeled in different ways, depending on the desired emphasis. Suppose we have a common connector, such as the Internet, through which pairwise interactions take place. Figure 6.5 shows how we could model this with the emphasis on the common connector, as on the

* The science of networks is a combined natural and artificial science because it abstracts patterns from both natural and artificial networks.

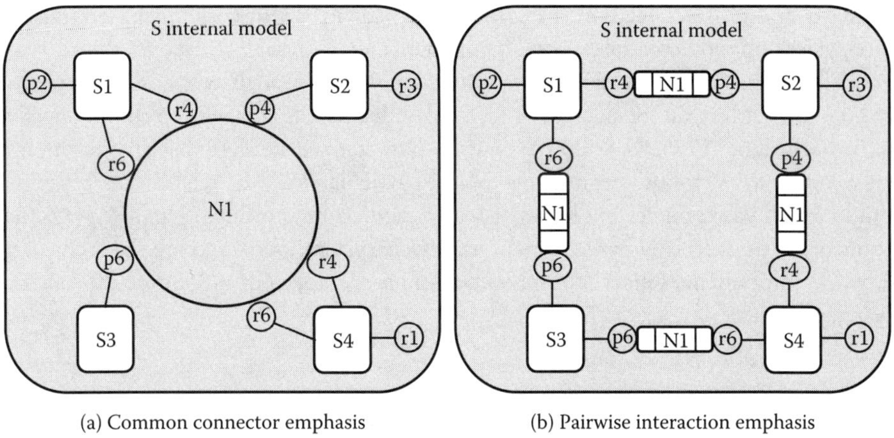

Figure 6.5 Two views of pairwise interactions via a common connector.

left, or with the emphasis on the pairwise interactions, as on the right. In both cases, systems S1 through S4 interact through one connector, N1.

One example of this configuration: S1 is a person accessing the Internet on their home computer, S4 is a neighbor down the street with similar access, S2 is the nearby pizza place both people order from, and S3 is an online bookseller they both use. The model on the right makes the interactions visually clearer. S1 and S4 independently interact with both the pizza place S2 and the bookseller S3. The two people, S1 and S4, do not interact with each other; and the two businesses, S2 and S3, do not interact with each other. One way these two models are used together is to define the common connector model as the physical system model, and the pairwise model as a view of certain interactions that use or overlay the physical model.

Note that both a component and a connector may be a composite of components and connectors. Examples: a network of cities as components and a transportation network as a connector. Each city has components and connectors internally — for example, components = buildings; connectors = streets, transporting vehicles, etc. And the transportation network has connectors and components — for example, connectors = roads, air routes, waterways; components = trucks, airplanes, boats, gateway cities. Thus, a truck can be a system in itself; or a component in a larger system, such as a fleet of trucks; or a component in a composite connector, such as a transportation network. A city can be a system, or a component in a composite connector such as a transportation network in which the city is a gateway between a source and destination. More extensive multilevel composition structures will be explored in the section entitled "Multilevel Composition Structures and Scales."

Some additional notes on connectors and components:

- Connectors as well as components can have ports. Examples: a portal to the Internet; a port to a limited access highway, called an interchange; a dock or ramp that provides boat access to a river or waterway.
- Connectors can be encapsulated. In the previous bullet, most of the examples are not accessible except via their ports.
- Services can be component services (tax preparation, Web services) or connector services (Internet, FedEx). This is a useful distinction that perhaps should be made more explicit in the service-oriented world.

The first-class stature of relations or connectors is such that, in component–connector based models of system relations on parts, both components and connectors may be regarded as parts. Some models emphasize components over connectors; other models emphasize connectors over components. Depending on the context or type of model, part might refer to components, or connectors, or both. Klir (1985, 4–8; 2001, 44–54) characterized models that are based on certain kinds of *components*, without committing to any particular kind of relation, as experimentally based, and models based on certain kinds of *relations*, without committing to any particular kind of component, as theoretically based. Relations reflect "systemhood" and are the focus of systems science, while components reflect "thinghood" and are the focus of more domain-specific traditional sciences such as physics or chemistry.

Composition Properties

We present three composition properties in this section: (1) exclusivity, (2) mutability, and (3) symmetry. These properties structure, and to some degree motivate, our discussion of composition models. The *exclusivity* of a part is the degree to which it is exclusively in one whole. The *mutability* of a part is the degree to which it changes in the composition process. The *symmetry* of a composition is the degree to which it combines parts to form a whole, as opposed to, for example, adding a part to a whole. We now discuss composition models associated with each of these properties, and then briefly discuss how the three relate.

Exclusivity

The exclusivity distinction is between a part that is in only one whole (exclusive) and a part that is, or can be, in multiple wholes (non-exclusive). One might think that a part can be in only one whole. How can an atom be in more than one molecule, a cell in more than one organ, a brick in more than one house, an engine in more than one car, a tree in more than one forest, or a planet in more than one solar system? If a part can exist in only one whole, we call it the contained composition model. In the most extreme or pure form of this model, the whole owns the parts; the parts do not exist separately from the whole; and if the whole is destroyed, the parts are also destroyed. In less extreme forms, a part cannot be part of multiple systems simultaneously, but could be part of one system at one time and part of another system at a different time. This is the consecutive model we describe at the end of this subsection.

However, in many cases, a part can be conceptualized as existing in multiple systems. In this case, which we call the shared model, a part exists separately and can be in multiple wholes; and if some wholes are destroyed, the part still remains. A common example of the shared model is a software component that is used by and is part of multiple software systems. Two dominant architectures of this type are layered architecture, where each layer has parts that support multiple parts in higher layers, and a service-oriented architecture (SOA), such as Web services that are accessible to multiple consumers. The same computer, CPU, operating system, file system, and middleware provide common services to multiple applications. Networks such as the Internet are shared parts of multiple systems. A more recent topic in service orientation is the concept of cloud computing, where computing power and other computing resources are available from data centers or computing centers on the Internet (Hayes 2008, Gruman and Knorr 2008). Variations on this theme — and this term — include on-demand computing, software as a service (SaaS), utility computing, platform as a service, Internet as platform, managed service provider, service commerce platform, and Internet integration, where the latter is integration of cloud-based services.

Shared models go beyond software and computer systems. Consider these physical service providers: the same agent — such as a book seller, car dealer, mail carrier, transport service (FedEx), doctor, tax preparer, etc. — serves multiple clients. In each case, the agent-client or server-client interaction can be designated as the internal model of a system. So if a doctor has 50 patients, it is entirely reasonable to designate that the doctor is part of 50 systems, plus several other systems, such as his colleagues and the office staff in his practice. The whole notion of common service is very strong and traditional in the physical world; common software services have nothing new on this concept. Outsourcing, such as accounting, HR, even IT, fits this model. So the basic shared composition model — separated, loosely coupled — is very similar for physical systems and software systems.

The shared model has two variants. One is where a shared part occurs once in each system. The examples given above are of this variety. In the shared model shown in Figure 6.6, part S14, as a part of systems S1 and S2, illustrates this.

The other variant is where a shared part occurs multiple times in a single system, illustrated in Figure 6.6 by the multiple appearances of part S14 in system S3. Most of the systems in which this occurs are models. Many target systems have multiple parts of the same type, that is, they have the same specification or model. For example, a house has many bricks that are similar. If we defined a model of a house that showed every brick, and we were using 1000 bricks of type S14 (a certain size, color, material, etc., maybe even from a certain vendor), our model would have part S14 repeated 1000 times. Those 1000 parts would be multiple occurrences of the same part in the model. The target system, the house, would also have 1000 parts of type S14. But in the target system, each of those 1000 bricks would be a distinct and unique entity; they would not be the same part. So in this case, the S14 brick would be a shared, multiple occurring part representing shared composition in the model of the house, and would represent contained composition in the target system (the house).

Contained and shared composition models apply not just to composition of components, but also to composition of connectors. A contained connector would be hardwired between interacting parties that form a closed set. A shared connector would be a common connector such as a shared network in bus topology that is part of multiple systems of interacting parties.

The shared model also brings together and enables unification of the concept of contract. A design by contract, an SOA service contract, an old-fashioned contract between two organizations or parties, and implicit contracts in the traditional service industry all align in the unified shared

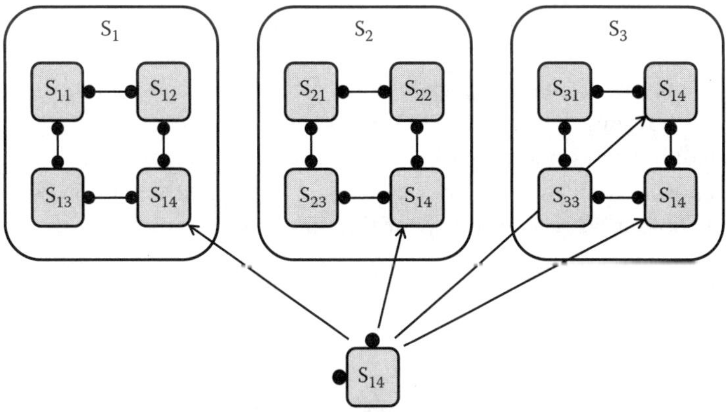

Figure 6.6 Shared composition model.

model. In addition, MOSES aligns and unifies both composition models—contained and shared—with the common contract specification approach, as described in more detail in Chapter 11.

There are intermediate models between fully contained and fully shared. One is a consecutive composition model, in which a part can only be in one system at a given time, but can be part of another system at another time. If it is difficult or rare to move a part from one system to another, it is closer to the contained side. Examples are moving an engine from one car to another, or using a brick in one house and later using it in another house, or removing a heart from one human and placing it in another human.

If it is easy or typical to move a part from one system to another, it is closer to the shared side. A common category of this is parts that can do multiple jobs but can only do one of them at a given time. Examples: a single computer processor supporting multiple programs via a time-slicing operating system, or a single accountant that can help multiple people prepare their tax returns but can only work on one at a given moment. In these cases, the difference between the consecutive model in which we allow a part to be contained in multiple systems but constrain to only one system at a time, and a parallel processing shared part that can be part of multiple systems becomes very small.

The foregoing discussion therefore represents the following composition model variants:

- *A pure contained model:* a component is part of only one system, and ceases to exist when the system ceases to exist.
- *A consecutive contained model:* a component is part of only one system at a time but may move to another system at a later time.
- *A consecutive view of a shared model:* a component is a part of multiple systems simultaneously but can only interact with parts of one system at a time, and this interaction constraint is part of the model.
- *A simultaneous view of a shared model:* a component is a part of multiple systems simultaneously; whether or not the component can only interact with parts of one system at a time is *not* part of this model.

Keep in mind that these variants and views represent points on a continuum that ranges from pure contained to simultaneous shared.

The UML (OMG 2007, 39) defines composition types that correspond approximately to our contained (pure or consecutive) and shared, which they refer to, respectively, as composite aggregation and shared aggregation. They describe the types this way.

> "Aggregation Kind is an enumeration of the following literal values:
> - None: indicates that the property has no aggregation.
> - Shared: indicates that the property has a shared aggregation.
> - Composite: indicates that the property is aggregated compositely, that is, the composite object has responsibility for the existence and storage of the composed objects (parts)."

Mutability

In the discussion thus far, we have used examples where parts are composed into wholes more or less intact. But in many cases, the parts change when they are composed with other parts. We call this a mutable composition model. Common terms for this type of composition are bonding, fusion, and merging. Prominent examples are chemical bonding and organizational mergers. The example in Figure 6.2 showed two molecules changing when they merged. Two hydrogen atoms and an oxygen atom were removed during composition and combined to produce water as a side effect of the composition.

When two organizations merge, they usually change or remove certain elements so they do not have unwanted duplication of common functions.

Symmetry

All the cases and models discussed so far are symmetric, in that the compositions were of parts into a whole. An asymmetric model also exists, where a whole adds a part; that is, a whole and a part are composed into a (revised) whole. Most of the composition discussion in this book is about symmetric composition. Examples of asymmetric composition include adding a small molecule (monomer) to a polymer chain in chemistry; or an organization hiring a new employee, a person joining a social club, or a network adding a node. There can be something of a continuum between asymmetric and symmetric composition. The pattern of company *mergers* and *acquisitions* reflects such a continuum. Generally, if two companies merge, we would call it symmetric composition, while if one company acquires another company, we would typically regard it as asymmetric.

Relations among the Three Categories

The mutability composition category is independent of the symmetry category. That is, both mutable and immutable composition can apply to symmetric and asymmetric models.

Immutable composition can apply to both shared and contained composition models. Mutable composition can also apply to contained composition. If a part is used only in one system, the part can be changed without affecting any other system, especially if the part is encapsulated.

Mutable composition does not, in general, apply to shared composition. A shared part must retain its separate existence to be available to other systems, so that if one system changed the part, the change would, in general, affect all other systems that used or included the part.

It is important to distinguish between mutability and adaptation or system change. A mutable part changes during composition. Adaptation involves system change over its operational life. A part can be both immutable and adaptable. Adaptation and other forms of change are discussed in Chapter 9.

Relations between Wholes and Parts

In this section we discuss a variety of relations that exist or can be defined between wholes and parts. We observe at the start that because SE in general and MOSES in particular support both physical systems and conceptual systems, we do not constrain either a whole or a part to be concrete or abstract.

The first section presents some basic whole–part concepts from systems science. Because the whole–part relation is where the concept of emergence arises, we briefly discuss in a sidebar the concept of emergence and the stance of MOSES toward this concept. Then we discuss a basic set and an extended set of whole–part relations.

The discussion of whole–part relations in this section and chapter is informal. There are formal treatments of this subject. Some of them are based on set theory, and others on mereology, or the theory of wholes and parts (see, e.g., Simons (1987)). Another formal approach is the theory of granular partitions (Bittner and Smith 2003). The latter has features that could be applied to the model space, especially composition and views. One such feature is the depiction of cells and their mapping to reality. This relation is analogous to the MOSES depiction of composition models and their correspondence with systems.

Before we proceed with the discussion, we need to remove a possible ambiguity for readers who tend to think in formal terms. In this chapter and book, part is meant to be proper part. Formally, a system can be part of itself, analogous to a set that can be a subset of itself. But a proper subset is not the whole set. A set cannot be a proper subset of itself; there must be elements of a set S that are not elements of a proper subset of S. Likewise, a proper part is not the whole system. A system cannot be a proper part of itself; given a proper part P of system S, there must be at least one part of S that is not P. Pursuant to this discussion, unless otherwise noted, everywhere in this book that we say part, we mean proper part. This convention holds for both contained and shared parts.

Basic Whole–Part Concepts

Multiple overlapping system distinctions can be defined, each of which can generate system taxonomies. Some common distinctions are open versus closed; natural versus artificial; or physical versus conceptual. Each distinction or taxonomy may be regarded as a view of the space-time region of all systems. (Views are discussed in Chapter 10.)

The characteristics of a system exhibit a common pattern in both a spatial and temporal view. In the spatial view, the pattern is the relation of characteristics or properties of a whole system to the characteristics of its parts. A system has a set of characteristics of interest, and each of its parts has a set of characteristics of interest. The relation of these sets is illustrated in Figure 6.7a. Some characteristics apply to both the whole and the part; others apply only to the whole, not to the part; and others only to the part, not to the whole.

Suppose entities A11 and A12 are parts of whole A1, and in turn A1 is part of whole A. Consider a property P. Suppose P is not a property of A11 or A12; P is a property of A1; and P is not a property of A. For example, A11 = tire, A12 = wheel, A1 = wheel assembly, A = car, and P = air pressure. We are interested in the air pressure of a tire assembled with a wheel, but not in the air pressure of either a tire or a wheel by itself, and not in the air pressure of a car. Thus, P is a property of only A1. In the relation of A1 with A11 and A12, P is a property of the whole only. In the relation of A1 with A, P is a property of the part only.

In the temporal view, the pattern is the relation of characteristics or properties of a system at time T1 to its characteristics at time T2. A system has a set of characteristics of interest at any given time, and the set may change over time. The relation of these sets across time is illustrated in Figure 6.7b. Some characteristics apply at both T1 and T2; others apply only at T1, not at T2; and others only at T2, not at T1. These can be referred to in the temporal relation, respectively, as persistent, earlier, and later characteristics.

Analogous to the spatial view, the designation of persistent, earlier, and later are not inherent but rather are context dependent, where the context is the relation in which the designation is made. For example, if a property P holds for system S at T1 and T2 but not at T3, then P is a persistent property of S in the T1–T2 relation but is not persistent in the T2–T3 relation.

Note that the diagrams in the figure are intended to show only the *existence* of these variations of characteristics, not the relative number or proportion of each characteristic.

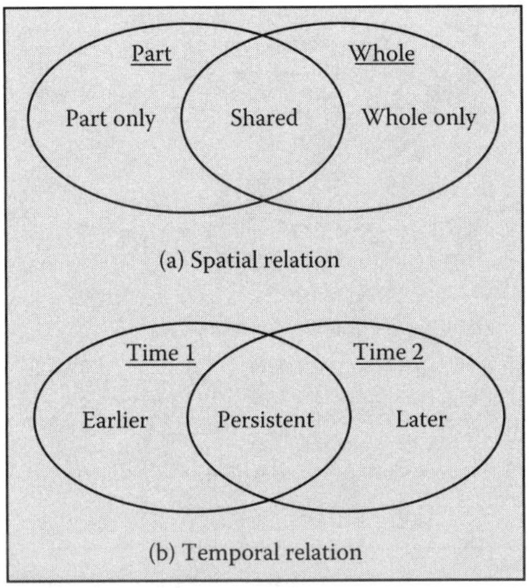

Figure 6.7 Relations of characteristics in space and time.

The Issue of Emergence

When we consider the relations between wholes and parts, we encounter the potential relation of emergence. The vernacular form of emergence is "The whole is more than the sum of its parts." A more formal definition of emergent properties, from Honderich (1995, 224), is: "A property of a complex system is said to be 'emergent' just in case, although it arises out of the properties and relations characterizing its simpler constituents, it is neither predictable from, nor reducible to, these lower-level characteristics." A variety of definitions of emergence exists. See Emmeche et al. (1997) and O'Connor and Wong (2006) for discussion of some of the variants.

In traditional systems engineering, emergence is often implicitly assumed but is rarely discussed as an explicit topic. Much of the complex systems literature includes emergence as a defining characteristic of complex systems. For example, Boccara (2004, 3) states that "The appearance of emergent properties is the single most distinguishing feature of complex systems." On the other hand, some researchers in complex systems call themselves reductionists — for example, Shalizi (2007).

The MOSES approach is to focus on and characterize the relations between wholes and parts. For both TS and CS, the whole often has characteristics that parts do not have, which may be called emergence. But parts may also have characteristics that the whole does not have, which we might call submergence. We explore these and other relations in the sections that follow. While we occasionally use the term "emergence," the primary MOSES depiction of the general situation for both TS and CS is to say that a whole and its parts have partially overlapping characteristics.

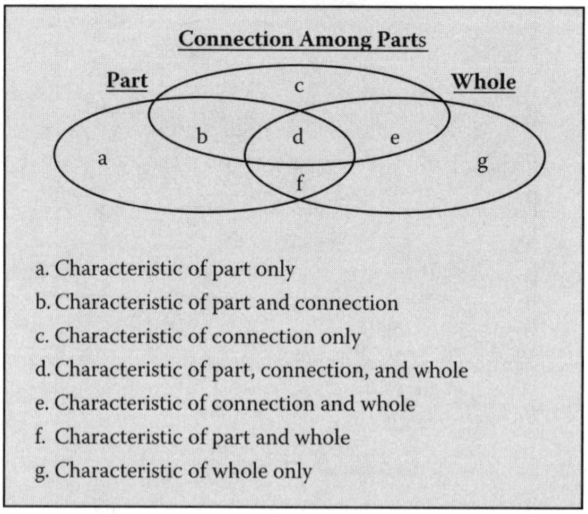

Figure 6.8 Differing characteristics among system views.

The more complete picture also includes the connection among parts, so that the views V1, V2, and V3 from Figure 6.1 can all exhibit these relations. We can say that the whole, connection among parts, and the parts have, in general, different sets of characteristics, as depicted in Figure 6.8. An example of (a), a characteristic of a part only, is a wheel assembly, if we consider it as part of a car. It has the characteristic of roundness, which the car does not have. An example of (g), a characteristic of the whole only, is the transportation function of a car as a whole, which is not a characteristic of any part or of the connection among parts. Examples of (c), a characteristic of the connection among parts only, include any topology of interacting nodes, such as parallel or serial or any of the network topologies depicted in Figure 6.4. Additional examples of the types in Figure 6.8 are given later in this chapter.

A concept closely related to the connection among parts is internal environment. The internal environment, for example, is the locus of homeostasis of a system — the process of maintaining internal state despite perturbations in the external environment.

In this chapter we only look at the whole–part relations, not at how they came into existence. That topic is addressed in Chapter 9, where we discuss temporal relations between wholes and parts, such as self-organization and evolution. We believe that this set of structural and temporal topics not only addresses the issues of emergence relevant to SE, but that it does so in a more direct and clearer way than would a discussion based on the traditional framework of emergence.

Span of Control

In traditional systems, the number of parts in a system is usually small, the order of magnitude ranging from tens to thousands. For hierarchical organizations, the number is described in terms of span of control. George Miller (1956) introduced the well-known norm of 7 ± 2, based on the average number of chunks of information a human can remember under certain conditions. But complex systems often have a much larger parts-to-whole ratio (i.e., a much flatter hierarchy), and the number of interconnections among parts also increases. An extreme example is neurons in the human nervous system. The number of neurons in the human brain is on the order of 10^{11}, and the number of interconnections or synapses among neurons is on the order of at least 10^{14} (Pakkenberg and Gunderson 1997; Pakkenberg et al. 2003).

A major reason for the large difference between traditional and complex systems is autonomy. That is, in traditional systems and organizations, the assumption is that one controller (or manager) controls the parts, so that controller has to keep track of everything. In complex systems, the parts can act on their own. No one is in charge of the neurons in the brain or the ants in an ant colony. Autonomous components are self-acting, self-organizing, and cooperating — they work it out among themselves. Consequently, many more of them can interact in such a system.

Defining the Set of Basic Relations

With the foregoing background discussion, we are ready to define the set of relations in the composition dimension. The full set is divided here into what we call a basic set and an extended set of whole–part relations. In this section we describe the basic set.

Two obvious relations between wholes and parts are composition and decomposition. These appear intuitively to be a conjugate pair (opposite views of the same relation), but in MOSES they are not. We follow the approach described in Meyer (1997, 40–43). There are four relations that form two conjugate pairs: decomposition-encapsulation, and composition-componentization. In addition, several other whole–part relations exist.

All relations in the basic set are explained in Table 6.1. We use a small example to illustrate the relations in the table. Suppose we have a system S with a set of parts consisting of components S1, S2, S3, and S4, and connectors N1, N2, N3, and N4, as shown in Figure 6.9. Define the external model of S as S_{ext} and the internal model of S, or the connection among the parts of S, as S_{int}, along with external models of each part, as shown in the figure.

Using these terms and this example, Table 6.1 identifies and explains the basic set of whole–part relations. Column 1 names the relation and maps it to the corresponding characteristic type (a, …, g)

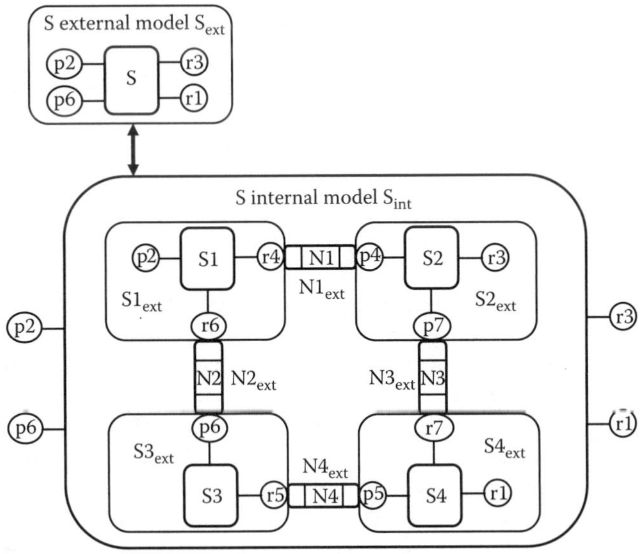

Figure 6.9 Example of basic whole–part relations.

Table 6.1 Basic Whole–Part Relations

Relation	*Description Using Entities in Figure 6.9*
Decomposition	S_{int} is a decomposition of S_{ext}. In engineering, decomposition is the traditional design problem, i.e., S_{int} represents a design or architecture of a system.
Encapsulation	S_{ext} is an encapsulation of S_{int} (conjugate of decomposition).
Composition	S_{int} is a composition of {S1,S2,S3,S4,N1,N2,N3,N4}
Componentization	{S1,S2,S3,S4,N1,N2,N3,N4} is a componentization of S_{int} (conjugate of composition).
Inclusion	Internal model of S (S_{int}) includes external model of each part ($S1_{ext}$, $S2_{ext}$, $S3_{ext}$, $S4_{ext}$, $N1_{ext}$, $N2_{ext}$, $N3_{ext}$, $N4_{ext}$); view of a whole as connection among parts includes view of each part as a whole; clear-box view of whole includes black-box view of each part.
Delegation (char C(d,e))	S_{ext} delegates characteristics (properties, behaviors, functions, services) to S_{int}. Refers to an external characteristic of the whole (S_{ext}) that is or becomes a characteristic of the connection among the parts (S_{int}).
Delegation (char C(b,d))	S_{int} delegates characteristics to {S1,S2,S3,S4,N1,N2,N3,N4}. Refers to a characteristic of the connection among the parts that is or becomes a characteristic of one or more parts.
Delegation (char C(d,f))	S_{ext} delegates characteristics to {S1,S2,S3,S4,N1,N2,N3,N4}. Refers to an external characteristic of the whole that is or becomes a characteristic of one or more parts.
Propagation (char C(d,e))	S_{int} propagates characteristics to S_{ext}. Refers to a characteristic of the connection among the parts that is or becomes a characteristic of the whole.
Propagation (char C(b,d))	{S1,S2,S3,S4,N1,N2,N3,N4} propagates characteristics to S_{int}. Refers to a characteristic of one or more parts that is or becomes a characteristic of the connection among the parts. A model in which all characteristics of S_{int} are propagated in this way, and are dependent on the parts, is sometimes called reductionism.
Propagation (char C(d,f))	{S1,S2,S3,S4,N1,N2,N3,N4} propagates characteristics to S_{ext}. Refers to a characteristic of one or more parts that is or becomes a characteristic of the whole. A model in which all characteristics of the whole are propagated in this way, and are dependent on the parts, is sometimes called reductionism.
Non-delegation (char C(f,g))	S_{ext} does not delegate to S_{int}. Refers to a characteristic of the whole that is not a characteristic of connection among parts. We call this a systemic characteristic.
Non-delegation (char C(c,e))	S_{int} does not delegate to {S1,S2,S3,S4,N1,N2,N3,N4}. Refers to a characteristic of connection among parts that is not a characteristic of parts. We call this a composite characteristic. If there is a dependence on parts, this relation corresponds to what is sometimes called a form of emergence.
Non-delegation (char C(g,e))	S_{ext} does not delegate to {S1,S2,S3,S4,N1,N2,N3,N4}. Refers to a characteristic of the whole that is not a characteristic of parts. We call this a systemic characteristic. If there is a dependence on parts, this relation corresponds to what is sometimes called a form of emergence.

Table 6.1 Basic Whole–Part Relations (Continued)

Relation	*Description Using Entities in Figure 6.9*
Non-propagation (char C(b,c))	S_{int} does not propagate to S_{ext}. Refers to a characteristic of connection among parts that is not a characteristic of the whole.
Non-propagation (char C(a,f))	{S1,S2,S3,S4,N1,N2,N3,N4} does not propagate to S_{int}. Refers to a characteristic of parts that is not a characteristic of connection among parts.
Non-propagation (char C(a,b))	{S1,S2,S3,S4,N1,N2,N3,N4} does not propagate to S_{ext}. Refers to a characteristic of parts that is not a characteristic of the whole.
Trace	Trace or mapping between characteristics of S_{ext} and S_{int}, or S_{int} and {S1,S2,S3,S4,N1,N2,N3,N4}, or S_{ext} and {S1,S2,S3,S4,N1,N2,N3,N4}.

in Figure 6.8. For example, the delegation relation from S_{ext} to S_{int} can occur for characteristics d and e in Figure 6.8.

All delegation, non-delegation, propagation, and non-propagation relations in the table can potentially be of two types: dependent or independent. For example, in the propagation of S_{int} to S_{ext}, the characteristics of S_{ext} may be dependent on, or independent of, the characteristics of S_{int}.

The connection among parts (S_{int}) clearly plays a central role in the set of views of a system. The relation represents both a decomposition of a whole and a composition of a set of parts.

The conjugation and directionality of these relations needs additional explanation. In this discussion we will simplify the three-part <whole – connection among parts – parts> structure and simply discuss whole and parts; the application to the full structure should be clear.

Characteristics in delegation and propagation, by definition, exist for both part and whole. Most, if not all, characteristics that participate in the delegation and propagation relations are not inherently directional. Whether they are modeled as delegation (whole to part) or propagation (part to whole) is a matter of choice or viewpoint. One heuristic for this set of characteristics is that a top-down engineering process typically views the relation as delegation, while a bottom-up process typically views the relation as propagation.

We use the property of mass to illustrate the difference in view. In a physical system, the whole system has a mass, each physical part has a mass, and the relation is that the mass of the whole is the sum of the mass of each part. Would we say that this property is propagated from the parts to the whole, or that it is delegated from the whole to the parts? In cases where mass is not a serious constraint, we would say the property is *propagated* because the mass of the whole comes directly from the mass of each part. On the other hand, suppose we are building a space vehicle where the overall mass is a significant constraint. In the design process, we use the total constrained mass as a budget that we apportion to the various parts. The design involves trade-offs; if we allow one part to increase mass, we must reduce the mass of another part. From this viewpoint, we would probably view mass as a *delegation* from the whole to the part. Allocation is a common term for this in cases, such as mass, where the value of the property can be subdivided.

A common example of a characteristic that forms one end of a conjugate relation is the category of required services and provided services, as shown in Figure 6.10. The conjugate nature of provided and required services is shown in the S internal model. For example, S1 requires service

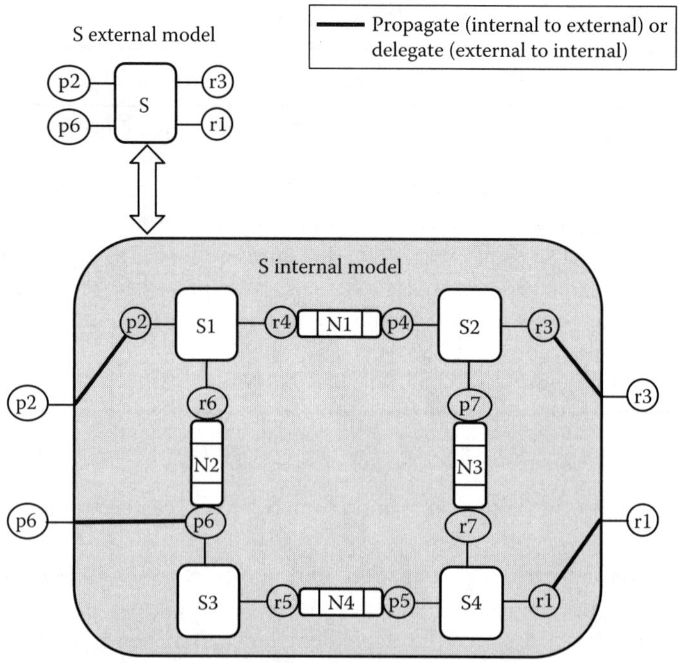

Figure 6.10 Conjugate propagation and delegation relations. (*Source*: Hybertson, D. Using a specification approach to facilitate component testing. In *Testing Commercial-off-the-Shelf Components and Systems*, S. Beydeda and V. Gruhn, Eds. New York: Springer-Verlag, p. 213–237, 2005. Used with permission of Springer Science + Business Media.)

4, S2 provides service 4, and in the S connection among parts, S1 obtains service 4 from S2. At the next higher composition level is where one end of each of several conjugates is exposed. The system S provides service 2 for any other system that requires it, and S requires service 3 from any other system that provides it. Examples of the delegation-propagation relation in the figure include the part S1 and the whole S provide service 2, and the part S2 and the whole S require service 3.

Whether we view these as propagation or delegation again depends on the situation, especially for provided services. If we are doing top-down engineering and specify that we need a system that provides service 2, and then an engineer designs it so that part S1 provides that service, we would likely view service 2 as a delegated characteristic from S to S1. On the other hand, if S1 were an existing system and we decided to compose S as a system of systems that took advantage of S1 to offer service 2, we would likely view service 2 as a propagated characteristic from S1 to S. The required services, such as service 3, are more often viewed as propagated than as delegated. While some required services are independent of the design, many depend on the design. Therefore, a full account of services required by a system — and system dependencies in general — must await completion of the design, which gives these characteristics more of a bottom-up propagation flavor.

The non-delegation and non-propagation relations, however, are inherently directional for all characteristics that participate. In non-delegation, a characteristic exists in the whole but not the parts; and in non-propagation, it exists in one or more parts but not the whole. In both cases, we have no symmetry and no choice in the relation designation.

Defining the Set of Extended Relations

Now we will address the extended set of relations — alternative, consecutive, and shared whole–part. Suppose we have a system S with three different internal models SI1, SI2, and SI3; an external model SE; and three different potential encompassing systems S1, S2, and S3, as shown in Figure 6.11.

This example is from the perspective of one whole, that is, an external view of system S. Using these terms and this example, Table 6.2 identifies and explains the extended set of whole–part relations.

For the consecutive and simultaneous relations, mutable composition does not generally apply, but neither is it precluded. It does raise the question of reversibility. For example, if S must be modified to be integrated into S1, can it be "unmodified" or returned to its original form if later removed from S1? And can it fit or be modified again to fit into S2?

In Figure 6.11, SE is a context-general external model of S. SE1, SE2, and SE3 are context-specific external models of S. The context-general and context-specific models correlate, respectively, with the lower and upper external models of S1 in Figure 6.1.

This distinction between context-general and context-specific external specs also illustrates a form of Gödel's incompleteness theorem. The application or analog of this theorem for systems engineering composition is something like this: The uses or roles of a system cannot be determined by or for the system separate from a context or environment. A larger environment is needed to define the uses and roles. Thus, the external model SE in Figure 6.11 does not define what roles S plays. The internal models S1, S2, and S3 define the respective roles of S in those environments.

There is a strong sense in which each system exists in its own world. More specifically, the external view and viewpoint of a given system share a world with the environment of the system, and that environment may share a world with a larger system of which it is a part; but the internal and external of a system are in many respects different worlds; and the same distinction repeats at each composition level and for each system at a given level. Differences in language, ontology, generalization, categorization, and characteristics can occur in each world. These different worlds are closely related to the concepts of non-propagating and non-delegating characteristics.

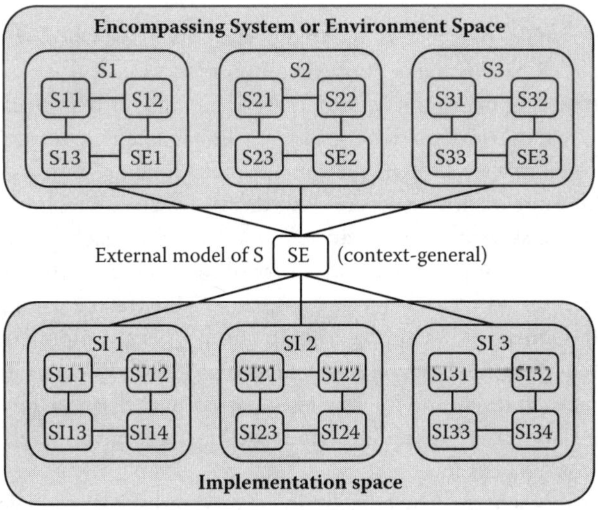

Figure 6.11 Example of extended whole–part relations.

Table 6.2 Extended Whole–Part Relations

Relation	*Description Using Entities in Figure 6.11*
The following relations can exist in *both* contained composition models and shared composition models:	
Alternative	One whole to many alternative potential sets of parts: S could be decomposed into many alternative relations on parts — such as SI1, SI2, or SI3.
Alternative	One whole to many alternative "actual" sets of parts: S could be decomposed into many alternative views, where each view has different components; e.g., SI1 might be a functional view, SI2 a physical deployment view, SI3 a data view.
Alternative	One part to many alternative potential wholes: S could be a part of many alternative systems, such as S1, S2, or S3.
Consecutive	One part to many wholes consecutively at different times, i.e., subject to the contained model constraint (but not the shared model constraint) that a part can only exist in one whole at a given point in time: S could consecutively be a part of multiple systems — first S1, then S2, then S3.
These relations can exist *only in shared* composition models:	
Simultaneous	One part to many wholes simultaneously.
Simultaneous	Network of many wholes to many parts.

Multilevel Composition Structures and Scales

A variety of multilevel composition structures exist, including repeating relations such as hierarchies and networks. This section discusses these structures and how they vary for the different composition models.

Multilevel Structures

We start with the basic whole–part relations, and especially the basic whole–part structure in which one whole has multiple parts. For the contained composition model, if we extend this to multiple levels, we generate a repeating one-to-many structure, a hierarchy or tree. Because we are discussing repeating wholes and parts, the primary structure is actually a holarchy, a hierarchy of holons. In holon terms, the key generator of this structure is the concept that a holon is both a part and a whole.

We can reveal a bit more of this structure if we use the inclusion relation from Table 6.1; that is, an internal model of a system includes an external model of each part. If we extend this to multiple levels, it produces an intertwining structure of internal and external models that toggles between internal and external at each level. Figure 6.12 illustrates how multiple levels can be generated. S_{444} is a component of system S_{44} and, in turn, S_{44} is a component of system S_4, which is, in turn, a component of S. This can be repeated as many levels as necessary. Both components and connectors exist at each level. The intertwining pattern was discussed by Davis (1990, 12-19) in terms of what versus how. The internal describes the 'how' at a given level, and that leads to the external or the 'what' of the next level.

We can go beyond this structure by considering the extended set of whole–part relations from the previous section and extrapolating to multiple levels. The multilevel structures described below are varying interpretations of the views in Figure 6.13.

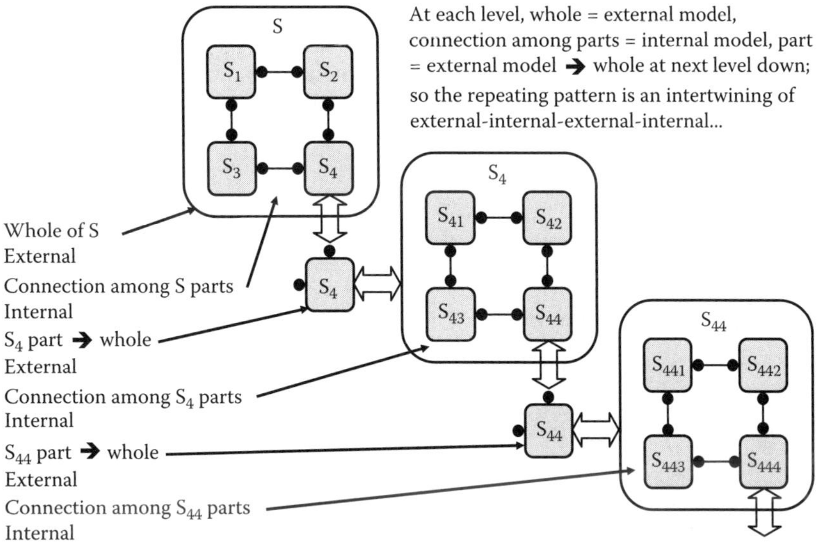

Figure 6.12 Repeating whole–part contained composition structure.

Composition structures generated by contained models and shared models include

- Hierarchy (repeating one to many relation) of one whole to many parts
- Hierarchy of one whole to many alternative potential sets of parts
- Hierarchy of one part to many alternative potential wholes
- Hierarchy of one part to many wholes at different times, that is, subject to the contained and consecutive model constraint that a part can only exist in one whole at a given point in time

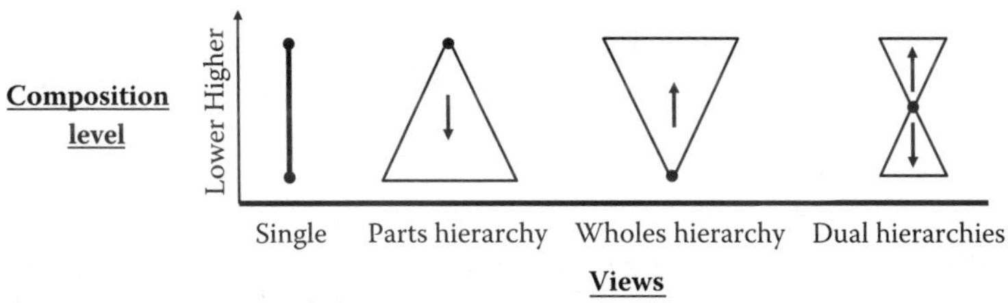

Interpreting these as alternatives:

- Single: One composition option, no alternatives
- Parts: From top down, multiple alternative sets of parts at each level—e.g., system can have multiple potential designs
- Wholes: From bottom up, multiple alternative encompassing wholes at each level—e.g., system can fit in multiple potential environments
- Dual: From any intermediate point, dual hierarchies or holarchies:
  - Down = Alternative sets of parts
  - Up = Alternative encompassing wholes

Figure 6.13 Views of extended relations in multilevel compositions.

Composition structures generated only by shared models include:

- Hierarchy of one part to many wholes simultaneously
- Network of many wholes to many parts, repeated to multiple levels
- Layered architectures

These structures, and especially the pervasive dual hierarchies, reveal the form of the many-to-many network structure of the composition dimension in the model space.

Multiple Levels of Living Systems

James Miller (1995) described in depth the architecture, functions, elements, and processes of living systems at multiple levels, ranging from a cell to a supra-national system. One might call this structure a living holarchy. For the purposes of this chapter, we consider the self-similarity of this structure from two viewpoints: (1) a set of 20 functions (Miller called them subsystems) in each system that are common across all levels, and (2) physical elements that vary across levels.

Miller described eight levels of composition: cell, organ, organism, group, organization, community, society, and supra-national system. Although he called these living systems, much of the focus is human-centric. He ascertained 20 functions that exist in all living systems at all these levels. The functions are divided into three groups as follows:

1. Two functions that process *both matter-energy and information:* Reproducer; Boundary
2. Eight functions that process *matter-energy:* Ingestor; Distributor; Converter; Producer; Matter-energy storage; Extruder; Motor; Supporter
3. Ten functions that process *information:* Input transducer, Internal transducer, Channel and net, Timer, Decoder, Associator, Memory, Decider, Encoder, Output transducer.

Miller discusses the physical elements that perform these functions at each level. For example, the ingest function is performed in an organ by an input artery; in an organism by mouth, nose, and skin; in a society by an immigration service. The boundary function is performed in a cell by plasma membrane; in an organism by skin; in an organization by guards at a building entrance for matter-energy, and by a network firewall for information.

There is another important perspective on the 20 functions — namely that they could serve as the elements of a universal architecture. This possibility, along with more description of the functions, is discussed in Chapter 7.

Another biological model of multiple levels is the framework for conflict and cooperation given by Futuyma (2005). The conflict and cooperation relations coexist in various combinations of attractive and repulsive forces in the formation of a higher-level entity from lower-level units, and this can recur across multiple composition levels.

Composition Scale Issue

MOSES considers repeating composition structures as open-ended in either direction, up or down. The same pattern can be repeated at as many levels as needed for the understanding and engineering of any system. The physical world is a giant composition hierarchy or holarchy that ranges from superstrings up to galaxies and the entire known universe. This is a large range of scales that represents a size difference of about 60 orders of magnitude, according to

Dyson (1989, 18). The scale where humans exist is approximately in the middle of this range. Traditionally, SE has operated primarily at this scale but it is now being extended both up and down. Nanotechnology takes engineering down to the molecular or atomic scale. The large range of scales is sometimes grouped into nano-scale, micro-scale, meso-scale (humans), and macro-scale levels. Because nano-devices will ultimately need to be used in the macro-world, SE will need to traverse and integrate all these scales, or at least from atomic up to several orders of magnitude above humans.

There has also been considerable discussion in SE about mega-systems and systems of systems, which require some degree of scale-up. Systems that cross organizational boundaries and even national boundaries, such as coalition forces, move into the complex systems arena and present significant challenges. However, in terms of levels of composition, SE is arguably being stretched less in the scale-up direction by these large systems than in the scale-down direction by nanotechnology. In any case, the self-similar composition structures and relations, and repeating patterns of MOSES facilitate both scale-up and scale-down. Those structures capture the similarities at all levels but we still need a way to capture the differences between levels. The MOSES mechanisms for capturing differences are described in Chapter 7.

The concept of scale is also connected to the MOSES subject of views. This connection is discussed in Chapter 10.

Membrane Characteristics

We now revisit the boundary or membrane issue that was discussed in Chapter 3. The reason is that composition brings this issue strongly into play. In particular, SE, including but not limited to the shared composition model, often requires explicit and non-trivial decisions on where system boundaries lie, as well as other characteristics.

The placement of a membrane or boundary may be obvious in some systems, but in many systems it needs to be defined by someone, usually by the observer who defined the system, as we noted in our Chapter 2 definition of system. A key characteristic of the membrane is its permeability, which may range from entirely transparent (no encapsulation) to entirely opaque (full or extreme encapsulation):

- Extreme encapsulation = boxology (see below)
- No encapsulation = invisible and non-utilitarian membranes (i.e., in a non-encapsulated system of several composition levels, the top-level system in effect consists of all the parts at the bottom level and the overall connection among them)
- Partial encapsulation, such as encapsulation of some parts and not others, or perhaps making a part visible but not accessible, or perhaps making system state visible
- Non-definitive membranes: fuzzy or uncertain membrane:
    - Fuzzy membrane: situation in which a part is partly in and partly out of a system, in a range of 0 (entirely out) to 1 (entirely in)
    - Uncertain membrane: situation in which it is unknown whether a part is in or out of a system, but in general a likelihood or probability can be estimated in a range of 0 (certainly out) to 1 (certainly in).

The need for fuzzy and uncertain boundary is mitigated by the fact that a system is what an observer says it is, and therefore the observer can define the boundary to include and exclude whatever he wants. Nevertheless, there are still situations in which fuzziness and uncertainty are useful in defining boundaries. Chapter 11 gives more details on the specification of fuzzy and uncertain boundaries and the relation between them.

Extreme Encapsulation: Boxology

The discussion in this section assumes open systems. It is also based primarily on a contained composition model. We describe extreme encapsulation in terms of boxes instead of systems, to emphasize that the concepts are very general. It also summarizes many of the points that have been made in this chapter. In the discussion that follows, box means what is often called black box. That is, a box that hides what is inside it.

- *Box*: something of interest as a whole; contains a set of interacting parts; has a defined boundary that separates what is inside from what is outside; hides both inside and outside from each other; and has points in the boundary, called *boundary points*, that can be interpreted as ports or access points or interface points. Stuff can go through the boundary at the boundary points, either from outside in or from inside out.*
- The *parts* of a box are themselves boxes and connectors through which they interact via their exposed boundary points.
- A box exhibits *characteristics*, which are separated by its boundary into external and internal characteristics:
  - The *external* (black box) characteristics of a box are externally visible at its exposed boundary points.
  - The *internal* characteristics of a box are internally visible in the connection among its parts (boxes) and their interaction.
  - An internal view of a box contains external views of its parts; this internal-external *intertwining relation* scales to an unlimited number of levels.
  - Parts that serve as ports or boundary points are partially or fully exposed; that is, their encapsulation is either non-existent or partial (see next section).
- Box *models* are also separated by the box boundary into two types: external models and internal models; both external and internal models have multiple *views*:
  - External models are typically called requirements, capabilities, or service models.
  - External models are separated into two types: *provided* capabilities or services, and *required* capabilities or services. Two boxes can interact if they have *conjugate* external models: capabilities or services provided by one box are required by the other box.
  - Internal models are typically called concept of operations, architecture, design, or implementation.
  - The relation between the two: an internal model of a box includes the external models of its parts.
  - Other relations between internal and external models exist, as described in this chapter in the section entitled "Relations Between Wholes and Parts," but those are less related to encapsulation.

* The exposed boundary points, or ports, of a system are in many cases also parts of the system (e.g., a control knob on an appliance, or the steering wheel in a car). When we discuss extreme encapsulation of such a system, we mean it hides all *internal* parts, not *ports*. Ports are, by definition, not (fully) encapsulated.

- *Boxes are nested*: a box contains all boxes below it. *Models of boxes are not nested*: a model is only one level deep.
- A box is a very general concept that can be interpreted as a natural or artificial system, a traditional or complex system, a component, a person, an organization or enterprise, a group of persons, a collection of any of the foregoing; and any composition level, from nano to macro.

Partial or No Encapsulation

What changes if a system is not encapsulated, that is, the system boundary is transparent and completely permeable? We could model this situation in at least two ways. One model is that the whole, the boundary, S_{ext} does not exist, except perhaps in some virtual or logical sense. The relation and the parts still exist, but not the boundary. In fact, if none of the parts at any level of the system are encapsulated, then the external view of the system collapses down the composition hierarchy chain to the bottom-level primitives such as nuts and bolts, atoms, or individual symbols. Another approach is that the model of the whole collapses to the internal relation; that is, $S_{ext} = S_{int}$.

How about partial encapsulation? What does this mean? A couple of examples help explore this issue. Is a glass house encapsulated? It is for some characteristics, not for others: yes with respect to air, water, wind, small physical objects that cannot penetrate or break the glass, and sound to some extent; and no with respect to visual access — the boundary is visually transparent, the internal elements in the house are visible. On the other hand, a house with no external walls but a regular roof is also partially encapsulated: yes, it is visually encapsulated from a viewpoint above the roof; no, it is not encapsulated from a viewpoint on the ground near the house. This latter example is analogous to other types of systems, such as information systems. They may have some parts whose characteristics — properties, services, behaviors — are open to use, and other parts whose characteristics are encapsulated and must be accessed only through the system boundary.

Consider another example: the Internet and the related structure of REST-based Web services. These are huge interconnected groups of entities and links with a deep nested or chained structure. Hinchcliffe (2008) claims that these structures violate encapsulation and call into question the established principles of encapsulation and information hiding. The MOSES interpretation of these structures is as follows. If we assume a contained composition model, then the structures look as if they break encapsulation and share their inner secrets with the world. But if we assume a shared composition model, then it appears that modularity and encapsulation are still alive and well. Each link or Web site is encapsulated but it can be shared in the context of multiple link chains.

Composition-Related Processes

Processes

Primary composition-related processes include compose, encapsulate, decompose, componentize, restructure, add, remove, and replace. These involve other subsidiary processes: match, fit, mutate, and integrate. The first four primary processes derive directly from the relations identified in the section entitled "Relations Between Wholes and Parts." The latter four primary processes are concerned with parts. Adding, removing, and replacing a part are intuitive processes. Restructuring involves changing the interaction pattern of what parts interact with other parts.

The processes transition a set of parts between these states: not composed, partially composed, and composed.

1. *Not composed:* Parts exist, are identified, are available and ready for composition, but are not composed; system does not exist as a whole.
2. *Composed:* Parts are in a relation; system can operate as a whole.
3. *Partially composed:* Parts are in a relation, but at least one part is not in the system. In this state, a system may be able to operate as a whole with little or no degradation, or with significant degradation, or not be able to operate as a whole.

Composition is a transition process from not composed to composed. Decomposition is a transition process from composed to not composed. Part replacement is a transition process from composed to partially composed to composed. Adding, removing, and restructuring start and end in a composed state, and during the process, may be in any of the three states. Adding, removing, and replacing are asymmetric composition models, because they involve a whole and a part.

Part replacement might involve simple plug out old, plug in new. Or it might involve mutating, making changes — external or internal — to the new part or to existing parts with which the new part is interacting. Or it might involve introducing adaptors as separate intermediary parts. The processes of {plug out and plug in one part} and {plug out old, plug in new + adaptor} both correspond to the immutable type of composition model, while mutating one or more parts corresponds to the mutable type.

Reversibility: In some cases, composition is reversible: the parts that were put together can be taken apart into their original state. In some cases, decomposition is reversible: the whole that was decomposed into parts can be composed again into the whole.

Composition process boundary: After composition, both wholes and parts can adapt as an ongoing change process. This is not part of the composition process. But in the mutable composition model, parts can also change and mutate during composition to fit into a whole. So when does composition mutation stop and adaptation begin? And where does temporal emergence fit? This may seem like a trivial issue, but it is important in adaptive and self-organizing systems. It is also needed to define the boundary between system composition and system change, which corresponds to the boundary between this chapter and Chapter 9. Our answer is that while we acknowledge this to be a fuzzy boundary, it is convenient to define it as occurring at the first point in time when the system is able to operate as a whole.

Composition also involves more specific processes of matching, fitting, and integrating parts together. An example is matching context-dependent external models with context-generalized external models. This process depends heavily on specifications of components and is therefore discussed further in Chapter 11.

Another composition variable is the sequence in which parts and wholes in a composition structure are created: whole, then parts (top down); parts, then wholes (bottom up); or simultaneous (e.g., a natural system such as a developing human). This variable gets into the engineering or actualization process, which is addressed in Chapter 13.

Agents

An additional element of the composition process is the concept of agent. A composition agent is an entity or condition that is instrumental in bringing about a composition. Some compositions only occur through the direct action of an external agent. Others occur spontaneously or through the choices of autonomous components.

Examples of Direct Agent Composition

In chemistry, catalysts and initiators are often used as agents in many compositions. An initiator helps start a composition and a catalyst increases its rate. A temperature range may induce a chemical composition. Agents in engineering include architects, designers, and builders who design and compose a variety of systems. Other agents include factory workers who produce products; robots that produce products; and software that assists in producing, or automatically produces, a design or produces other software. Most or all of these agents can be characterized as allopoietic systems, or systems of one type that produce a system of a different type.

Examples of Non-Agent Compositions or Indirect or Virtual Agents

Compositions of most natural systems, for example, formation of galaxies, solar systems, planets, mountain chains, oceans, rivers, hurricanes, organisms, social groups, molecules, or biological cells are of this type. Some of these could be characterized as autopoietic systems, or systems that maintain, organize, and regenerate themselves.

Partial or Mixed Direct/Indirect Agents

Some compositions are in between, where an agent has a partial or indirect role.

In recent years, the Internet has provided new opportunities for information system composition, especially in the context of open-source software, SOA, Web services, and Web 2.0. Available software services from third parties can be composed via public interfaces and made available on Web sites. The resulting composition is called a mashup. Many of these include geospatial maps as one part of the composition — for example, HousingMaps (2007) combines Google maps with Craigslist to enable searches for available housing in a selected area. Wood et al. (2007) discuss a geovisualization mashup. Areas of application include enterprise services (Zou and Pavlovski 2007) and Web service development (Jin and Lee 2008).

More generally, this category of mixed composition is closely related to the dominant process of collective actualization introduced in Chapter 1 and discussed further in Chapter 13. Traditionally, SE puts together large components or subsystems, each of which was built by a separate vendor or contractor, and in many cases were preexisting (e.g., COTS). That pattern is indicated by the M (machine) part in Figure 6.14. But the A (agent) part exhibits the same pattern: The human is a preexisting part, coming to this role in system S with a set of capabilities acquired through growth, education, and experience. So in the sense of putting preexisting parts into a system, this example is not very different from a familiar SE process.

The significant difference comes in the contribution of the autonomous system A to the formation, integration, and organization of the hybrid system S. In the unified SE of the future that addresses both machines and autonomous agents, we have argued that the dominant type of systems will be hybrid. The composition process of hybrid systems is a key part of the overall collective actualization process. Table 1.3 showed the system row with SE and SA boxes as part of collective actualization. Mapping that to Figure 6.14, the SE part of Table 1.3 is the role the systems engineer plays in Figure 6.14 in putting together system S, which involves bringing in machine M and some

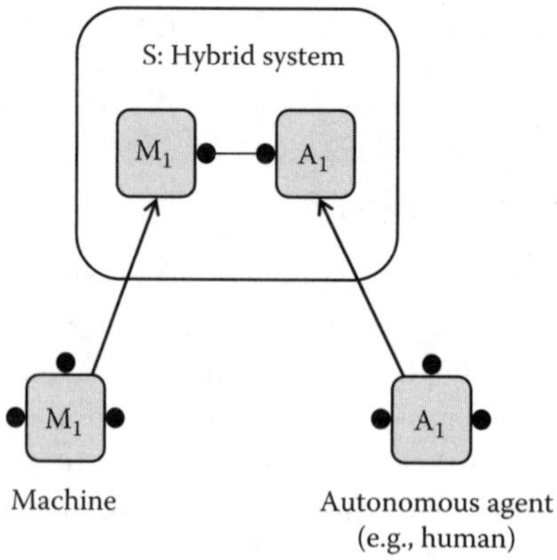

Figure 6.14 Mixed composition of hybrid systems.

portion of the integration. The SA part of Table 1.3 is the role the agent A plays in Figure 6.14 in putting together system S, which involves a portion of the integration, ongoing change, and self-management.

A summary of the three types of composition in this section in the context of Table 1.3 actualization is as follows:

1. Direct agent: all SE, no SA
2. Non-agent: no SE, all SA
3. Partial or mixed agent: some SE, some SA

References

Allen, R. and D. Garlan. 1997. A formal basis for architectural connection. *ACM Trans. Software Eng. and Methodol.*, 6(3):213–249.

Atay, F, and J. Jost. 2004. On the emergence of complex systems on the basis of the coordination of complex behaviors of their elements: synchronization and complexity. *Complexity,* 10(1).17–22 September. Published Online: 25 October 2004. DOI: 10.1002/cplx.20045. (Accessed January 11, 2009.)

ATIS. 2008. *ATIS Telecom Glossary 2007.* Washington, D.C.: Alliance for Telecommunications Industry Solutions. http://www.atis.org/glossary/definition.aspx?id=3516. (Accessed January 11, 2009.)

Barabási, A.-L. and R. Albert. 1999. Emergence of scaling in random networks. *Science*, 286:509–512.

Bittner, T. and B. Smith. 2003. A theory of granular partitions. In *Foundations of Geographic Information Science*, M. Duckham, M.F. Goodchild, and M.F. Worboys, Eds. London: Taylor & Francis Books, p. 117–151.

Boccara, N. 2004. *Modeling Complex Systems.* New York: Springer-Verlag.

Davis, A. 1990. *Software Requirements: Analysis and Specification.* Englewood Cliffs, NJ: Prentice-Hall.

Dyson, F. 1989. *Infinite in All Directions.* New York: Harper and Row, Perennial Library edition.

Emmeche, C., S. Køppe, and F. Stjernfelt. 1997. Explaining Emergence: Towards an Ontology of Levels. *Journal for General Philosophy of Science* 28: 83-119 (1997). http://www.nbi.dk/~emmeche/co Publ/97e.EKS/emerg.html. (Accessed January 11, 2009.)

Futuyma, D. 2005. *Evolution.* Sunderland, MA: Sinauer Associates.

Gruman, G. and E. Knorr. 2008. What cloud computing really means. *Infoworld.* April 7. http://www.infoworld.com/article/08/04/07/15FE-cloud-computing-reality_1.html. (Accessed January 11, 2009.)

Hayes, B. 2008. Cloud Computing. *Commun. ACM.* 51(7):9–11.

Hinchcliffe, D. 2008. 12 Things You Should Know about REST and WOA. http://hinchcliffe.org/archive/2008/04/08/16627.aspx. (Accessed January 11, 2009.)

Honderich. T. 1995. *The Oxford Companion to Philosophy.* New York: Oxford University Press.

HousingMaps. 2007. http://www.housingmaps.com/. (Accessed January 11, 2009.)

Hybertson, D. 2005. *Testing Commercial–off-the-Shelf Components and Systems,* S. Beydeda and V. Gruhn, Eds. New York: Springer-Verlag. p. 213–237.

Jin, H-J. and H-C. Lee. 2008. Web Services Development Methodology Using the Mashup Technology. *Int. Conf. Smart Manufacturing Application,* 9–11 April. p. 559–562.

Kirchner, G. 2001. Self-Instructional on Biological Molecules: Organic Compounds. http://homepages.ius.edu/GKIRCHNE/biomolec.htm. (Accessed January 11, 2009.)

Klir, G. 1985. *Architecture of Systems Problem Solving.* New York: Plenum Press.

Klir, G. 2001. *Facets of Systems Science, 2nd ed.* New York: Kluwer Academic/Plenum Publishers.

Meyer, B. 1997. *Object-Oriented Software Construction, 2nd ed.* Upper Saddle River, NJ: Prentice Hall.

Miller, G. 1956. The magical number seven, plus or minus two: some limits on our capacity for processing information. *Psycholog. Rev., 63:81–97.*

Miller, J. 1995. *Living Systems.* Liwot, CO: University Press of Colorado. (Originally published 1978.)

Newman, Mitch., A.-L. Barabási, and D.J. Watts. 2006. *The Structure and Dynamics of Networks.* Princeton, NJ: Princeton University Press.

O'Connor, T. and H. Wong. 2006. Emergent Properties. Stanford Encyclopedia of Philosophy. http://plato.stanford.edu/entries/properties-emergent/. (Accessed January 11, 2009.)

OMG (Object Management Group). 2007. Unified Modeling Language (OMG UML), Superstructure, V2.1.2. OMG Document Number: formal/2007-11-02 Standard document URL: http://www.omg.org/spec/UML/2.1.2/Superstructure/PDF. (Accessed January 11, 2009.)

Pakkenberg B. and H. Gundersen. 1997. Neocortical neuron number in humans: effect of sex and age. *J. Comp. Neurol.,* 384:312–320.

Pakkenberg, B., D. Pelvig, L. Marner, M.J. Bundgaard, H.J.G. Gundersen, J.R. Nyengaard, and L. Regeur. 2003. Aging and the human neocortex. *Exp. Gerontol.,* 38:95–99.

Shalizi, C. 2007. Emergent Properties. http://www.cscs.umich.edu/~crshalizi/notebooks/emergent-properties.html. (Accessed January 11, 2009.)

Shaw, M. and D. Garlan. 1996. *Software Architecture: Perspectives on an Emerging Discipline.* Upper Saddle River, NJ: Prentice Hall.

Simons, P. 1987. *Parts: A Study in Ontology.* New York: Oxford University Press. (Paperback edition, 2000.)

Wood, J., J. Dykes, A. Slingsby, and K. Clarke. 2007. Interactive visual exploration of a large spatio-temporal dataset: reflections on a geovisualization mashup. *IEEE Trans. Visualization Computer Graphics.* 13(6):1176–1183.

Zou, J. and C. Pavlovski. 2007. Towards accountable enterprise mashup services. *IEEE Int. Conf. e-Business Eng.,* 24–26 October, p. 205–212.

Chapter 7

Commonization

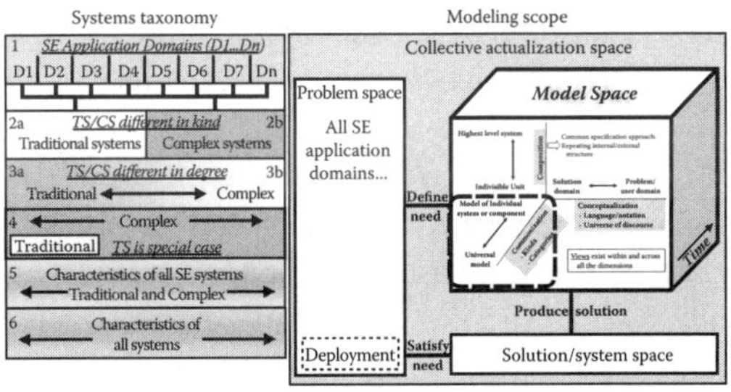

Commonization is the story of similarities and differences, of patterns among multiple systems. In a collection of systems, what is common to all and what varies? And which of the commonalities and variabilities do we want to model? By focusing on models of collections of systems rather than individual systems, we can capture patterns in a given domain or discipline, and those patterns represent knowledge. Commonization is the model space dimension that enables knowledge to be accumulated in the model space, shared, and used in the application of SE. The region of a model is important in this dimension.

Basic Concepts: Similarities and Differences

Much of the depiction of engineering models in Chapter 6 reflects how we often think of such models: as models of one specific system — the one we are building or analyzing. If we are building system S, we typically write a requirements spec R that we view as the requirements of S, and we describe an architecture A that we view as the architecture of S.

But in any engineering discipline, as the discipline matures, certain common patterns emerge as common problems and common solutions. In automotive engineering, while cars have evolved over time, the basic function of transporting a few people from point A to point B has remained the same. And while engineers and designers have continually reinvented the car, the general design, the basic set of interacting parts, has been fairly stable. The same situation exists in other engineering disciplines. General patterns or models have traditionally been defined in engineering handbooks. They emphasize what is common for all systems in the domain. Each system that is engineered is a realization of models that are specialized from the general models.

Even in the nascent domain of bio-nanosystems, there are efforts to define standard parts to be used as building blocks, called BioBricks, as mentioned in Chapter 1.

Recall the Chapter 4 discussion of correspondence set, which is the join structure for separate, loosely coupled models and systems. The correspondence set pattern in this chapter is that M is an *individual* model while S is a *set* of systems that correspond to that model. More specifically, this chapter addresses the intersection of two ideas: (1) the abstraction aspect of models and (2) the similarities and differences that exist among a set of systems.

Model abstraction: A model of a system focuses on certain characteristics and ignores the other characteristics of the system. Thus, any given model partitions system characteristics into two categories: those that matter and those that do not matter. As soon as we abstract in this way, it introduces the possibility that the model applies to more than one system. That is, the model might capture what is common to a set of systems. And because abstraction is inherent in every model, groups of corresponding systems are an inherent potential of every model.

Similarities and differences: In general, the members of a given set of systems are similar in some ways and differ in other ways. Furthermore, if the set is defined by a model, then "some ways" in which the systems are similar are defined by the model. That is, if a model defines characteristics A, B, and C, then the systems all exhibit characteristics A, B, and C. A homogeneous set has more similarities or more characteristics in common, and a heterogeneous set has more differences or more variation. This fundamental and universal distinction is the basis of the ability to sort out and categorize the "furniture" or the mosaic of the world, and therefore the basis of much of our knowledge of the world.

Intersection: If we overlay the model abstraction distinction on the similarity-difference distinction, we obtain four categories, as shown in Figure 7.1: similarities that matter, differences that matter, similarities that do not matter, and differences that do not matter. Ideally, a model addresses or specifies everything that matters and ignores or leaves open everything that does not matter, for a given purpose.

It is important to note that the column distinction in Figure 7.1 — between similarities and differences — depends on the systems involved, not on the model involved. And that the row distinctions — between what does and does not matter — depends on the purpose of the model involved.

One might think that a model only defines similarities among corresponding systems, not differences. Does a model not describe a set of characteristics that hold for all of the corresponding

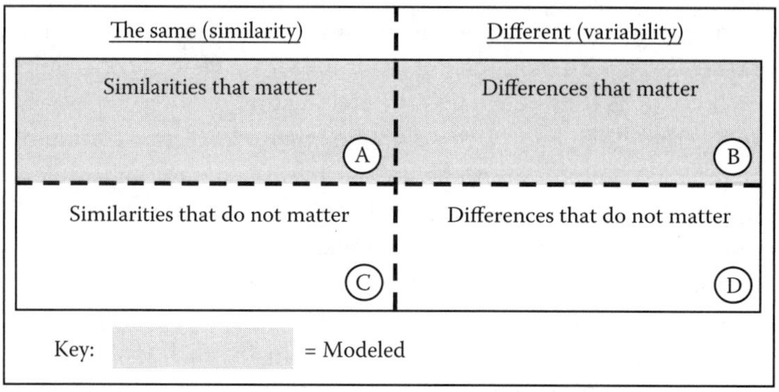

Figure 7.1 Modeling similarities and differences.

systems? But a model includes variables as well as constants. The variables define differences or variability that matters and is within the purview of the model. An example of the areas in Figure 7.1 is a model of linear momentum of physical objects. We use the classical model: Momentum = Mass × Velocity.* Consider two systems: a red apple 8 cm in diameter with a mass of 100 grams and a velocity of 2 ft/sec, and a red ball 12 cm in diameter with a mass of 150 grams and a velocity of 3 ft/sec.

- A: Both are physical objects and therefore have properties of mass and velocity. Those similarities clearly matter to momentum.
- B: The two have different values of mass (100 versus 150 grams) and velocity (2 ft/sec versus 3 ft/sec). Those differences affect the momentum.
- C: Both have red color. This similarity is immaterial to momentum.
- D: They have different diameters (8 cm versus 12 cm), and one is natural while the other is artificial. These differences are immaterial to momentum.

A and B are included in the model. In this case, both A and B are represented in the variables mass and velocity. The similarity (A) is partially implicit: only systems that have properties of mass and velocity are in the region of this momentum model. The difference (B) is more explicit: the range of values for mass and velocity is represented in those two variables.

C and D are excluded from the model. The fact that color (C) and diameter (D) do not matter to the momentum model is implicit in their absence from the model. In general, the exclusion of C and D from a model can be explicit or implicit in the model. *Implicit*, as in the example above: characteristics that do not matter do not appear in the model, and the implication is that those characteristics have no effect on the model. *Explicit:* The model states the characteristic but only for the purpose of making explicit that the characteristic is unconstrained in the model, that is, not of interest in the model. Sometimes it helps to be explicit to counter assumptions. Many of the examples of decoupling in Chapter 4 were of this sort. The decoupling allows us to explicitly separate characteristics that matter (as in A and B in Figure 7.1) from those that do not (as in C and D). For example, by decoupling control from hierarchy, we can define a model that separates the one-to-many relation, which matters to hierarchy, from a control relation, which is explicitly unconstrained in the general hierarchy model.

Expressing similarities and differences in common models is the primary mechanism humans use to capture and share knowledge. Common models include scientific theories, complex systems models, and traditional engineering artifacts such as conventional designs in a given domain. Common models are used extensively in all areas of engineering. In older traditional engineering disciplines, common models often take the form of conventional designs or common system properties expressed in engineering handbooks. Software engineering makes extensive use of common models.† Systems engineering makes use of many of these same types of common models, including large-scale architectures, but there has been less community codification of SE common models.

---

* For purposes of this discussion, we address the speed aspect of velocity but ignore the direction aspect.

† Examples include data types, classes in class hierarchies, component and connector types, metamodels, generics, templates, reference models, architecture styles, domain and product line architectures, and design patterns.

Schindel (2005a, b) has described an SE method in terms of model-based SE (MBSE) and patterns-based SE (PBSE). MBSE recognizes the importance of describing SE artifacts, such as requirements and design, as specific models. PBSE extends this by defining patterns, which in MOSES are general models.

Klir (2001, 214–215) has identified a number of areas in which systems science (and the study of complex systems) has extended our capabilities for modeling through generalizations of mathematical theories. These generalizations include

- From quantitative to qualitative theories
- From functions to relations
- From ordinary geometry to fractal geometry
- From linear to nonlinear theories
- From classical set theory and logic to fuzzy set theory and logic
- From intolerance of inconsistencies to the logic of inconsistency
- From single objective to multiple objective criteria optimization

Both the original and extended element of each pair in this list fit comfortably in the commonization dimension of the MOSES model space. The extended elements are part of the complex system expansion of SE.

Category and Kind Structures

Two varieties of commonization exist. While they have multiple names in the literature, they are often called *categories* and *kinds*. To understand the distinction, we use the concept of members of a set. A category is a set whose members are viewed as individuals, each of which is an instance of the category. A kind is a set whose members are viewed as sets, each of which is a sub-kind of the kind. Why do we use the phrases "are viewed as individuals" and "are viewed as sets" instead of "are individuals" and "are sets," respectively? Because, as we will see shortly, category, individual, kind, and sub-kind are all roles that a single entity can assume in a given context — they are not inherent properties of an entity.

Category and kind have the corresponding relations *instance-of* and *kind-of*. These relations form two separate hierarchies that may be called categorization and generalization. In software object-oriented parlance, "kind-of" is a class hierarchy relation (i.e., the relation of a subclass with its parent class), and "instance-of" is the relation of an individual object with the class from which it was instantiated.

Terms used for the elements in a kind-of relation include

- Kind – sub-kind
- Super-kind – kind
- Class – subclass
- Type – subtype
- Parent – child*

* This is a modeling relation, not a biological one. Often used in conjunction with class. Biologically, a child is not a kind of parent. But in modeling terms, a child class is a kind of a parent class.

- General model – specific model
- Generalization – specialization

Terms used for the elements in an instance-of relation include

- Category – instance
- Class – object
- Universal – particular
- Model – individual
- Metamodel – model

The easiest way to explain these two structures and how they relate is by example. Figure 7.2 illustrates the distinction between the hierarchies, and also the intersection of the two, with a partial sketch of the two-dimensional biological taxonomy. A given entity can have multiple roles in these hierarchies. In Figure 7.2, for example, Dog is a *category* of Lassie, an *instance* of species, a *generalization* of Collie, and a *specialization* of Mammal. Using these terms, a class in object-oriented modeling is a *kind* in the context of its class hierarchy, and it is a *category* in the context of its instantiated objects.

This distinction is discussed in Borgida and Brachman (2007, 353), where they use the terms "individual" and "concept" for what we here call "instance" and "category." The role of an entity,

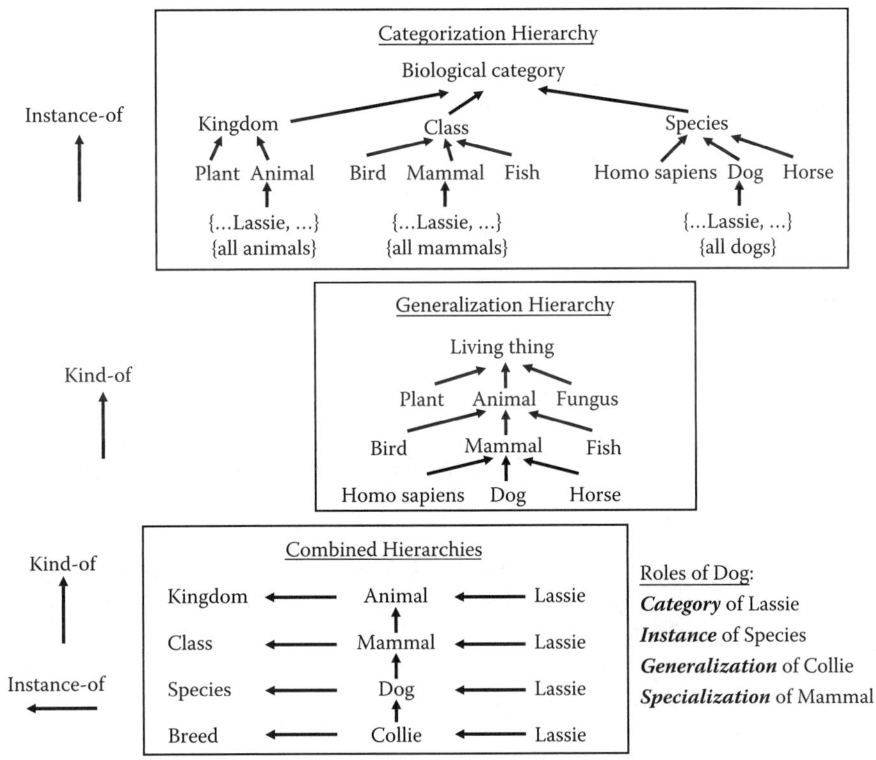

Figure 7.2 Intersection of kinds and instances using biological taxonomy.

Table 7.1 Dog Roles and Relations in Two Domains

Dog Role/Relation	*In Context*	*Domain*
1 Category	of Lassie	Biological taxonomy
2 Instance	of Species	Biological taxonomy
3 Generalization	of Collie	Biological taxonomy
4 Specialization	of Mammal	Biological taxonomy
5 Instance	of Category	Modeling domain
6 Instance	of Instance	Modeling domain
7 Instance	of Generalization	Modeling domain
8 Instance	of Specialization	Modeling domain

in terms of category or instance, may vary by the operation performed on it. For example, consider a book, say Dickens' *Great Expectations*. In the operation of writing the book, *Great Expectations* is an instance of the category book. In the operation of buying or reading it, *Great Expectations* is a category and my copy is an instance of it.

Kind-of is a transitive relation. In the Figure 7.2 generalization hierarchy, because dog is a kind of mammal and mammal is a kind of animal, then dog is a kind of animal. Instance-of is not a transitive relation. In the categorization hierarchy, Lassie is an instance of dog, and dog is an instance of species, but it is not true that Lassie is an instance of species.

The four roles of Dog shown in Figure 7.2 are defined in the context of the biological taxonomy. But Dog also has the role of instance in four other relations in the modeling domain. Because Dog is, in Figure 7.2, a category, an instance, a specialization, and a generalization, in a modeling sense Dog is an instance of all those modeling categories. All eight roles/relations are shown in Table 7.1.

The second group (5–8) represents a commonization modeling domain of discourse. The first group (1–4) represents a content domain, in this case biological taxonomy, that is represented in the commonization modeling domain.

Based on this discussion, we can say that a category is a model role whose targets are instance roles. A kind is a model role whose targets are kind roles of generalization (parent) or specialization (child).

Figure 7.3 shows a similar approach to categorization by Atkinson and Kühne (Atkinson and Kühne 2001; Atkinson and Kühne 2003; and Kühne 2005). They use different terms; instead of "biological domain" and "modeling domain," they use the terms "ontological instance" and "linguistic instance":

- Lassie is an ontological instance of Collie; Collie is an ontological instance of Breed.
- Lassie is a linguistic instance of Object; Collie is a linguistic instance of Class; breed is a linguistic instance of Metaclass.

Our treatment is similar, but differs from theirs in several ways:

- They use a predefined set of absolute modeling categories based on UML: object, class, metaclass; we use two intertwining relative modeling categories: instance and category. This makes it much more scalable, and it shows the repeating pattern.

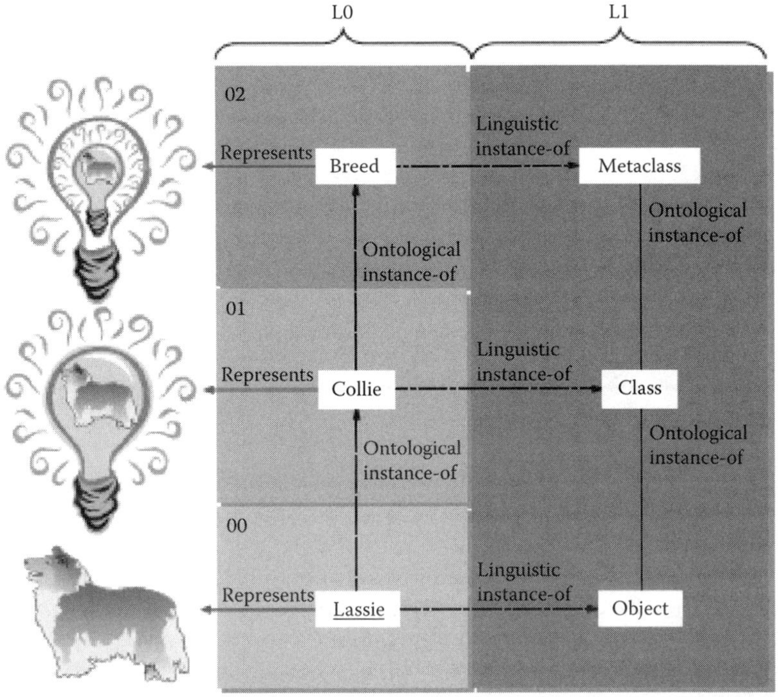

Figure 7.3 Ontological metamodeling view. (*Source:* Atkinson, C. and T. Kühne. Model-driven development: a metamodeling foundation. *IEEE Software*, 20(5):36–41. Published by IEEE Computer Society, 2003. Used with permission.)

- Their use of absolute categories obscures the concept of context-dependent role.
- Our context-dependent role also applies to generalization in addition to categorization; they show the two-dimensional generalization versus categorization of the biological taxonomy, but they do not show the roles, the fact that each element plays multiple modeling roles.

Figure 7.4 shows some relations between categorization and generalization, and between correspondence and categorization. Assume A and B are systems, and M1 and M2 are models, where M2 is a specialization of M1. The circles indicate the regions of M1 and M2; that is, all the systems that correspond respectively to M1 and M2. A corresponds to M1 but not to M2, while B corresponds both to M1 and M2. Figure 7.4 shows four examples of models (or model names) for both M1 and M2. The examples are from domains associated with SE: aeronautical engineering, information systems/software engineering, biology, and mathematics. Each illustrates this relation between correspondence and categorization: A system that corresponds to a model (and therefore is in the model region) is an instance of the category defined by that model.

- In the relation of A to M1 and M2, A is an instance of the category defined by M1 but is not an instance of the category defined by M2. For example, A is a specific individual aircraft of type Airbus 300, or is a specific database management system of type object-oriented DBMS, or is a specific mammal of species dog such as Lassie, or is a specific second-order differential equation such as $3x - dy/dx - d^2y/dx^2 = 0$. Each example given is clearly in the M1 region but outside the M2 region.

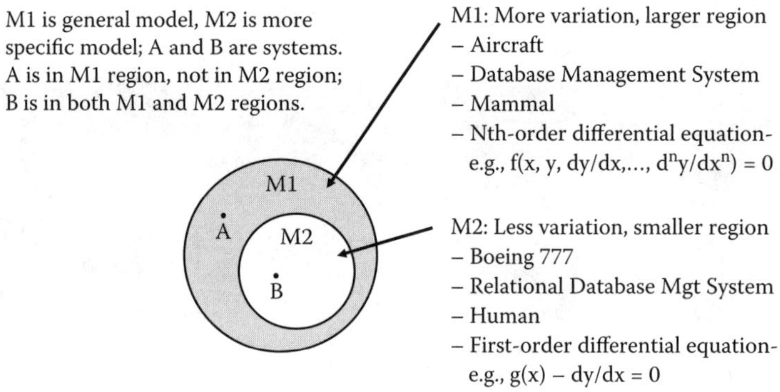

Figure 7.4 Variation allowed in models.

- In the relation of B to M1 and M2, B is an instance of the category defined by M1, and is an instance of the category defined by M2. For example, B is a specific individual aircraft of type Boeing 777, or is a specific database management system of type relational DBMS, or is a specific mammal of species human, or is a specific first-order differential equation such as $3x - dy/dx = 0$.

In sum, a more general model such as M1 defines a category that has more instances, and a region that has more potential corresponding systems, than does a more specific model such as M2. Furthermore, if the more specific model is a specialization of the more general model, as M2 is a specialization of M1, then all systems in the region of the more specific model are also in the region of the more general model.

We now consider the overlap of selected categories of nonlinear dynamical systems, specifically, chaotic systems, logistic maps, and the Lorenz attractor. All these categories are both nonlinear and deterministic.

There is a general model of chaotic systems. All chaotic systems are in the region of this model. The model can be expressed in a number of ways. The expression we use here follows (Boccara 2004, 153). It is based on the behavior trajectory of a system, expressed as a set S (of behaviors) mapped onto itself.

If S is a set, the mapping $f: S \rightarrow S$ is chaotic on S if:

1. Unpredictable: f has sensitive dependence on initial conditions;
2. Nondecomposable: f is topologically transitive (cannot be broken into separated subsets);
3. Regular: f has periodic points that are dense in S.

There is a general dynamical system model called the logistic map (Martelli 1999, 11; May 1976), originally designed for modeling population trends. It specifies the characteristics of a class of systems, in MOSES terms a region of systems, that correspond to this model. Mathematically, the logistic map, a nonlinear difference equation, is written as

$$x_{n+1} = ax_n (1 - x_n)$$

where, for population modeling

- x_n is a number between zero and one, and represents the ratio of existing population to sustainable population at year n, and hence x_0 represents the initial population ratio (at year 0) and x_{n+1} is the population ratio at year n + 1.
- a is a positive number, and represents a net growth rate.

A subset of these systems, for certain values and ranges of the parameter *a*, are chaotic.

A general model called the Lorenz equations specifies a region of systems with a variety of characteristics, including both chaotic and non-chaotic. The equations are (Boccara 2004, 174)

$$dx/dt = \sigma (y - x)$$

$$dy/dt = x (\rho - z) - y$$

$$dz/dt = xy - \beta z$$

where the parameters of this set of equations — σ, β, and ρ — are all greater than zero. The parameters can be interpreted as simplifications of a model of the physical system in which a box is filled with a gas. σ is the ratio of the fluid viscosity of the gas to its thermal conductivity; ρ represents the difference in temperature between the top and bottom of the gas in the box, and β is the width to height ratio of the box. At least one of these systems, in which $\sigma = 10$, $\beta = 8/3$, and $\rho = 28$, is chaotic. This is the famous Lorenz "butterfly" strange attractor (see Figure 7.5 below — from Wikipedia 2008). The logistic map and Lorenz equations regions are disjoint but both regions have an intersection

Figure 7.5 Plotted trajectory of Lorenz strange attractor. (Source: Wikipedia 2008.)

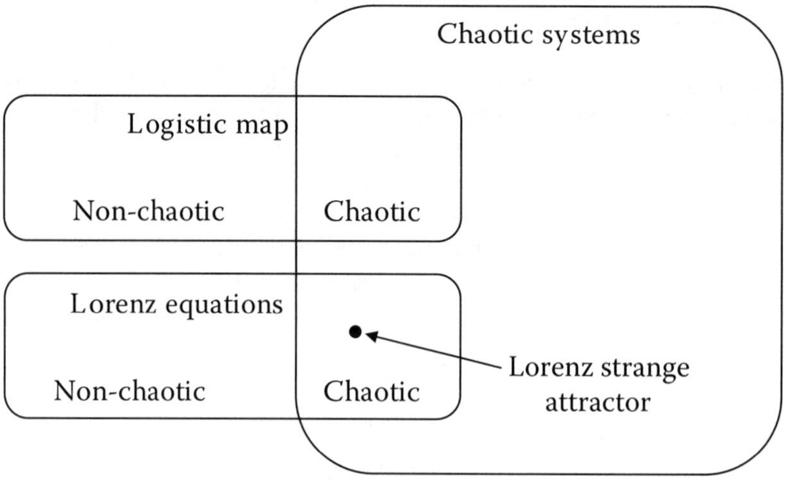

Figure 7.6 Overlap of chaos, logistic map, and Lorenz equations.

with the chaos model region that consists of multiple chaotic systems. The Lorenz strange attractor intersection is a point, representing the chaotic system with the parameters discussed above. The overlap is illustrated in Figure 7.6.

Intension and Extension

The terms "intension" and "extension," as terms of logic, are encountered in some of the disciplines supporting SE. This sidebar briefly describes these concepts and how they relate to SE and MOSES.

Intension and extension are two different ways of defining a term. An intensional definition defines one or more properties. Those properties may apply to any number of individuals, from zero to many. An extensional definition defines a set of individual things by enumerating the individuals. Either of the two definitions can exist without the other. When both definitions exist for a term, then the extension (or the extensional definition) is the set of individuals whose properties match those of the intensional definition.

A few examples will illustrate the difference. We will define a function, a model, and two specific concepts: the morning star and the evening star.

1. *Function:* intensional definition: $y = f(x) = x^2$ in the domain of positive integers. Extensional definition: $(x,y) = \{(1,1), (2,4), (3,9), \ldots\}$.
2. *Model:* intensional definition: for stimulus T, response is R; response time ≤ 2 sec; extensional definition: {S1, S2, S3, S4} (assuming S1, …, S4 are all systems that satisfy the model and no other systems satisfy it).
3. *Morning star:* intensional definition: bright star seen in the morning; extensional definition: {Venus}
4. *Evening star:* intensional definition: bright star seen in the evening; extensional definition: {Venus}

Often, one set of properties (intensional definition) corresponds to multiple individual things (extensional definition), as in the first two examples given. On the other hand, many times a given individual thing corresponds to multiple intensional definitions. In examples 3 and 4, Venus is both the morning star and the evening star, although these terms have different intensional meanings. Taken together, these examples illustrate that the relation between intensional definitions and individual things is many-to-many.

How do these concepts relate to SE? In MOSES and SE in general, models typically have multiple corresponding systems. In some cases, a model is defined to correspond to one system only — for example, a biometric model intended to uniquely identify one individual human. In other situations, one system satisfies multiple models. One such situation is where a set of models describes the same system, for example, a requirements model of system S1, an architecture description of S1, and a design model of S1. A second situation is when a given system plays different roles in different environments, for example, a system has multiple features, and different environments use different features. Another example of the second situation is when a systems engineer plays multiple roles on a project such as team leader and chief designer. Team leader and chief designer are two different categories, two different intensions, but the extension is the same for both.

More generally, in MOSES, the relation between intension and extension is the same as the relation between category and instance. This has implications:

- Intensional and extensional definitions are at different levels of the categorization hierarchy. Intension defines the properties of a *category* of things, and extension enumerates the *instances* of that category — which is one level down.
- Intensional definition defines a model, and extensional definition defines the region of the model.
- Both provide pieces of information that the other does not provide. An extensional definition of a model provides information on what systems are in the region but it does not provide explicit information on the system characteristics of the model. An intensional definition of a model is the opposite; it provides information on the system characteristics of the model but not what systems are in the region.

Another distinction of the intension–extension duality is time invariance versus time variance. Let's use the notation $R_I(W)$ to mean an intensional definition of a region in defined world W, and $R_E(W)$ to mean the analogous extensional definition. Now consider this example: the employee database of ACME company is a region R that results from the model "is employee" applied to the defined world W = "ACME company." Then $R_I(W) = \{S|S = \text{ACME employee}\}$. This is abstract with respect to time; that is, it does not specify a point in time or a time period, and it does not change over time.

$R_E(W)$, on the other hand, is the set of ACME employees as recorded in the employee database at any point in time, and it changes as employees are hired by or leave ACME. The concept of an employee database ($R_I(W)$), which is stable, and the current set of employees ($R_E(W)$) in that database, which changes, are both useful. And the relation between the two is equally important. If all we see is a list of names, we do not know what it represents. If someone tells us that "This is the list of current ACME employees," then we know what the list means.

This approach of abstracting time is a very common use of the intension – extension distinction.

Intension is the dominant way models are defined in MOSES. One goal is to encourage defining a set of characteristics and constraints on them; for example, give characteristics of a platform instead of simply saying "Windows XP." Many things are defined extensionally in TSE (platforms, connectors, components, etc.); this is one reason the U.S. Department of Defense (DoD) changed its acquisition approach from defining "requirements" to defining "capabilities": DoD customers (i.e., DoD managers and organizations that were requesting solutions) were defining requirements extensionally, in terms of "I need System S" or "I need Product P" instead of intensionally, such as "I need a system that performs these functions and has these properties…" In addition, intensionally defined models significantly help define a body of knowledge, as opposed to a collection of individual systems. General models are essential to defining and using knowledge. An appropriate combination of general intensional models that define the properties and rules, and specific extensional regions that serve as examples, is a good basis of knowledge.

Commonization Relations and Multilevel Structures

The intension–extension relation illustrates the more general many-to-many relation between category and individual.

The same is true for kinds. A kind can have multiple sub-kinds. Aircraft can be specialized into rotary and fixed-wing aircraft. A kind can also be generalized in multiple ways. Aircraft can be generalized along with space vehicles into vehicles that fly above the Earth, or it can be generalized along with cars, buses, trains, and passenger ships into transportation vehicles. In general, a kind that is defined by a certain number of properties or characteristics can be generalized by removing some of those characteristics, and specialized by adding to those characteristics. For example, the kind "Boeing commercial aircraft" is defined by the characteristics of commercial aircraft plus manufacturer. It can be generalized to commercial aircraft by removing the manufacturer characteristic from the model; and it can be specialized to Boeing 777 aircraft by adding the characteristics that distinguish the 777 from other Boeing commercial aircraft.

These many-to-many commonization relations are analogous to the single and dual composition hierarchies described in Chapter 6.

- Repeating specializations, and repeating instantiations — both are analogous to the parts hierarchy in Figure 6.13 (Chapter 6).
- Repeating generalizations, and repeating categorizations — both are analogous to the wholes hierarchy in Figure 6.13.
- From a given model, specializations emanating downward and generalizations emanating upward are dual hierarchies analogous to those in Figure 6.13, as are instantiations emanating downward and categorizations emanating upward.

Categorization is the relation that involves what many call metamodels. A metamodel is a category of which a model is an instance. Some modeling structures define four such levels: Object is an instance of a Model, which is an instance of a Metamodel, which is an instance of a Meta-metamodel. A corresponding example from Figure 7.2 is: Lassie (object) is an instance of Dog (model), which is an instance of Species (metamodel), which is an instance of Biological Category (meta-metamodel). As originally defined, the Object Management Group's (OMG) Meta Object Facility (MOF) had this four-layer architecture. In MOSES, all levels are models — model is a relative term. As described above, each model is defined and distinguished by its roles in the categorization and generalization hierarchies. MOSES does not restrict to four levels. Starting with MOF 2.0 (OMG 2006), MOF has also switched to this more flexible and extensible architecture.

Generalization allows architectures of varying levels of generality to be defined. Specific architectures can be defined that include extensive detail and admit a small class of corresponding systems. General architectures can be defined that include little detail and admit a large class of corresponding systems. Examples of the latter are reference architectures or reference models.

Under the constraint that each level only specializes its parent, and does not modify what it inherits from its parent, and there is only single inheritance (i.e., a kind inherits from a single parent), then a kind hierarchy has the following pattern: the top (most general, most inclusive) kind in the hierarchy is relatively heterogeneous, and each successive level below that becomes relatively more homogeneous. Equivalently, the top of the hierarchy has the fewest constraints, and the bottom of the hierarchy has the most constraints. The top has the largest region or population of systems; and the bottom has the smallest region. Inheritance is discussed further in Chapter 9.

Categorization is, arguably, the more fundamental of the two commonization types. Kind or generalization is critical and useful but it depends on categorization. Kind adds a relation (generalization) between categories. Categories could exist without kinds but kinds could not exist without categories.

Gödel's incompleteness theorem also applies, by analogy, to categorization. The characterization or role of a model is incomplete within the model itself. A model can only be defined as a category or an instance in context: It is a category in its relation with an instance; it is an instance in its relation with a category.

Example Tool: The Generic Modeling Environment

The Generic Modeling Environment (GME) is a toolkit for creating domain-specific modeling environments through the use of metamodels that specify the modeling language of the application domain (GME 2008; Ledeczi et al. 2001). It is based on UML and the Object Constraint Language (OCL). At the most general level, the modeling constructs include hierarchy, aspects, sets, references, and constraints. Other first-class objects include models, atoms, and connections, which express relations. A new modeling language can be defined using the MetaGME framework and its metamodeling language.

GME has been used to define domain-specific modeling frameworks such as an environment for modeling physical systems (Manders et al. 2006) and the modeling language for integrated modular avionics design (MIMAD) (Gamatié et al. 2006).

Leverage

MOSES recognizes a trade-off between general and specific solutions, and addresses it through a principle we call *leverage*. In the context of the overall model space, an important goal is to develop an efficient set of "reusable" models that will satisfy specific problems and result in specific systems that operate in given environments. The reusable goal is achieved if the model captures what is common across multiple systems. Satisfying a specific problem is achieved if the model addresses what is different or unique about that specific problem. Therefore, the reusable goal and the specific satisfaction goal are in conflict. Several approaches have been proposed to satisfy the common and unique constraints. These include

- *Custom solution approach:* Create custom solutions to each problem in each domain (i.e., the radical design approach). In general, this is unnecessarily and prohibitively expensive. It emphasizes unique aspects but does not exploit the common aspects of problems.
- *General solution approach:* Define a one-size-fits-all solution that captures what is common across all problems in all SE application domains. In software engineering, this approach is reflected in higher-level general languages, to extend automation. This approach is highly reusable but requires significant augmentation and specialization for each individual solution. In fact, most of each solution must be custom tailored; so although this approach looks very different from the custom solution approach, the effect is very similar.
- *Application domain approach:* Define a general solution within an application domain that captures what is common to all systems in that domain. The approach exploits the fact that problems and solutions and systems within a domain are typically much more homogeneous than those across domains. Therefore, the general solution captures a significant portion of problem variability within the domain.
- *Enterprise approach:* This criterion would result in maintaining enterprise-specific concepts throughout. This addresses enterprise-specific issues but duplicates domain and more general aspects of problems.

Each of these criteria makes a contribution to problem solving, but we now develop a more complete and balanced solution. The criterion that we propose is maximum leverage. Leverage of a model, seen as a solution, is defined as the degree to which it satisfies the following two criteria (which are, in general, conflicting): (1) number of problem situations to which it applies, and (2) extent to which it provides the complete solution needed for the specific problem situations to which it applies. Essentially, leverage is the product of these two criteria. These criteria represent a general trade-off in problem solving: generality versus power (illustrated in Figure 7.7). Note that leverage is possible primarily in mature or normal design, as opposed to radical design, because it takes time and experience to determine what is common and what is unique among the problems and solutions in a given domain. Generality refers to the number of problems a solution helps solve, the first criterion; power refers to the degree to which a solution helps solve a given problem, the second criterion. Ideally, one solution would completely solve all problems (upper right of box in Figure 7.7). However, there is a limit even when a discipline reaches maturity; the trade-off still exists. In an immature discipline, the limit is not yet reached, and leverage is even more restricted. The current and ultimate trade-offs are shown by the diagonal dashed lines in the figure. Thus, two models of equal leverage may differ in that one may provide a small part of the solution for a large number of problems (shown as the area labeled "Leverage of general solution" in the figure), while the second may provide most of the solution for a small number of problems (shown as "Leverage of specific solution" in the figure). The first criterion reflects the perspective of the person with the solution—the producer or provider: I have a solution that will give some help in solving everyone's problems (one size fits all). The second criterion reflects the perspective of the person with the problem—the consumer or client: I want a solution that completely solves my specific problem.

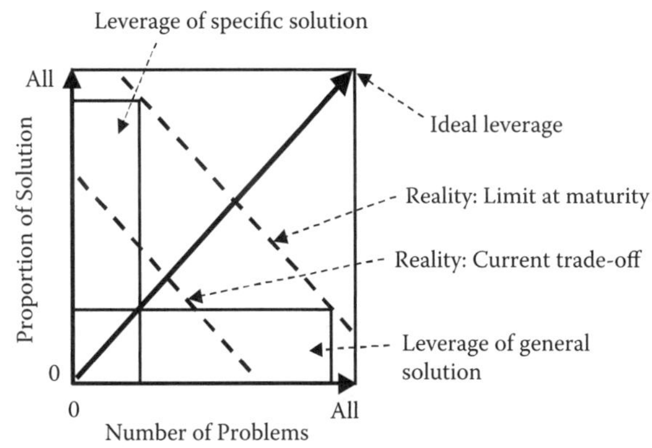

Figure 7.7 Leverage metric. (*Source:* Hybertson, D. A uniform component modeling space. *Informatica,* 25(4):475–482. Published by Slovenian Society Informatika, 2001. Used with permission.)

Leverage thus exploits commonality and balances it with the other side of problem solving: *power of solution*. It fosters a multi-model strategy that is driven by the goal of finding models that best exploit the full commonization spectrum of the model space. That is, it exploits and combines the strengths of each of the four approaches above. The way to achieve maximum leverage is to factor out the common and domain-specific parts of a set of problems. Then one can find a general solution for the common part and a more specialized solution for the domain-specific part, and integrate them into more powerful solutions than either part separately.

The four approaches described above can be positioned on Figure 7.7. The first two represent the extreme ends of the limit spectrum, and achieve little leverage. The custom solution approach is represented by a very tall narrow rectangle at the left of the box. Each custom solution solves all of one specific problem, so the top of the rectangle is at proportion "all" and the right side is at 1 problem. The general solution one-size-fits-all approach is represented by a very short, wide rectangle at the bottom of the box. The general solution solves a tiny fraction of every specific problem, so the top of the rectangle is just above 0 proportion and the right side is at "all" problems. The application domain and enterprise solutions are in between these extremes.

The concept of layered architecture is an exemplar of the leverage concept; that is, it is an architecture style designed to achieve maximum leverage. For example, a layered software architecture (Jacobson, Griss, and Jonsson 1997) factors out commonality and variability from the perspective of problem domains. The upper layers are more problem or application domain specific. The middle layers contain components of intermediate generality from supporting domains, such as database management, distributed object computing, transaction processing, and messaging. The lower layers contain the most general elements that apply to all problem domains, such as services provided by operating systems or networks. Additional interpretations of layered architectures are discussed in Chapter 10.

There is a space-versus-time aspect to commonality and leverage. Much of the motivation for Parnas' information hiding (Parnas 1972) was related to the time dimension: identify and separate the parts of a system that will change or evolve over time from the parts that will remain stable, and hide or encapsulate the changing parts. But the space dimension is perhaps even more significant because it captures the composite systems space and supports more general reuse: Identify and separate the parts that are common to multiple systems from the parts that are unique to each system, and exploit the common parts.

Achieving leverage is critical to any component-based domain in terms of market viability. A component developer must produce a component that simultaneously satisfies enough of the specific needs ("proportion of solution") of a sufficient number ("number of problems") of individual system composers, to establish a market. A component-based architecture can be expected to have a combination of problem domain-specific components and general components. A layered or tiered architecture, as discussed above, supports this concept.

Within the problem domain context, another leverage mechanism is enterprise-specific models, such as product line architectures. Enterprises can exploit the domain-based knowledge foundation and specialize domain models for their own purposes. It is much more difficult to create enterprise models, components, and architectures without this more general foundation. For example, General Motors would not have been in a good position to create a new Saturn line of cars without the solid foundation of a shared knowledge of automotive engineering that they could specialize. In the software arena, it became much easier for an enterprise to build a new compiler after a standard architecture with standard components emerged and was understood in the compiler community.

Limits of Generality: Is a Universal Architecture Possible?

What is the practical limit of generality of a model? The previous section on leverage would suggest that the limit is a model that applies to all problems but offers only modest help in solving any specific problem. But what if we could push that leverage limit line a step or two toward the upper right ideal, where a model applies to most problems and offers substantial help in solving each one of them?

Suppose we define the set of all systems of interest to SE — that is, all systems in the taxonomy in Figure 2.3. Now consider internal models of these systems, such as architectures. Is there one internal model, one architecture, that would apply to all SE systems? We could define a model that has these constraints: number of components is at least two, and they have a variable number of connections or interactions among them. That model would apply to all SE systems — in fact to all systems — but it would not be of much use in solving (i.e., engineering) any specific system.

But there is a more substantial candidate we could consider as an internal model that applies to all SE systems and gives us some heft, something useful for all these systems. That candidate is the set of 20 functions that Miller (1995) defines as being common to all living systems at all levels, from cell to supranational society. To enable discussion of the universal possibilities of this set of functions, we list and summarize them based on Miller (1995, xix). Functions 1 and 2 process both matter-energy and information. Functions 3 through 10 process matter-energy, while functions 11 through 20 process information.

1. *Reproducer:* Carries out the instructions in the genetic information of a system and mobilizes matter, energy, and information to produce one or more similar systems.
2. *Boundary:* At the perimeter of a system, holds together the components that make up the system, and excludes or permits entry.
3. *Ingestor:* Brings matter-energy across the system boundary from the environment.
4. *Distributor:* Carries inputs from outside the system or outputs from its subsystems to each component.
5. *Converter:* Changes certain inputs to the system into forms more useful for the special processes of that system.

6. *Producer:* Forms stable associations that endure for significant periods among matter-energy inputs to the system or outputs from its converter, the materials synthesized being for growth, damage repair, or replacement of the components of the system, or for providing energy for moving outputs to its suprasystem.
7. *Matter-energy storage:* Places matter or energy at some location in the system, retains it over time, and retrieves it.
8. *Extruder:* Transmits matter-energy out of the system.
9. *Motor:* Moves the system or parts of it in relation to part or all of its environment.
10. *Supporter:* Maintains the proper spatial relationships among components of the system so that they can interact.
11. *Input transducer:* Sensory; brings markers bearing information into the system, changing them to other matter-energy forms.
12. *Internal transducer:* Sensory; receives, from subsystems or components within the system, markers bearing information, changing them to other matter-energy forms.
13. *Channel and net:* Composed of a single route in physical space, or multiple interconnected routes, over which markers bearing information are transmitted.
14. *Timer:* Transmits information to the decider about time-related states of the environment or of components of the system.
15. *Decoder:* Alters the code of information input to it through the input transducer or internal transducer into a "private" code.
16. *Associator:* Carries out the first stage of the learning process, forming enduring associations among items of information.
17. *Memory:* Carries out the second stage of the learning process, storing information, and then retrieving it.
18. *Decider:* Executive receives information inputs from all other subsystems and transmits to them guidance, coordination, and control of the system.
19. *Encoder:* Alters the code of information input to it from other information processing subsystems.
20. *Output transducer:* Puts out markers bearing information from the system, and changes markers within the system, into other matter-energy forms.

The 20 functions do not form a composition fractal. That is, they do not represent a pattern in which each function subdivides into the same 20 functions. The 20 simply exist repeatedly at every level, in every system, which suggests that the set of functions is both highly scaleable and broadly based. If they hold for all living systems from cell to supranational system, that naturally raises the question of whether they hold for all systems of interest to SE. How does the scope of SE systems compare to living systems? Miller's depiction of living systems is that they are divided into functions that process matter and energy versus those that process information. A first possibility then is that the 20 functions cover three types of systems: (1) systems that process both matter-energy and information, which use all 20 functions; (2) information systems, which use the 12 functions that process information; and (3) systems that process only matter-energy, which use the 10 functions that process matter-energy. These three categories would seem to cover the scope of SE systems.

However, a closer look at information processing functions shows that they are defined in the context of a physical matter-energy system, where some of the functions transduce stuff between *matter-energy* and *information* forms. To support a pure information system, these functions would need to be modified or replaced with corresponding functions that simply process information — an eminently doable task based on the work Miller has done. A partial analysis suggests the following changes for a pure information system:

- *Function 1/Reproducer:* Does this apply to information systems? Note "produce one or more similar systems." This suggests the function is autopoietic but programmers are allopoietic in their role of producing code. Even software code generators would appear to be allopoietic — they do not reproduce themselves; they produce other different code. Self-modifying software might come closest to the reproducer function as it stands. This function requires further analysis but would probably need to be modified a bit to be fully consistent with information systems.
- *Functions 12/Internal transducer and 15/Decoder:* Information systems would not need the Internal transducer function because all they have is information. They just need the Decoder, analogous to the matter-energy Converter.
- *Functions 11/Input transducer and 20/Output transducer:* These would need to be changed to be just simple ingestors and extruders, analogous to the matter-energy functions so named.

The matter-energy functions would also need to be analyzed to see if they were adequate for machines as well as living systems. If we are considering this as a potential universal architecture, we also want to allow for two variations in specific cases: some of these functions might not exist in some classes of systems; and some additional functions not covered in the 20 might exist in some classes of systems. Nevertheless, it is still an interesting and potentially useful possibility that the 20 functions form the elements of a universal functional architecture of all SE systems. We do not pursue the matter further in this book but we conjecture that, within the caveats listed above, the 20 functions do form the elements of such an architecture, and that the possibility is worth investigating further.

Another "universal" model candidate is the set of common criteria in the domain of security (CCRA 2006, 2007a, b). This would be an external model, relating more to requirements, and it would not be as complete as the universal functional architecture because it only covers the security requirements, or the security view. Nevertheless, it is another potentially useful model that has significant generality.

Knowledge, Body of Knowledge (BoK)

In a mature engineering discipline or domain, knowledge of the general class or classes of problems and systems that solve those problems is accumulated and used in the engineering of new systems. From the perspective of MOSES, that knowledge is captured and represented in general models that include similarities and differences that matter in the domain. As mentioned earlier, these models are in the form of conventional designs, patterns, common requirements, reference architectures, standards, etc. Models can represent relatively specific knowledge or general knowledge. In particular, specific models that have many details and small regions represent narrow deep knowledge, while general models that have few details and large regions represent broad shallow knowledge. Both types are useful; they represent the ends of the leverage spectrum as discussed in the section entitled "Leverage." General models that capture a body of knowledge form the contents of the knowledge space portion of the model space, as described in Chapter 5.

What is a BoK?

A BoK is a codified set of information about some profession that is generally accepted by the community of practice in that profession. The set of information is typically organized or categorized in a way that is understood by the practitioners. *The Guide to the Software Engineering Body of Knowledge* (Abran et al. 2004, viii) states:

> "Every profession is based on a body of knowledge and recommended practices, although they are not always defined in a precise manner. In many cases, these are formally documented, usually in a form that permits them to be used for such purposes as accreditation of academic programs, development of education and training programs, certification of specialists, or professional licensing. Generally, a professional society or related [organization] maintains custody of such a [BoK]. In cases where no such formality exists, the body of knowledge and recommended practices are 'generally recognized' by practitioners and may be codified in a variety of ways for different uses."

A guide to a BoK may contain part of the BoK but does not contain the complete BoK. Instead, a guide provides a structure that organizes and summarizes key aspects of the BoK and contains a set of references to the BoK.

Knowledge Categories

Because SE is an engineering discipline, one might ask what kinds of knowledge are contained in an engineering BoK? In a mature engineering discipline, most engineering involves "normal" or "routine" design. The engineering content in any given domain embodies the essential knowledge of the problems and solutions of that domain. This knowledge is an evolving combination of practical knowledge and theory- or science-based knowledge. The growth of such knowledge reduces uncertainty and risk in engineering.

A set of categories of knowledge is taken from Vincenti's book on engineering knowledge (Vincenti 1990), especially Chapter 7, "The Anatomy of Design Knowledge." This describes mature engineering knowledge, and is a good model for our depiction of an SE BoK. Table 7.2 describes Vincenti's knowledge categories.

Does this notion of BoK apply to complex systems? The answer is yes. Some claims are made that while traditional systems follow patterns, each complex system is unique and continually changing in response to its particular environment, and thus does not follow a common script or design. The implication would be that no BoK is possible for complex systems. We agree with the premise of this argument but not with the conclusion. While complex systems in general may experience more uniqueness and more change than traditional systems, we believe the following. First, both traditional systems (at least analog or physical ones) and complex systems are unique; each one is a little bit different. Second, both traditional and complex systems follow patterns. We can write a book about the common patterns of bridge design; but we can also write a book about common patterns of human design, such as anatomy. Even if we consider psychological aspects of the brain, we have common patterns of perception, personality types, and disorders. Counterintuitively, even chaos exhibits patterns. Attractors and repellers certainly represent patterns of behavior. Strange attractors also represent patterns; they are chaotic attractors that are

Table 7.2 Vincenti's Engineering Knowledge Categories

Category	Contents
1. Fundamental design concepts (standard models or design patterns)	– Operational principle: how the system/component works; – Normal configuration of the system or component: general shape and arrangement, characteristic parts, general design of the system or component
2. Criteria and specifications (system properties)	– Salient properties of the system or component; – Specified values for those properties – Applicable generally (e.g., regulated limits)
3. Theoretical tools (for design analysis)	– Mathematical tools, no physical content; – Mathematical tools, based on scientific theory; – Mathematical tools, based on engineering theory; – Mathematical tools, based on assumptions; – Intellectual concepts (variety)
4. Quantitative data (measurement and experimentation)	Data to input to the mathematical analysis tools: – Empirical data (applicable generally) – Theoretical data
5. Practical considerations	Rules of thumb based on experience or convention
6. Design instrumentalities	Process of design

fractals (Boccara 2004, 167). The most compelling argument for the possibility of a complex systems science BoK is that it already exists. This book has cited representative pieces of this BoK, which has been accumulating for several decades. Multiple disciplines have contributed to the CS BoK, including physics, biology, psychology, sociology, organizational theory, and systems theory.

Value of BoK Guide

In general, the value of a guide to a BoK is twofold. First, it points to the primary elements of knowledge in a discipline or field of study. Second, it provides an organizing framework that integrates the elements and gives a comprehension of the whole. A guide to a mature BoK reflects known elements and a common organizing framework. A guide to an immature BoK, such as that for CSE, does not have a complete set of knowledge elements or a common organizing framework. In this case, a framework can be proposed, and will have some empty knowledge slots. But if the framework is well-conceived, it can help shape the thinking about the field and its development, in terms of both practice and research. Such a framework can pre-integrate the field and identify missing elements, and show where those elements fit in the larger picture. The BoK guide can then be an enabler or facilitator of good practice and further development of the field.

The SE BoK is a mix of many mature disciplines and a few immature ones. The maturity of the overall SE BoK is discussed in the section entitled "Current State of SE BoK." One goal of the design of the model space is to define a structure that applies to all the SE disciplines and domains, exploits the knowledge of mature disciplines, and helps shape the knowledge of immature disciplines. The model space does not claim to be at all complete in terms of an organizing structure of the SE body of knowledge; more structure can be and needs to be added for that purpose. The model space does claim to provide important elements of an organizing structure of SE models, and claims that the body of SE knowledge can be captured in models that populate that structure. Stated differently, the model space is capable of holding the complete SE BoK, and partially structuring it as such, and provides the tools for additional BoK structure as needed.

Contributors to SE BoK

Figure 7.8 shows how the BoKs of various disciplines contribute to an envisioned SE BoK. Various sciences contribute to a number of engineering disciplines, and both the sciences and engineering disciplines contribute to the SE BoK. The sciences shown here are a subset of the more complete list of disciplines shown in Table 2.2.

The discussion in this section is on SE BoK rather than SES BoK because the ultimate goal is to have a body of knowledge that supports SE. The relation between SES and SE in this context is that all the disciplines that support or contribute to SES, as identified in Table 2.2, may be said to contribute to the SES BoK. Those are represented by the dashed boxes in Figure 7.8. The disciplines that contribute to the SE BoK include the SES BoK (dashed-line boxes) and the engineering or "actualization" disciplines (solid-line boxes).

Current State of SE BoK

Chapter 5 assessed the readiness of SE knowledge to be incorporated into the model space. This section briefly assesses the maturity and completeness of SE knowledge.

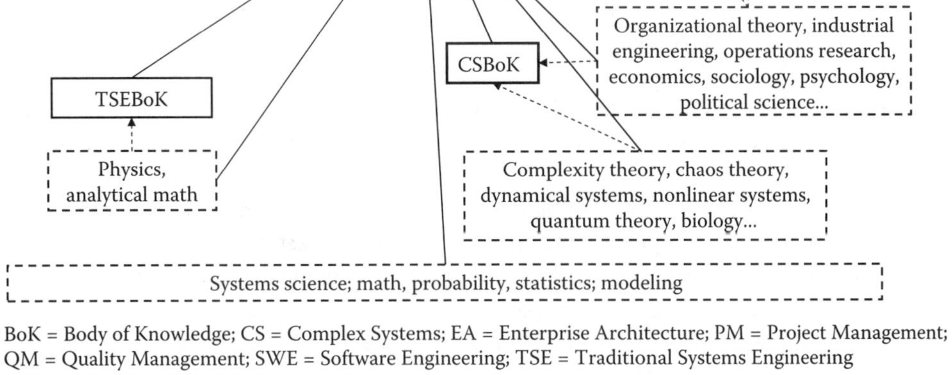

Figure 7.8 Contributing elements to SE BoK.

SE BoK is less mature than traditional engineering disciplines. One challenge is the extreme multidisciplinary nature of SE, which means the knowledge is highly distributed. Another challenge is the current expansion to CSE and to a larger range of scales. The knowledge in these areas to date has limited visibility to SE practitioners, and has not been incorporated or integrated into the SE BoK. It also is to some extent immature (e.g., nanotechnology).

In addition, within TSE, there is a significant difference between the kind of information in traditional engineering handbooks and that in SE handbooks or "body of knowledge" compendiums such as INCOSE's *Systems Engineering Handbook* (INCOSE 2007) and their *Guide to the Systems Engineering Body of Knowledge* (G2SEBoK) (INCOSE 2004). The difference is basically that the traditional disciplines emphasize properties and conventional designs of the systems or products they are engineering, while the latter emphasize the process of engineering. That is, a traditional engineering body of knowledge gives the message "These are the conventional designs and design parameters for this class of systems" while the TSE body of knowledge gives the message "These are the process steps for creating a design for all SE systems."

The MOSES model space has a place for models of actualization processes but the goal is to be oriented more like the traditional engineering handbook type of models: general designs and properties of systems engineered by SE. This goal reflects the belief that the difference in emphasis between traditional engineering (models of produced systems) and SE (models of the engineering process) is not a reflection of essential differences between the disciplines, but is rather a reflection of the maturity levels. When SE becomes more mature, it will focus more on system models and less on engineering process models.

In sum, considerable knowledge in the SE set of disciplines does exist. But parts of it are immature, and it has not been collected or integrated. From the viewpoint of the SE of the future, much work is required to find and collect the SE knowledge in all of its disciplines and integrate that into one composite SE BoK, in preparation for the equally challenging step of bringing the knowledge into a form suitable for the model space. One primary goal of MOSES and this book is to provide a foundation that facilitates the integration of these areas, and to provide a knowledge container — in the form of the model space — that will facilitate their unification into one coherent SE BoK of the future.

Summary of Commonization-Related Processes

The basic processes in the commonization dimension are inheritance, top down, bottom up, categorization, instantiation, generalization, and specialization. We discussed some of these earlier but summarize each of them here.

Inheritance: Process by which a sub-kind obtains all the characteristics of its parent kind. Can also have variations, such as multiple inheritance — sub-kind inherits from multiple parent kinds.

Top down: Process in which the higher-level entity is defined first, and then the lower-level entity is defined. Examples are instantiation and specialization.

Bottom up: Process in which the lower-level entity is defined first, and then the higher-level entity is defined. Examples are categorization and generalization.

Categorization: Process in which a category is defined by abstraction from the characteristics of one or more individuals.

Instantiation: Process in which an individual is defined from the characteristics of a defined category.

Generalization: Process in which a kind is defined by abstraction from the characteristics of one or more sub-kinds.

Specialization: Process in which the characteristics of a kind are added to or specialized.

In some cases, multiple processes are used together. For example, the full definition of a sub-kind uses inheritance and then specialization.

References

Abran, A., J.W. Moore, P. Bourque, and R. Dupuis (Eds.). 2004. *Guide to the Software Engineering Body of Knowledge*, 2004 Version. Los Alamitos, CA: IEEE Computer Society Press.

Atkinson, C. and T. Kühne. 2001. The essence of multi-level metamodeling. In *Proceedings of the 4th Int. Conf. on the Unified Modeling Language*, M. Gogolla and C. Kobryn (Eds.), Lecture Notes in Computer Science, 2185:19–33.

Atkinson, C. and T. Kühne. 2003. Model-driven development: a metamodeling foundation. *IEEE Software*, 20(5):36–41.

Boccara, N. 2004. *Modeling Complex Systems*. New York: Springer-Verlag.

Borgida, A. and R. Brachman. 2007. Conceptual modeling with description logics. In *The Description Logic Handbook: Theory, Implementation, and Applications*, Baader, F., D. Calvanese, D. McGuiness, D. Nardi, and P. Patel-Schneider, Eds. Cambridge, U.K.: Cambridge University Press.

CCRA. 2006. *Common Criteria for Information Technology Security Evaluation. Part 1: Introduction and general model.* Version 3.1 Revision 1. Common Criteria Recognition Agreement. September. CCMB-2006-09-001. http://www.commoncriteriaportal.org/thecc.html. (Accessed January 11, 2009.)

CCRA. 2007a. *Common Criteria for Information Technology Security Evaluation. Part 2: Security functional components September.* Version 3.1 Revision 1. Common Criteria Recognition Agreement. September. CCMB-2007-09-002. http://www.commoncriteriaportal.org/thecc.html. (Accessed January 11, 2009.)

CCRA. 2007b. *Common Criteria for Information Technology Security Evaluation. Part 3: Security assurance components. September.* Version 3.1 Revision 1. Common Criteria Recognition Agreement. September. CCMB-2007-09-003. http://www.commoncriteriaportal.org/thecc.html. (Accessed January 11, 2009.)

Gamatié, A., C. Brunette, R. Delamare, T. Gautier, and J.-P. Talpin. 2006. A modeling paradigm for integrated modular avionics design. *Proc. 32nd EUROMICRO Conf. on Software Engineering and Advanced Applications* (EUROMICRO-SEAA'06), p. 134–143.

GME. 2008. The Generic Modeling Environment. http://www.isis.vanderbilt.edu/Projects/gme/. (Accessed January 11, 2009.)

Hybertson, D. 2001. A uniform component modeling space. *Informatica*, 25(4):475–482.

INCOSE. 2004. Guide to the Systems Engineering Body of Knowledge — G2SEBoK. International Council on Systems Engineering. http://g2sebok.incose.org/ (Accessed January 11, 2009.)

INCOSE. 2007. *INCOSE Systems Engineering Handbook*, v3.1. Seattle, WA: INCOSE (International Council on Systems Engineering).

Jacobson, I., M. Griss, and P. Jonsson. 1997. *Software Reuse: Architecture, Process and Organization for Business Success*. New York: ACM Press/Addison-Wesley.

Klir, G. 2001. *Facets of Systems Science, 2nd ed.* New York: Kluwer Academic/Plenum Publishers.

Kühne, T. 2005. Understanding metamodeling. Tutorial, *Int. Conf. Software Eng.*, 12–25 May, St. Louis, MO.

Ledeczi, A., M. Maroti, A. Bakay, G. Karsai, J. Garrett, C. Thomason, G. Nordstrom, J. Sprinkle, and P. Volgyesi. 2001. The generic modeling environment. In *Proc. IEEE Workshop on Intelligent Signal Processing.* http://www.isis.vanderbilt.edu/sites/default/files/GME2000Overview.pdf (Accessed January 11, 2009.)

Manders, E.-J., G. Biswas, N. Mahadevan, and G. Karsai. 2006. Component-oriented modeling of hybrid dynamic systems using the generic modeling environment. *Fourth and Third International Workshop on Model-Based Development of Computer-Based Systems and Model-Based Methodologies for Pervasive and Embedded Software.*

Martelli, M. 1999. *Introduction to Discrete Dynamical Systems and Chaos.* New York: Wiley.

May, R. 1976. Simple mathematical models with very complicated dynamics. *Nature,* Vol. 261, 459-467, June 10.

Miller, J. 1995. *Living Systems.* Niwot, CO: University Press of Colorado. (Originally published 1978.)

OMG. 2006. Meta Object Facility (MOF) Core Specification, Version 2.0 (formal/06-01-01). January. http://www.omg.org/docs/formal/06-01-01.pdf. (Accessed January 11, 2009.)

Parnas, D. 1972. On the criteria to be used in decomposing systems into modules, *Commun. ACM,* 15(12):1053–1058.

Schindel, B. 2005a. Requirements statements are transfer functions: an insight from model-based systems engineering. *Proc. INCOSE 2005 Symp.*, July.

Schindel, B. 2005b. Pattern-based systems engineering: an extension of model-based SE. *INCOSE 2005 Symp.,* Technical Information Exchange Session, July.

Vincenti, W. 1990. *What Engineers Know and How They Know It: Analytical Studies from Aeronautical History.* Baltimore, MD: The Johns Hopkins University Press.

Wikipedia. 2008. Lorenz attractor. http://en.wikipedia.org/wiki/Lorenz_attractor. (Accessed January 11, 2009.)

Chapter 8

Conceptualization

Coauthored by Leo Obrst

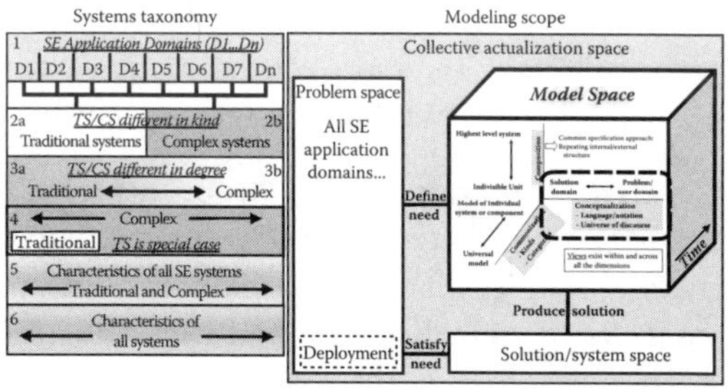

In previous chapters we have said that systems engineering (SE) defines models that are internal and external, address multiple levels of composition, are general or specific, and define categories or individuals. But how do we state or represent such models, what do the models mean, and how do we communicate about models in such a clear way that we achieve the shared understanding of specifications? The "we" here refers to all agent interactions, in any combination of humans and machines.

That is the focus of this chapter. The conceptualization dimension of the model space is concerned with the content, languages, and meaning of models.

The conceptualization spectrum includes languages (how and ultimately what to represent) and concepts or referents from the problem domain, which ranges over subject areas such as aerospace, transportation, construction, and organizations. It also includes considerations such as the design, analysis, or modeling constructs that will be necessary to solve those modeling problems in the solution space, to the degree of verisimilitude necessary. The heterogeneous subject matter highlights the complicating fact that the model space of SE is fundamentally multilingual and multi-ontological, and must implicitly deal with the necessary mapping from the representation of the problem to the representation of the solution space.

The discussion in this chapter uses concepts presented in Chapters 6 and 7. The multilevel, whole–part structures in Chapter 6 are used to show the structural context for much of the multiplicity of ontologies and languages in SE. The category hierarchy concept in Chapter 7 is discussed in this chapter in terms of how domain category hierarchies are generated and used.

Concepts and Definitions

The conceptualization dimension of the model space embodies a number of related concepts that need to be understood before proceeding into the substance of the chapter. The terms we use for key concepts are domain, universe of discourse (UD), conceptualization, ontology, language, and semantics. Conceptualization is a key concept that we chose as the overall name of the chapter.

We define the terms and how they relate to each other and to the previously defined concepts of model and specification. The definitions given here are an amalgam of definitions from several sources, including Daconta, Obrst, and Smith (2003); Guarino (1998), Honderich (1995); and Masolo et al. (2003).

Domain: Application domain or area of knowledge (or subject area) associated with SE. In the present context, the salient characteristic is that a domain has a community of agents (people, software, etc.) that share some degree of knowledge of the domain — with more sharing in a mature domain.

Note: Domains identified in the aircraft example given below in the section entitled "Set-up Example" include commercial aircraft systems and all the internal and supporting domains discussed in Chapters 5 and 6—avionics, etc. This list illustrates that the scope of domains and the relations between them are not simple. For example, domains can overlap, and one domain such as SE can be a composite of other domains. A domain can also be very general, as a context for general models. The most general domains associated with SE are probably systems science and mathematics because their models abstract away the differences between other domains.

Universe of discourse (UD): Set of entities or objects that are deemed to exist in a domain. These can be physical, such as airframe and engine entities, or conceptual, such as software and mathematical entities. They can be individual instances, or categories. In a logical sense, these objects are those that can be quantified over. Quantifiers are simply operators that bind specific kinds of variables. If we say "All women are human," that is different from "All humans are women," or "All humans work at IBM," or "Some humans are women," or "Some humans work at IBM." Those "some" and "all" constructs are quantifiers; they are operators that bind variables that range over different sub-domains or categories such as humans, men, women, etc. Some of these statements with quantifiers are appropriate because they represent the meaning of the real world; e.g., "All humans are mortal". Some are not: "All women work at IBM." So the universe of discourse involves both the instances and the categories of those instances that are appropriate for the model we are creating.

Conceptualization: An informal view of conceptualization is discussed first, followed by discussion of a more formal view.

> **Informal:** In this view, conceptualization means that a person has an informal or naïve conception of the part of the world that he/she is interested in modeling. From that implicit ("in the head") notion of the portion of interest of the world, one tries to lay down some tenets or statements about what that portion of the world means to the person. An example is familial relations. Initially, one might consider parents, fathers, mothers, children, sons, and daughters. Parents have children. Fathers and mothers are specializations or particular roles of Parent. Children are sons or daughters, i.e., specializations or roles of Child. When we talk about "roles" we are focusing in particular on roles in relationships, some of which are permanent and some of which are not. A father will always be a father (and in the father role) to his child. A child will always be in the child role in the relationship to his/her father. So far, so good. But

consider other semantic notions about the real world of the family, some of which are so obvious to humans that they are not even considered. Can a parent be a parent of himself? Can a parent be a child of himself? Can you be your own grandparent? In the real world (or this very simple snapshot of the real world), of course these relationships do not hold.

Formal: The informal concepts described above can be refined while still implicit. A more formal conceptualization of a domain typically involves consideration of more aspects of the domain, to include any of the following: the universe of discourse of a domain, the relations among the entities of that domain; the properties and property values of the entities; the functions of and processes involving the entities; and constraints on and rules involving the entities. The categorizing of domain entities is a common element of more formal conceptualization.

Note: Once a category has been established by the process of conceptualization, that category is deemed to exist in the domain; this is the reason a UD includes categories as well as instances.

Ontology: Description or model of a conceptualization and the definition of terms used in the description. These are the same entities and relations in the conceptualization, but they are modeled in a modeling language, and given more-or-less precise logical characterization in that language. If the modeling or content language is based on a formal logic, then the resulting model (ontology) is more likely to accurately reflect the conceptualization of the user. That is, it is semantically correct or semantically describes the portion of the world that the modeler intended to describe.

The distinction between conceptualization and ontology is therefore that conceptualization is partially implicit in the minds of one or more people in the domain; while ontology is fully explicit, and captured in some form that can be communicated.* The traditional philosophic notion of ontology is closer to the definition here of conceptualization, but the traditional notion does not make a distinction between implicit and explicit. The definitions here are more or less consistent with the notions of conceptualization and ontology as developed and used in the engineering community,† which takes a more formal approach in order to enable computers to manipulate and interpret semantics and ontologies.

There are other distinctions between ontology as used in traditional philosophy and as used in engineering. Poli and Obrst (2008) explore these distinctions. They use the term "ontology_t" to mean ontology as technical or engineering product, sometimes called "little o ontology"; and "ontology_c" to mean ontology in the sense of philosophical categorial analysis, sometimes called "big O ontology." Ontology and science work hand-in-hand. Science tries to find out what is real, and generalizations about what is real, and finally theories about what is real. Ontology organizes and classifies the results from that which science discovers about reality. Ontology_t is an engineering product that uses ontology_c, but at the same time it provides domain modeling to information technology. On the other hand, ontology_t may create categories about common-sense realms about which neither science nor ontology_c has as yet anything to say.

Language: System of symbols and rules for composition, structure, and interpretation that is used to communicate between different parties. In general, the symbols can be visual, audio, or tactile. Visual symbols can be text, graphical, or mathematical/logical. All languages have both a

* This is why Gruber (1993), for example, initially defined an ontology as "a specification of a conceptualization." He thought that information technologists in their usage generally understood the definition of "specification" from the practice of specifying designs, schemas, etc.

† We say more or less consistent because the notions of conceptualization and ontology are only partially mature and still evolving in the engineering community.

syntax and a semantics. The syntax specifies the symbols and formation rules that put the symbols together in valid ways. The semantics (see below) specifies the meaning of the symbols and the syntactically valid expressions built up from the symbols.

Note: A large variety of languages exist, including natural (e.g., English), mathematical, logical, simulation, programming, computer CPU, script, markup, modeling, business processing, architecture, design, requirements, network, functional, multi-paradigm, data, knowledge representation, general purpose, domain specific, etc. The foregoing categories are not mutually exclusive — for example, a programming language can be a functional language. A set of SE-related language categories is presented and discussed in the section entitled "Sample of Languages that Support SE."

Semantics: Interpretation or meaning of signs, especially of words and sentences of languages. In formal languages and formally described natural languages, semantics is expressed as interpretations of the symbols (expressions) in the model, where the model is either a formal model of set-theoretic objects (e.g., relations, functions, partially ordered sets, lattices, etc.) which act as "stand-ins" for the real world objects, or just a model of the world (where trees, people, etc., are identified). Those objects are the denotations of the symbols (expressions). The set of interpretations of the symbols constitutes the meaning/semantics, and is typically called "model-theoretic semantics."

Semantics has intensional and extensional forms, consistent with the definition of these terms in Chapter 7. That is, intensional semantics is expressed by defining properties or characteristics; while extensional semantics is expressed by enumerating instances (that have those properties or characteristics) in some world or UD.

Related to semantics is the concept of pragmatics. The focus of pragmatics is on how words and sentences are used in a given context. Semantics therefore correlate with a context-general external model of a system, as described in Chapter 7, which states the behavior of the system separate from any given environment; and pragmatics correlates with a context-specific external model of a system, which defines how the system is used in a specific environment — what its role is. In actuality, the boundary between semantics and pragmatics is fuzzy. The two determine a spectrum, with one bleeding into the other.

Defining Categories in an Ontology

The process of defining categories in a domain is an important element of conceptualization and ontology. Individuals or instances exist in any domain. For example, in the aircraft domain, there are individual aircraft that we can see and fly, their components, their crews, etc. Categories such as fixed-wing aircraft, commercial aircraft, or pilots are defined when categorizing the domain. After the categories are established, they become part of the domain, the UD.

There are two recent trends in the arena of defining classification systems and, more generally, defining ontologies. Both are to some degree in response to the challenges and opportunities of the Internet. One trend is bottom up, and the other is top down. The bottom-up trend is publicly shared tagging by users of Web sites such as Flickr and Del.icio.us to form folksonomies. The large number of tags categorize the material on the Web site, and the links associated with the tags support improved search capability. However, the model that they form is weak, often without any taxonomic structure. Folksonomies represent another example of mass collaboration, a topic that was discussed in Chapter 1. The use of the folksonomies approach in some SE domains is being explored — for example, see Nauman and Hussain (2007) and Lee and Yong (2007).

Gruninger et al. (2008) report on the Ontology Summit 2007, which examined distinctions among ontology, taxonomy, and folksonomy.*

The top-down trend is to increase emphasis on explicitly defining and to varying degrees formalizing ontologies. A key driver of this trend is the difficulty of communicating across multiple communities and domains, especially in the context of the Internet. The Semantic Web is a focal point for many of the efforts. Because SE, and especially the SE of the future, has similar multi-domain difficulties, SE should leverage the work being done in this area. After the introduction of an example in the section entitled "Set-up Example," the following two sections discuss further the multi-domain challenges and some of the ontology work being done. The MOSES orientation is that shared understanding of conceptualizations among the agents of a community is the primary goal, and ontologies can help achieve that goal.

A project that could be characterized as a bridge or a middle ground between top down and bottom up is called myOntology (Siorpaes et al. 2008). This project is producing an open ontology environment intended for Semantic Web-based e-Commerce. The bottom-up part is the Wiki-based community collaboration aspect. The top-down part includes an underlying meta-model based on a lightweight subset of OWL DL (Web Ontology Language based on Description Logics), and algorithms that run in the background to check inconsistencies.

In actual practice, of course, when ontologists develop an ontology, they do so according to particular use cases (refined to scenarios) that the ontology is being developed to address. As part of this process, the ontologist looks at any existing data or knowledge source whose semantics (as ascertained via analysis) the proposed ontology will have to include. This is largely a bottom-up analysis. But the ontologist also has to look at the refined scenarios, develop competency questions (questions that the resulting ontology should answer), and then develop a model of the semantics that addresses the requirements of the use cases. In addition, those competency questions also act as criteria for when the ontology is ready, that is, when those questions can be adequately answered at the right level of granularity, as judged by a domain expert. So the usual ontology development process in practice includes bottom-up and top-down notions. Using solely one or the other will result in a less than adequate ontology.

Set-up Example

A brief example is introduced in this section to set up and illustrate the discussion in later sections. Reconsider the models introduced in Figure 6.1 for a system of interest (SOI), and the basic categories of internal and external models. Figure 8.1 recasts these into three types: internal, external, and environment. Suppose we have a systems engineer who is engineering system S1, say a commercial aircraft. A partial list of its environment includes passengers, human pilots, air traffic control, runways, air pressure, force of gravity, fuel source, and other aircraft and vehicles in its vicinity. These are notionally represented by S2 and S3 in the figure (clearly, an aircraft has more than two environment elements with which it interacts).

S1 includes several parts, each of which is designed by other engineers. To give a simplifying example, let's say the engineers and parts are a software engineer designing the avionics software S11, a mechanical engineer designing the hydraulics system S12, a propulsion engineer designing the power plant S13, and an aeronautical engineer designing the airframe structure S14. The SOI for the systems engineer is S1, while each subsystem is the SOI for the engineer designing it; so the SOI for the software engineer is S11. Figure 8.1 shows a decomposition of models and views from

* Ontology Summit 2007: http://ontolog.cim3.net/cgi-bin/wiki.pl?OntologySummit2007.

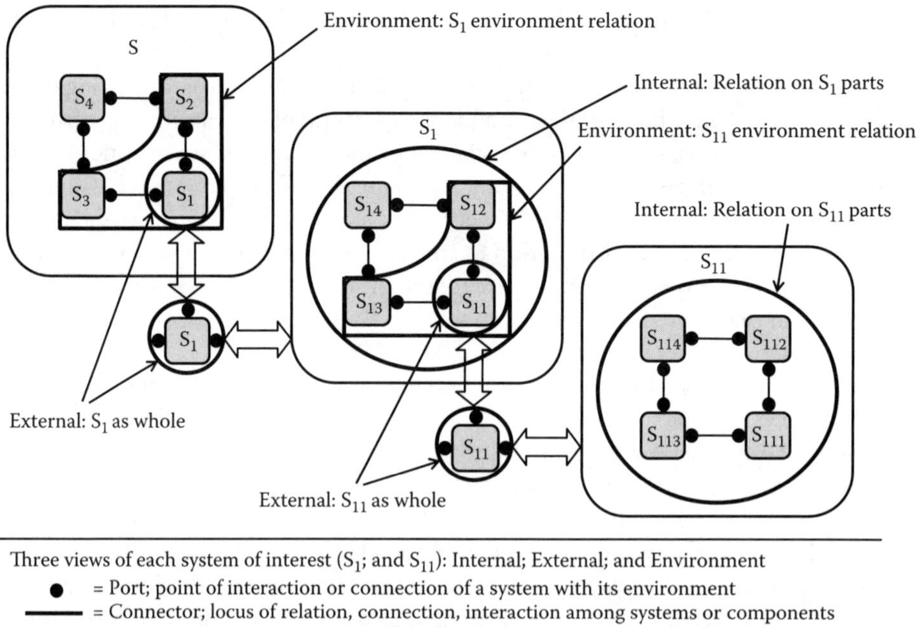

Figure 8.1 SE context for multiple domains, languages, and ontologies.

the viewpoint first of S1 (the aircraft) in its environment and its composition, and then S11 (the software) in its environment and its composition.

The four primary models of interest for each SOI are: (1) an environment model of the interaction of the SOI with its environment (a type of internal model); (2) an external model of the SOI; (3) an internal model of the SOI; and (4) an external model of each part of the SOI.

The systems engineer models his/her SOI, S1, with these models:

1. Environment model of the interaction of S1 with its environment, S2 and S3
2. External model of S1: what it provides, what it requires
3. Internal model of S1: its parts (S11, S12, S13, S14) and their interactions
4. External model of each part (S11, S12, S13, S14)

The software engineer models his/her SOI, S11, with these models:

1. Environment model of the interaction of S11 with its environment, S12 and S13 (which form part but not all of S1, according to the figure)
2. External model of S11: What it provides, what it requires
3. Internal model of S11: Its parts (S111, S112, S113, S114) and their interactions. For avionics software, these might be communications, navigation, flight control, and collision avoidance, for example.
4. External model of each part (S111, S112, S113, S114).

The overlap of these sets of models is that the external model of S11, and the model of S11 interacting with its environment S12 and S13, are part of the internal model of S1. The systems engineer develops an understanding of S1 providers and consumers in defining the external model of S1 in its S environment. The systems engineer is the architect of the internal model of S1. From the perspective of the software engineer of S11, S12 and S23 are the providers/consumers of S11 in its S1 environment, and the software engineer negotiates the external spec of S11 with the systems engineer as the definer of the S11 environment.

Challenge of Multiple Domains

As discussed above, a system can exist in multiple domains, especially if we consider its full decomposition. Its external environment might be one domain, such as commercial air transportation. Not only can its internal parts be different from the external domain, but the part domains also can differ from each other. In the aircraft example above, at the system internal level, we saw a wide variety of domains. Each domain can and typically does have its own universe of discourse, conceptualization, ontology, and language(s). In SE, this is especially true of high-level subsystems — in this example avionics, power plant, etc. This presents an interesting challenge to systems engineers, to bring together and integrate parts from a variety of domains. The Rosetta Stone example (Baez and Stay 2008) referenced in Chapter 2 illustrated the difficulties of communicating among disciplines even in the simplified case when the concepts are the same.

We can use the example in the previous section to discuss possible differences in the UD and languages being used in these various models. First, the three boxes in Figure 8.1 labeled S, S1, and S11 could each have a different conceptualization and use different languages. The external languages and conceptualization of S1 could be different from the internal ones of S1, and likewise the external languages and conceptualization of S11 could differ from the internal. The services provided and required by S11 are in the languages of internal S1, and those are the languages on which the systems engineer and software engineer base their discussions and negotiation of S11 requirements. The languages of internal S11, however, could be different — in this case, software design languages such as UML and one or more programming languages. Sensors could detect various aspects of the S1 environment or state, feed certain status information to the software system (in this case S11), and the software system in turn could feed commands to aircraft actuators. Further examples of interaction are given in the next section.

The Figure 8.1 example and discussion above reflect the multidisciplinary nature of SE. We noted in Chapter 1 that SE is multidisciplinary in several ways. We now expand and interpret the earlier itemization in the context of languages and ontologies. The items that follow illustrate the extent of the challenge:

- The scope of systems that SE engineers includes multiple "primary" problem domains (aerospace, defense, transportation, construction, organizations, etc.), each with its own set of domain languages and ontologies. Expansion to complex systems and scaling only adds to the existing TSE domains. SE and domain integration would be simplified if they all used the same language, but that seems unlikely to happen.
- Individual systems that SE engineers involve multiple "supporting" problem domains in their subsystems or components (communications, information systems, mechanical, electrical, control systems, humans, etc.). Each of these supporting domains has one or more domain languages and ontologies.
- As discussed above with Figure 8.1, what goes on inside a system is often very different from what goes on outside the system, and languages and concepts of internal models of a system can be very different from those of external or environment models of the

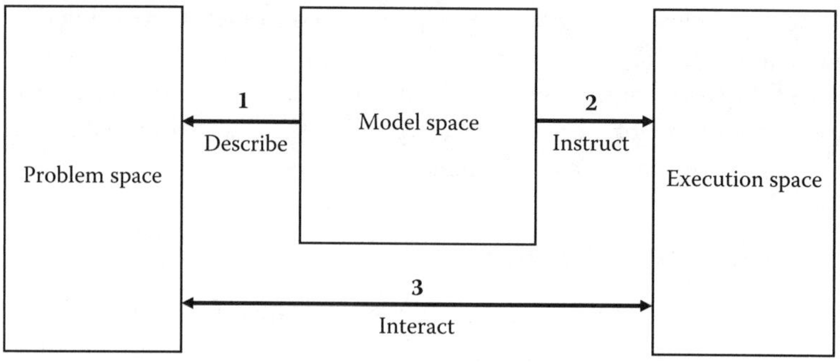

1. Describe: Model space entities describe problem space entities and behavior
2. Instruct: Models are interpreted by execution space entities as instructions and data
3. Interact: Problem space entities interact with execution space entities
   - Interaction is governed by models and problem space data
   - Entities operate within their own universe of discourse: problem vs. computer

Figure 8.2 Semantic range: architecture of software engineering.

same system. One of the dominant themes of MOSES is the emphasis on the distinction between internal and external views or models of a system. Even different views of the same internal or external model can use different languages.

- Within the internal view of a system, each component may have a different UD and language, as illustrated with the aircraft example. In addition, in the expanded SE, systems can include humans, groups, and organizations, which represent different domains.
- Related to the internal–external difference, domain semantics of a system range from the problem environment space to the furthest reaches of the model space and solution spaces. This range defines the scope of the conceptualization spectrum. For the general case, it is illustrated in Chapter 5, Figure 5.6, and for software, in Figure 8.2 in this chapter.

Greater distribution and heterogeneity of information systems, as they have evolved from self-contained monolithic homogeneous systems to decoupled modular distributed heterogeneous global systems, have added to the multi-domain multi-ontology trend (see Obrst 2006). The global reach brings natural language and culture differences prominently into play, as an additional dimension in the multilingual and multicultural challenge. Trying to reach a common understanding in this environment requires much greater attention to explicit semantics, ontologies, and languages, and they need to be machine-readable and machine-interpretable. The goal when SE is mature is not to have one universe of discourse, conceptualization, ontology, and language, but rather to recognize and support the individual domains and ontologies, and their integration into a coherent whole in support of the expanded SE. How does SE meet this challenge? Some ideas are presented in the next section.

Response to the Challenge

Mitigation

One significant element of the response to the challenge is the mitigation achieved by separating the internal world of a system from its external world by means of its boundary reinforced by encapsulation. That is, the composition structure that fosters the challenge of different languages inside and outside a system is also part of the solution. To the degree that the internal and external are separated, it does not matter if they speak different languages. However, that is not the complete solution because SE deals primarily with open systems, in which stuff goes in and out of the system. So some interaction — for example, translation or transformation — must occur between the languages of the internal world and those of the external, and between the conceptualizations.

This interaction is part of the design of a system. Systems engineers (1) use technology to design and develop a solution or system whose internals the system user — whether machine or human — does not in general know or care about, and (2) "translate" between the internals and externals of the system in a way that enables the user to understand, use, and interact with the system. In many cases, this translation is implicit or built into the system. For example, a bridge builder does not need to translate anything to a user; the bridge simply needs to present a flat surface to someone wanting to drive or walk across the bridge. But in most cases, some communication is needed, which means that translation is needed. The whole field of human-machine interface has focused on the communication and interaction between humans and machines, and there are many lessons, patterns, and constraints that help improve this interaction.

Perhaps the most extreme difference between internal and external UD and languages occurs in computer systems that interact with humans. A large conceptual distance exists between the domain concepts of a computer processor (load, store, etc.) and those of banking, geospatial systems, and aircraft control. The software engineer is required both to translate between languages and to transform between very different UDs. In fact, software engineering could be defined as the discipline that designs systems that enable successful communication between two agents, neither of whom has any idea what the other is talking about (the two parties being a problem domain specialist and a computer processor). The situation is illustrated in Figure 8.2. One key solution approach is to embed the symbols of the problem domain language inside the computer processor language. Under certain conditions that the computer understands, it prints out "Insert card"; under other conditions, it prints out "Deposit or Withdrawal?" These symbols are meaningless to the computer but the human user understands what they mean.

Two additional mechanisms are mentioned here. They do not directly support the user-computer communication because the mechanisms are hidden from both; however, they do help improve the likelihood that the software is correct, which does facilitate user-computer communication.

- The use of high-level programming languages such as Java, and translators/transformers (compilers) of software from high-level languages and their concepts to computer processor languages and their concepts. Both the high-level language and its UD are insulated from those of the computer, and are closer to human language, which means they are closer to the problem domains that humans understand.
- The use of domain specific languages, which are even closer to the languages and UDs of the specific problem domains for which they are written.

Ontology Efforts and Tools

A significant body of work is being done in the engineering and formalization of ontologies. The origins of this work go back at least several decades. The application of formal logic to languages and semantics has been a significant element of this genealogy. An example is Carnap (1958). The study of logics is still active and is an important basis for ontologies and modeling in SE because of their high degree of expressiveness — see the next section on ontology spectrum.

Another element of the genealogy is the artificial intelligence community, especially in the area of knowledge representation techniques such as semantic networks (Barr and Feigenbaum 1981). In semantic networks, the nodes typically represent objects or concepts, and the arcs represent relations between the objects or concepts. The set of relations is open-ended but common relations include instance-of, subclass-of, and part-of. These relations are common not just in semantic networks, but also in all semantic, conceptual, and ontology modeling. The first two of these are the relations in the MOSES commonization dimension, and the third is the relation in the MOSES composition relation.

A third important and more recent movement is the vision of the Semantic Web and the languages and standards leading to it. The goal of the Semantic Web is to capture more types of context and semantic information for Web data and thereby make it more understandable and processable by computers (Daconta, Obrst, and Smith 2003). Semantics and ontology engineering, including ontology design and analysis tools, are major elements in this effort. A sequence of machine-processable languages has been used in the Semantic Web effort, many of them developed prior to or separate from the advent of the Semantic Web idea. The languages include XML (Extensible Markup Language), RDF (Resource Description Framework), and OWL (Web Ontology Language), as well as various logics.

One of the most difficult issues to solve is the incompatibility between models based on ontologies from different domains. This poses a serious challenge for SE, where systems from different domains and ontologies must interact. SE needs ontology matching, mapping, integration, interoperability, sharing, and merging across domains. We now discuss some of the approaches being tried.

Upper Ontologies

One approach is the distinction between domains of different levels of generality. At the most general level are upper-level ontologies, of which arguably there would be a single one that would apply to what is common across all domains. Candidates for the upper level include Cyc, Basic Formal Ontology (BFO), Descriptive Ontology for Linguistic and Cognitive Engineering (DOLCE), General Formal Ontology (GFO), WordNet, and Suggested Upper Merged Ontology (SUMO).

- Cyc (Lenat and Guha 1990; Lenat 1995) is perhaps the earliest general ontology effort and one of the most comprehensive. It includes not only an upper ontology, but also domain ontologies, a large number of basic facts, and a set of tools, including the ability to infer new knowledge from existing facts. It is broadly targeted, attempting to cover the scope of human common-sense information. It is proprietary but its open-source version, OpenCyc (OpenCyc 2008), contains the full Cyc upper ontology.
- BFO (BFO 2008) is an upper ontology that is based on Barry Smith's "realist" approach to ontology (Smith 2004). It focuses on the physical world, and is intended to support domain ontologies for physical science research (physics, chemistry, biology).
- DOLCE (Masolo et al. 2003) is an extensional ontology; that is, its universe of discourse includes only individuals, not categories. It is not intended to be a universal ontology, but rather the first in a library of foundational ontologies. Its focus is on natural language and human common sense.
- GFO (Herre et al. 2006) is a foundational ontology that emphasizes the formal approach. It is a component of the Integrated System of Foundational Ontologies (ISFO).

It comprises three category levels: a basic level consisting of all relevant GFO categories such as objects, processes, properties, relations; a meta-level or abstract core level containing meta-categories; and an abstract top level containing only the meta-meta-categories of set theory and category theory.

- WordNet (2008) is a lexical database of English words in which nouns, verbs, adjectives, and adverbs are grouped into sets of synonyms (synsets). Synsets are linked by semantic and lexical relations. As an ontology, it is general but it has a limited categorical structure.
- SUMO (2008) is an upper ontology sponsored by IEEE. The total package includes the upper ontology; a mid-level ontology of entities such as communications, finance, and geography; and a set of domain ontologies. It also links SUMO with WordNet to associate natural language words with its categories.

Domain Ontologies

There are fewer mid-level ontologies than upper level but there are a great number of domain ontologies. One area that has pursued ontologies vigorously is biomedicine. A whole catalog of open biomedical ontologies (OBO) is being maintained at the OBO foundry (OBO 2008). The evolving ontologies that are foundry candidates are categorized into domains that include anatomy, biochemistry, biological function, environment, health, phenotype, proteins, and taxonomy. The OBO and the foundry are described in Smith et al. (2007).

One element of the OBO set of ontologies provides a good illustration of a concept that integrates MOSES ideas in composition, commonization, and conceptualization. The integrating concept is the use of ontology to show the similarities and differences of certain characteristics across different composition levels. The OBO element is the biological process ontology, specifically the development node part of this ontology; the illustration is the structure for development terms involving four composition levels: cells, tissues, organs, and organisms. Here are the structure and terms as shown in (GO 2008), where [p] means part of the overall process and [i] means "is-a" or specialization of the overall process.

This is the structure for development terms involving tissues, organs, and organisms. [X] means tissue, organ, or organism; [Y] means cell.

```
[X] development
[p] [X] morphogenesis
---[p] [X] formation
------[p] [Y] cell differentiation
---[p] [X] structural organization
[p] [X] maturation
```

This is the structure for development terms involving cells.

```
cellular process
[i] cell differentiation
---[p] cell fate commitment
------[p] cell fate specification
------[p] cell fate determination
```

```
---[p] cell development
------[p] cellular morphogenesis during differentiation
------[p] cell maturation
```

The thing to notice about this example is that three levels of composition are self-similar with regard to development, and the fourth level is different. The development of tissues, organs, and organisms all consist of morphogenesis, formation, etc., while cell development follows a different process.

At a broader level, the similarities and differences expressed in this example illustrate a principle of MOSES: modeling and engineering should allow system similarities and differences to assume their natural place in models. Too often we either try to force every system to be different, or force all systems to be the same. The OBO development example is consistent with the MOSES principle: it captured the similarities at the tissue, organ, and organism level, but allowed the cell to be different, and captured the different cell development process. This principle applies not only to composition levels, as in this example, but to all types of similarities and differences, as discussed in Chapter 7.

Another example of a domain ontology of interest to SE is GUMO — General User Model Ontology (Heckmann et al. 2005). It is a taxonomy of user dimensions, and each dimension is a category that has a number of identified elements. A few entries from the ontology are:

- *Basic User Dimensions:* contact information, demographics, ability and proficiency, personality, emotional state, etc.
- *Characteristics:* talkative, assertive, dominant, quiet, reserved, shy, etc.
- *Personality:* extravert, introvert, thinking, feeling, etc.

We include this model for several reasons. One is to show efforts to bring humans more fully into the SE arena and define more complete models of them, as part of the expansion to CS. A second reason is to show an example of an ontology in the expanded SE. The GUMO model extends beyond traditional models of humans as users. This model is discussed further in Chapter 10.

Bridging and Mapping Ontologies

An area important to SE is the work that addresses multiple domains and ontologies: bridging, integrating, mapping, merging, and composing. We briefly discuss a few examples. The OntoMap project (Kiryakov, Simov, and Dimitrov 2001) is an effort to provide a general upper-level ontology that serves as a common locus for mapping ontologies (primarily upper level) to it, and thereby to each other. It uses the general relations within its ontology, plus a few cross-ontology relations, to define each mapping. If the ontologies are formally defined, this process should be automatable.

The CROSI project (Capturing, Representing and Operationalising Semantic Integration) (Kalfoglou et al. 2005) investigated semantic integration technologies, including an extensive survey of existing technologies, for the purpose of ontology mapping. They also developed a prototype ontology mapping system, the CROSI Mapping System (CMS), that demonstrates mapping techniques.

Work on composing ontologies has been done by Kaushik, Wijesekera, and Ammann (2006). Their work is in the context of a Semantic Web-based environment in which multiple ontologies are proposed by distributed collaborative communities. They have defined a type system for the RDF, and an algebra for composing ontologies based on structural comparisons. They have also compared their approach with existing approaches.

More narrowly focused language mapping work has been done by Matulevičius, Heymans, and Opdahl (2006). Their scope is goal-oriented languages, their interoperability with each other and with other modeling languages. The approach is to map languages onto the same ontology, which is based on the Bunge-Wand-Weber (BWW) representation model (Wand and Weber 1993 and 1995). They illustrate it using a goal-oriented language called KAOS. They are also applying the method to develop a new version of the Unified Enterprise Modelling Language (UEML).

Storytelling

Storytelling was mentioned in Chapters 1 and 3 as a means of communicating that is increasingly being used in SE. A preliminary ontology model of storytelling has been defined (Nakasone and Ishizuka 2006). The goal is to generalize modeling of storytelling to a broader level. It is based on relations from the Rhetorical Structure Theory. The storytelling classes in the model include concept, event, relation, act, scene, agent, and role. The Model defines relations among these classes.

Automation

Many of the projects listed above have developed tools that move toward automating the various processes. Recall that automation is one of the key elements of SE as envisioned by MOSES. Model orientation fosters the automation of engineering systems. The work to make ontologies explicit and formal, with the goal of automating the analysis and use of ontologies, fosters the reconciliation and integration of multiple ontologies from disparate domains. That is a key enabler of the SE of the future.

Spectrum of Semantics and Ontologies Used in SE Models

From the viewpoint of semantics and ontologies, a variety of methods and approaches have been developed over the past few decades in the database, software engineering, and knowledge representation communities, many of which were noted above. There are debates concerning which of these methods may properly be called ontologies. Several different authors have independently defined an ontology spectrum or semantic spectrum that reflects a range of ontology-related characteristics. Each author used slightly different terms to characterize the range:

- Daconta, Obrst, and Smith (2003) define a range from less to more expressive, where the less expressive is associated with weak semantics and the more expressive with strong semantics.
- Smith and Welty (2001) define a range from low to high complexity, where the high end is associated with automated reasoning methods, but not the low end.
- McGuinness (2003) defines a range from low to high specification detail and expressiveness, within the scope of ontologies emerging on the Web.
- McComb (2004) defines a range of semantic precision, from a simple data catalog at the low end to general logical constraints at the high end.
- Kalfoglou et al. (2005) define a range of semantic intensity, where the low end is characterized as semantically poor and the high end as semantically rich.

The semantic spectrum presented in Figure 8.3 is a slight modification of that presented in Daconta, Obrst, and Smith (2003, 157). It portrays a range of semantic models that potentially

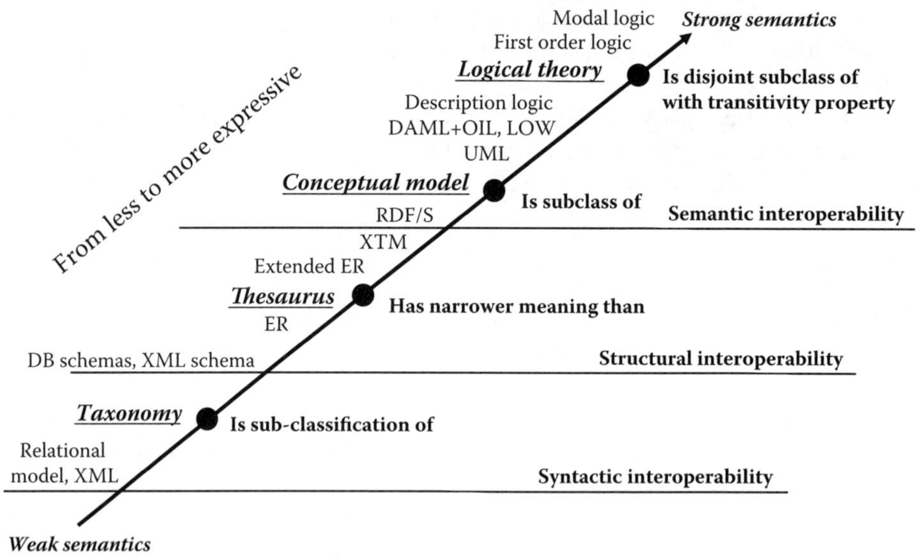

Figure 8.3 Semantic spectrum (*Source:* Daconta, M., L. Obrst, and K. Smith. *The Semantic Web: A Guide to the Future of XML, Web Services, and Knowledge Management*. Indianapolis, IN: Wiley Publishing, 2003. Used with permission.)

characterize ontology. The examples are organized around four constructs: taxonomy, thesaurus, conceptual model, and logical theory (or logical domain theory). A taxonomy is taken to mean terms or concepts with minimal hierarchic or parent/child structure. A thesaurus includes terms, synonyms, and term taxonomies based on the broader than/narrower than relation and association relation. A conceptual model structures concepts in a subclass hierarchy with generalized relations, properties, attributes, and instances. A logical model or theory comprises elements of a conceptual model focused on real-world semantics and extended with axioms and rules, and is represented in a logical language enabling machine semantic interpretation. A conceptual model can be considered a weak ontology; a logical theory can be considered a strong ontology. The interoperability labels on the right indicate roughly the expressiveness of the model required to address each level of interoperability. XML is sufficient for syntactic interoperability, XML Schema enables structural interoperability, but a minimum of RDF is necessary for semantic interoperability.

From this analysis, Daconta, Obrst, and Smith state that an ontology includes classes (general things), instances (individual things), relations among things, properties and property values of things, functions of and processes involving those things, and constraints on and rules involving those things. In MOSES, what is important is not the debate about what constitutes ontology or whether all of the items in Figure 8.3 represent an ontology. What is important is that those items can represent SE models, including domain ontologies. This usage is consistent with the MOSES concept that general models capture and represent SE knowledge, and collectively form a body of knowledge of SE. We also note the similarity of this list of ontology items with the specification items identified in Chapter 11.

Sample of Languages That Support SE

Daconta, Obrst, and Smith (2003, 188) describe the benefits of using representation languages that are more clear, precise, and unambiguous than natural language. Although natural language can be used to define models, it is not the first choice. Numerous languages and representation approaches are available in SE. We list and briefly describe some of them here. The language categories represent an informal and ad hoc set, and are not presented in any particular order. The categories are also not mutually exclusive; some languages belong to more than one category.

Simulation Languages

Continuous System Simulation Language (CSSL) is an evolving set of languages that was initially defined in 1967 and is designed to help engineers and scientists model the dynamic behavior of physical phenomena. For brief examples, see Huntsinger (1988). One member of this set is the European Space Agency (ESA) Simulation Language (ESL) (Crosbie et al. 1985). Another member is the Advanced Continuous Simulation Language (ACSL), which is targeted to time-dependent nonlinear differential equations.

Dymola also simulates dynamic behavior of physical systems but it is designed to follow the conservation principles of energy and momentum and not rely strictly on state-space models (Raczynski 2006).

Neural Simulation Language (NSL) is designed for large-scale general neural networks (Weitzenfeld, Arbib, and Alexander 2002).

System Modeling and Design Languages

The Specification and Description Language (SDL), defined as ITU standard Z.100, has evolved over a 20-year period. It is a wide-spectrum language from system requirements to implementation, and includes both graphical and textual elements. It is designed to support formal models and is suitable for real-time systems. According to SDL (2008), when SDL is used for systems engineering, it is usually in combination with other languages, including Message Sequence Charts (MSC), ASN.1, TTCN, and UML.

UML (OMG 2007a, b) is also used extensively in software engineering modeling and design, and is the basis for other profiles and languages. An extension of UML called SysML (OMG 2007c) has been defined to address more systems-oriented models such as block diagrams. Another extension of UML that addresses real-time systems is MARTE — Modeling and Analysis of Real-Time and Embedded systems (Demathieu et al. 2008; OMG 2008). MARTE supports modeling and design of both hardware and software aspects of these systems.

The Architecture Analysis and Design Language (AADL) was defined by the Society of Automotive Engineers (SAE) to be a standard for model based systems engineering. It supports formal modeling and systems performance analysis (Feiler, Gluch, and Hudak 2006). MARTE is also compatible with AADL (Faugere et al. 2007).

In the hardware arena, two languages have been used extensively: VHSIC Hardware Description Language (VHDL) and Verilog HDL. Both contain hardware modeling constructs, such as time concepts and parallel processing, that are not found in typical software programming or modeling languages. Both also have variants for analog and mixed-signal applications, called respectively VHDL-AMS and Verilog-AMS. In 2000, the two organizations VHDL International and Open Verilog International joined to become Accellera (Accellera 2008).

In 2007, the IEEE initiated a new standard called P1699: Rosetta System Design Language (IHS 2007). The goal is to provide "a system-level approach for developing, composing and analyzing specification models from multiple design domains, each with their own vocabulary and semantics." This should be of interest to SE with the multi-domain challenges discussed earlier in this chapter.

Architecture Description Languages (ADLs)

In the 1990s, a coalescence of research activity produced a variety of architecture description languages, including Rapide, C2, ACME, Wright, Aesop, and Unicon. A significant portion of this work was funded by the Defense Advanced Research Projects Agency (DARPA), and the primary focus was software systems. A summary of this work, as well as an application, is presented in Petty, McKenzie, and Qingwen (2002). A more recent unification and extension of this work is presented in Dashofy, Van der Hoek, and Taylor (2005). The ADL research was the source of the general component-connector model that can be applied to all systems, not just software systems, and which is the basic structure of the binary interaction model in Chapter 6.

Software Languages

A large number of programming languages such as Java and Visual Basic have been developed and used over the past half century. Many of the early languages, including the venerable Fortran, and even COBOL, are still in use, although they have evolved to some degree.

There are also many scripting languages such as Perl and Python. Modeling languages we mentioned earlier such as UML are also used extensively in software engineering.

Business Process Languages

WS-BPEL (Web Services Business Process Execution Language) is an orchestration language for specifying business process behavior based on Web Services (OASIS 2007). It is based on XML.

BPMN (Business Process Modeling Notation) is a graphical modeling language for business processes oriented toward workflows. The next version (V2.0) intends to incorporate the metamodel, graphical notation, and interchange format into one package, and the name is therefore changed to the Business Process Model and Notation (BPMN 2007).

The XML Process Definition Language (XPDL) is an interchange language for BPMN.

YAWL (Yet Another Workflow Language) was designed to support a basic set of workflow patterns or views: control perspective, data perspective, resource perspective, and exception-handling perspective. It also provides some support for human tasks.

Human business processes have been getting more attention recently. An extension of BPEL called BPEL4People has been defined that focuses on human tasks (Agrawal et al. 2007). A framework for defining human services and unifying human and software services is described in Schall, Truong, and Dustdar (2008).

Holmes et al. (2008) have defined a view-based, model-driven framework that supports both human and automated business processes. It provides a higher level of abstraction that allows both a human view and a technical view of business processes. This approach is consistent with MOSES.

Requirements/External Specification Languages

A large variety of system requirements or (external) specification languages have been defined and used over the past several decades, many of them in software engineering. They range from natural language to formal languages, and from general purpose to domain specific. Formal languages are typically based on the language of mathematics and logic, while informal languages typically use a combination of text and graphics.

Some languages are designed for what we call an external model only, not describing the internal system, while others include both internal and external. Formal languages include VDM, Z, Lotos, and OCL. UML is sometimes used as a requirements language, more informal than formal. SDL, discussed earlier in the system design languages, is also used to describe the boundary between the system and its environment, which puts SDL into this category as well.

The language T, according to Sunik (2005), is an open non-executable language that can be used to model a wide variety of systems, including software, hardware, technical appliances, biology systems, and knowledge representation systems.

On the other hand, examples of domain-specific languages include the Geographic Markup Language GML to specify geographic information systems, and a number of languages for agent systems. Agent languages include the Cognitive Agents Specification Language (CASL) (Shapiro et al. 2002), the Agent Modeling Language (AML) (Trencansky and Cervenka 2005), a specification language for coordinated objects (Ciobanu and Lucanu 2005), and the Autonomic System Specification Language (ASSL) (Vassev and Paquet 2007). In addition, a variety of logics are used to model and specify agents as rational entities. These include BDI (belief, desire, intention) logics (Rao and Georgeff 1991), and a dynamic framework based on BDI called KARO (knowledge, abilities, results, opportunities) (van Linder, van der Hoek, and Meyer 1998; Meyer, van der Hoek, and van Linder 1999; Hustad et al. 2001). Agent languages and ontologies are important for the future of SE because their constructs focus on autonomous behavior, which can be applied to both people and software agents.

XML, Markup Languages

XML (Bray et al. 2006) is a text language designed to be used on the Internet and to be readable by both humans and computers in a variety of domains. It is a subset or profile of the Standard Generalized Markup Language (SGML; ISO 8879). It is used to exchange data among machines but arguably its most significant role is as a metalanguage or specification language for more specialized markup languages. A large number of languages have been based on XML. We have mentioned a few of these earlier in this section. To give a sense of the variety of languages based on XML, here is a small sample:

- Graphical languages such as Scalable Vector Graphics (SVG) (W3C 2005)
- The Mathematical Markup Language MathML (MathML 2008)
- Electronic Business using Extensible Markup Language (ebXML) for global business transactions on the Internet (ebXML 2008)
- Human Markup Language (HumanML) for describing human characteristics and bridging cultural gaps (Peltz and Thunga 2005)

- Service Modeling Language (SML) for modeling services on the Web and their constraints and policies (SML 2008)
- The Architecture Description Markup Language (ADML), an XML-based extension of the ACME architecture description language (ADML 2008)
- Voice Extensible Markup Language (VoiceXML) to support speech-enabled applications (Voice 2008).

In addition, a number of related standards have been defined, such as Extensible Stylesheet Language (XSL), XML Schema, and XML Remote Procedure Call (XML-RPC). A growing number of Web Services standards are also built on XML, including Simple Object Access Protocol (SOAP); Universal Description, Discovery, and Integration (UDDI); Web Services Description Language (WSDL); and a host of other standards for addressing, policy, choreography, coordination, security, etc. Some standards, such as RDF and OWL, are part of the Semantic Web effort to expand from human-to-machine interaction in Web 1.0 to more machine-to-machine interaction in Web 2.0.

Knowledge Representation/Ontology Languages

RDF and OWL can also be categorized as knowledge representation (KR) or ontology languages. One of the early KR languages for exchanging knowledge was KIF, the Knowledge Interchange Format (KIF 2008). KR and ontology languages range from natural language dictionaries to description or high-order logics. This spectrum of ontology representation strengths and categories was discussed in the previous section.

Mathematics and Sciences, and Other Supporting Disciplines

Many of the SE supporting disciplines identified in Chapter 2 could be regarded as being or having their own languages and universes of discourse. Some of these, such as logics, have been discussed here. The number and variety of these disciplines and languages, especially in the future SE, are in fact a major driver for using ontology to improve communication among them in the engineering of SE systems, and a major impetus for this conceptualization dimension of the MOSES model space.

Natural Language

Natural languages, such as English, have shortcomings as modeling or engineering languages, including imprecision and ambiguity, although it is conceivable that in the future, as we get more into engineering complex systems these will be regarded as features rather than shortcomings. In any case, natural languages and translations between them will likely receive more attention in the future SE with its greater emphasis on human agents, its global reach, and its use of natural language mechanisms such as narrative and storytelling. In addition, models in many domains use constrained subsets of natural language such as domain glossaries and text descriptions of systems.

Language Leverage

The concept of leverage introduced in Chapter 7 applies to languages, which range from general purpose to specific to a narrow domain. Figure 8.4 shows the leverage trade-off of general and specialized languages. Relations and Operations in the Conceptualization Dimension

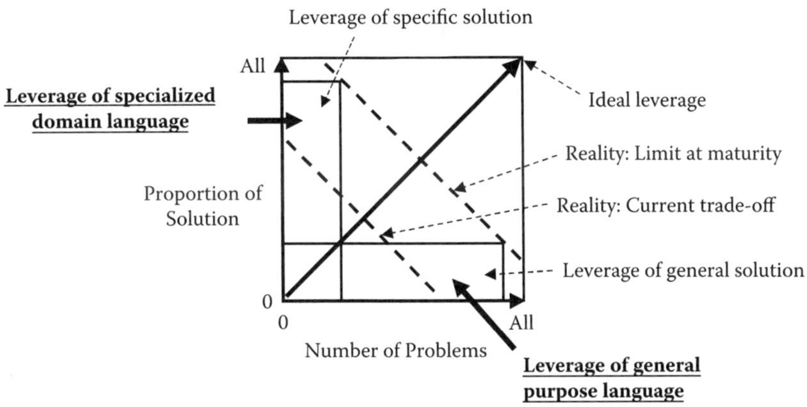

Figure 8.4 General and specialized language leverage.

Relations of Conceptualization Dimension

Relations between Models

Relations between models in this dimension include translation of a model from one language to another, interpretation and abstraction of a model, and mapping concepts between different ontologies (Daconta, Obrst, and Smith 2003, 218–220). An example of interpretation is applying semantics to a formal abstract model to produce a domain model, while abstraction is the conjugate — that is, removing the semantics from a domain model and producing a formal abstract model.

These relations are many-to-many. Many languages can be translated to many other languages, and many ontologies can be mapped to many other ontologies. A formal abstract model can be interpreted into multiple domain models, and a domain model can be abstracted into multiple formal abstract models.

Relation to Commonization Dimension

The relation between upper ontologies, mid-level ontologies, and domain ontologies maps directly onto the generalization spectrum from most general to most specific model. In addition, the category-to-instance relation is common to most general ontologies. An example from Jeffrey (1967, 228–230) is the following pattern in higher-order logic, which is consistent with the MOSES categorization hierarchy pattern (specific examples have been added):

- Instances (such as blue) have categories (such as color) at level 1.
- The categories at level 1 (such as color) become instances of level 2 categories (such as property).
- The categories at level 2 become instances of level 3 categories.
- etc.

At a given level *n*, the structure is called *n*th-order logic.

Relation to Composition Dimension

External and internal models of a system can be different worlds, different ontologies, and different languages, as this chapter has discussed extensively. This pattern extends up and down the composition spectrum.

We have positioned conceptualization as a dimension in the model space. Each dimension represents a spectrum. What distinguishes one point on the conceptualization spectrum from another?

The answer is that we propose that different models in this dimension are based on UDs that have varying degrees of similarity, and this similarity can potentially be measured by a conceptual distance function. Gärdenfors (2004, 17–21 and 110) provides an approach to this, in which he defines distance in a conceptual space and defines similarity as an exponentially decaying function of that distance. Consider the distance between the universes of discourse of two models — say the problem space and the internal solution — is measured by the number of transformations required in the model space to reach from one to the other. If we simply said the number of transformations required to reach from one to the other, the distance could always be defined as one unit, from external to internal.

However, this is where the relation with the composition dimension comes into play and limits the reach of a given transformation. There is at least one transformation between internal and external models. We typically have multiple levels of composition, and there is one transformation for each level. Therefore, the conceptual distance between models in the conceptualization dimension could be reflected in the number of levels of composition of the solution. How far down would we go in a physical system? Answer: To the lowest level that agents must deal with. As we have seen in previous chapters, with the extended scales of SE, this could be very many levels. In any case, if this notion has merit, the composition dimension could help measure the conceptual or semantic distance across the conceptualization spectrum.

References

Accellera. 2008. Accellera home page. http://www.accellera.org/home. (Accessed January 11, 2009.).

ADML. 2008. The Open Group Architecture Description Markup Language (ADML). http://www.opengroup.org/architecture/adml/adml_home.htm. (Accessed January 11, 2009.)

Agrawal, A., M. Amend, M. Das, M. Ford, C. Keller, M. Kloppmann, D. König, F. Leymann, R. Müller, G. Pfau, K. Plösser, R. Rangaswamy, A. Rickayzen, M. Rowley, P. Schmidt, I. Trickovic, A. Yiu, and M. Zeller. 2007. WS-BPEL Extension for People (BPEL4People), Version 1.0. June. http://download.boulder.ibm.com/ibmdl/pub/software/dw/specs/ws-bpel4people/BPEL4People_v1.pdf (Accessed January 11, 2009.)

Baez, J. and M. Stay. 2009. Physics, Topology, Logic and Computation: A Rosetta Stone. To appear in B. Coecke, Ed., *New Structures for Physics.* Springer. http://math.ucr.edu/home/baez/rosetta.pdf (Accessed January 11, 2009.)

Barr, A. and E. Feigenbaum. 1981. *The Handbook of Artificial Intelligence, Vol. 1.* Los Altos, CA: William Kaufmann.

BFO. 2008. Basic Formal Ontology. Institute for Formal Ontology and Medical Information Science, Saarland University, Saarbrücken, Germany. http://www.ifomis.org/bfo/home. (Accessed January 11, 2009.)

BPMN. 2007. Business Process Model and Notation (BPMN) 2.0 RFP. OMG. http://www.omg.org/cgi-bin/doc?bmi/2007-6-5. (Accessed January 11, 2009.)

Bray, T., J. Paoli, C.M. Sperberg-McQueen, E. Maler, F. Yergeau, and J. Cowan, Eds. 2006. XML 1.1 (second edition), W3C Recommendation, 16 August 2006. http://www.w3.org/TR/2006/REC-xml11-20060816/. (Accessed January 11, 2009.)

Carnap, R. 1958. *Introduction to Symbolic Logic and Its Applications*. New York: Dover.

Ciobanu, G. and D. Lucanu. 2005. A Specification Language for Coordinated Objects. In *Proceedings of the 2005 Conference on Specification and Verification of Component-Based Systems* (Lisbon, Portugal, September 05 - 06, 2005). SAVCBS '05. New York: ACM, 23-30. Available via online ACM SIGSOFT *Software Engineering Notes*, 31(2), March 2006. http://portal.acm.org/citation.cfm?id=1118537.1123062&coll=ACM&dl=ACM&idx=J728&part=newsletter&WantType=Newsletters&title=ACM%20SIGSOFT%20Software%20Engineering%20Notes&CFID=15943256&CFTOKEN=93064975 (Accessed January 11, 2009.)

Crosbie, R., S. Javey, J. Hay, and J. Pearce. 1985. ESL — A new Continuous System Simulation Language. *Simulation*, 44(5):242–246.

Daconta, M., L. Obrst, and K. Smith. 2003. The Semantic Web: A Guide to the Future of XML, Web Services, and Knowledge Management. Indianapolis, IN: Wiley Publishing.

Dashofy, E., A. van der Hoek, and R.N. Taylor. 2005. A Comprehensive Approach for the Development of Modular Software Architecture Description Languages. *ACM Trans. Software Eng. Methodol.*, 14(2): 199–245.

Demathieu, S., F. Thomas, C. André, S. Gérard, and F. Terrier. 2008. First Experiments Using the UML Profile for MARTE. 11th IEEE Int. Symp. on Object Oriented Real-Time Distributed Computing (ISORC). p. 50–57.

ebXML. 2008. ebXML. http://www.ebxml.org/ (Accessed January 11, 2009.)

Faugere, M., T. Bourbeau, R. De Simone, and S. Gerard. 2007. MARTE: Also an UML Profile for Modeling AADL Applications. 12th IEEE Int. Conf. Engineering Complex Computer Systems, p. 359–364.

Feiler, P., D.P. Gluch, and J.J. Hudak, 2006. The Architecture Analysis & Design Language (AADL): An Introduction, Technical Note, CMU/SEI-2006-TN-011. http://www.sei.cmu.edu/pub/documents/06.reports/pdf/06tn011.pdf. (Accessed January 11, 2009.)

Gärdenfors, P. 2004. Conceptual Spaces: The Geometry of Thought. Cambridge, MA: Bradford/MIT Press.

GO. 2008. Biological Process Ontology Guidelines. Gene Ontology Consortium. http://www.geneontology.org/GO.process.guidelines.shtml. (Accessed January 11, 2009.)

Gruber, T. 1993. A Translation Approach to Portable Ontology Specifications. Knowledge Acquisition, 5(2):199–220.

Gruninger, M., O. Bodenreider, F. Olken, L. Obrst, and P. Yim. 2008. The 2007 Ontology Summit Joint Communiqué. Ontology, Taxonomy, Folksonomy: Understanding the Distinctions. J. Appl. Ontol., 3(3):191–200.

Guarino, N. 1998. Formal Ontology and Information Systems. In Formal Ontology in Information Systems. Proceedings of FOIS'98, N. Guarino, Ed. Trento, Italy, 6–8 June 1998. Amsterdam, IOS Press, p. 3–15.

Heckmann, D., T. Schwartz, B. Brandherm, M. Schmitz, and M. von Wilamowitz-Moellendorff. 2005. GUMO — The General User Model Ontology, in Proc. UM 2005: Int. Conf. User Modeling, July 24–30, Edinburgh, U.K.

Herre, H., B. Heller, P. Burek, R. Hoehndorf, F. Loebe, and H. Michalek. 2006. General Formal Ontology (GFO): A Foundational Ontology Integrating Objects and Processes. Part I: Basic Principles. Research Group Ontologies in Medicine (Onto-Med), University of Leipzig. Version 1.0, Onto-Med Report Nr. 8, 01.07.2006. http://www.onto-med.de/Archiv/ontomed2002/en/publications/scientific-reports/om-report-no8.pdf (Accessed January 11, 2009.)

Holmes, T., H. Tran, U. Zdun, and S. Dustdar. 2008. Modeling Human Aspects of Business Processes — A View-Based, Model-Driven Approach. http://www.infosys.tuwien.ac.at/staff/tholmes/publications/human_vbm.pdf. (Accessed January 11, 2009.)

Honderich. T. 1995. The Oxford Companion to Philosophy. New York: Oxford University Press.

Huntsinger, R. 1988. Continuous System Simulation Languages (CSSL's). Proc. 1988 Winter Simulation Conf.

Hustadt, U., C. Dixon, R.A. Schmidt, M. Fisher, J.-J. C. Meyer, and W. van der Hoek. 2001. Reasoning about agents in the KARO framework. In *Proc. 8th Int. Symp. Temporal Representation and Reasoning (TIME'01),* C. Bettini and A. Montanari, Eds., p. 206–213. IEEE Computer Society. http://citeseer.ist.psu.edu/hustadt01reasoning.html (Accessed January 11, 2009.)

IHS. 2007. IEEE Starts Standard for Rosetta System Design Language — IEEE P1699. http://electronics.ihs.com/news/ieee-rosetta-language.htm. (Accessed January 11, 2009.)

Jeffrey, R. 1967. Formal Logic: Its Scope and Limits. New York: McGraw-Hill.

Kalfoglou, Y., B. Hu, D. Reynolds, and N. Shadbolt. 2005. *Capturing, Representing and Operationalising Semantic Integration (CROSI) project - final report. http://eprints.ecs.soton.ac.uk/11717/.* (Accessed January 11, 2009.)

Kaushik, S., D. Wijesekera, and P. Ammann. 2006. An Algebra for Ontologies. *Int. Conf. Formal Ontology in Information Systems (FOIS 2006).* Baltimore, MD, 9–11 November. (Accessed January 11, 2009.)

KIF. 2008. Knowledge Interchange Format. draft proposed American National Standard (dpANS). NCITS. T2/98-004. http://logic.stanford.edu/kif/dpans.html. (Accessed January 11, 2009.)

Kiryakov, A., K. Simov, and M. Dimitrov. 2001. OntoMap: Portal for Upper-Level Ontologies. Proc. Formal Ontology in Information Systems, FOIS 2001, 17–19 October 2001, Ogunquit, ME. http://www.ontotext.com/publications/fois01.pdf. (Accessed January 11, 2009.)

Lee, S. and H.-S. Yong. 2007. Component Based Approach to Handle Synonym and Polysemy in Folksonomy. 7th IEEE Int. Conf. Computer and Information Technol., CIT 2007. 16–19 October 2007, p. 200–205.

Lenat, D. 1995. CYC: a large-scale investment in knowledge infrastructure. Commun. ACM, 38(11): 33–38.

Lenat, D. and R. Guha. 1990. Building Large Knowledge-Based Systems: Representation and Inference in the Cyc Project. Boston, MA: Addison-Wesley.

Masolo, C., S. Borgo, A. Gangemi, N. Guarino, and A. Oltramari. 2003. WonderWeb Deliverable D18: Ontology Library (final). Laboratory For Applied Ontology — ISTC-CNR, Trento, Italy. http://www.loa-cnr.it/Papers/D18.pdf (Accessed January 11, 2009.).

MathML. 2008. Mathematical Markup Language (MathML) Version 3.0. http://www.w3.org/TR/MathML3/. (Accessed January 11, 2009.)

Matulevičius, R., P. Heymans, and A.L. Opdahl. 2006. Ontological Analysis of KAOS Using Separation of Reference. Proc. 11th CAiSE'06 Int. Workshop on Exploring Modeling Methods in Systems Analysis and Design (EMMSAD'06), Luxembourg, June 2006, p. 395–406.

McComb, D. 2004. Semantics in Business Systems: The Savvy Manager's Guide: the Discipline Underlying Web-Services, Business Rules, and the Semantic Web. San Francisco: Morgan Kaufmann.

McGuinness, D. 2003. Ontologies Come of Age. In *Spinning the Semantic Web: Bringing the World Wide Web to Its Full Potential,* D. Fensel, J. Hendler, H. Lieberman, and W. Wahlster, Eds. Cambridge, MA: MIT Press.

Meyer, J.-J. Ch., W. van der Hoek, and B. van Linder. (1999). A logical approach to the dynamics of commitments. *Artificial Intelligence,* 113:1–40. http://citeseer.ist.psu.edu/ch99logical.html. (Accessed January 11, 2009.)

Nakasone, A. and M. Ishizuka. 2006. Storytelling Ontology Model using RST. *Proc. the IEEE/WIC/ACM Int. Con. Intelligent Agent Technology,* p. 163–169.

Nauman, M. and F. Hussain. 2007. Common Sense and Folksonomy: Engineering an Intelligent Search System. *Int. Conf. Information and Emerging Technologies.* ICIET 2007. 6–7 July, p. 1–6.

OASIS. 2007. Web Services Business Process Execution Language Version 2.0. OASIS Standard. http://docs.oasis-open.org/wsbpel/2.0/wsbpel-v2.0.pdf. (Accessed January 11, 2009.)

OBO. 2008. The Open Biomedical Ontologies. http://www.obofoundry.org/. (Accessed January 11, 2009.)

Obrst, L. 2006. The Ontology Spectrum and Semantic Models. Presented at *Ontology for the Intelligence Community: Towards Effective Exploitation and Integration of Intelligence Resources.* Columbia, MD. 30 November–1 December. URL: http://ncor.us/oic/obrst.ppt (Accessed January 11, 2009.)

OMG (Object Management Group). 2007a. Unified Modeling Language (OMG UML), Superstructure, V2.1.2. OMG Document Number: formal/2007-11-02 Standard document URL: http://www.omg.org/spec/UML/2.1.2/Superstructure/PDF. (Accessed January 11, 2009.)

OMG (Object Management Group). 2007b. Unified Modeling Language (OMG UML), Infrastructure, V2.1.2. OMG Document Number: formal/2007-11-04 Standard document URL: http://www.omg.org/spec/UML/2.1.2/Infrastructure/PDF. (Accessed January 11, 2009.)

OMG (Object Management Group). 2007c. Systems Modeling Language (OMG SysML™), V1.0. OMG Document Number: formal/2007-09-01 Standard document URL: http://www.omg.org/spec/SysML/1.0/PDF. (Accessed January 11, 2009.)

OMG. 2008. The official OMG MARTE Web site. http://www.omgmarte.org/. (Accessed January 11, 2009.)

OpenCyc. 2008. OpenCyc: Formalized Common Knowledge. http://www.opencyc.org/. (Accessed January 11, 2009.)

Peltz, Jay and R.K. Thunga. 2005. HumanML: The Vision. http://www.oasis-open.org/committees/download.php/13625/HumanMLReport-WhitePaper.pdf (Accessed January 11, 2009.)

Petty, M.D., F.D. McKenzie, X. Qingwen. 2002. Using a software architecture description language to model the architecture and run-time performance of a federate. *Proc. Sixth IEEE Int. Workshop on Distributed Simulation and Real-Time Applications*, p. 85–92.

Poli, R. and L Obrst. 2009. *The Interplay Between Ontology as Categorial Analysis and Ontology as Technology.* Chapter 9 in Part One: Ontology as Technology in the book: *TAO — Theory and Applications of Ontology, Volume 2: The Information-Science Stance*, M. Healy, A. Kameas, and R. Poli, Eds. Forthcoming, Springer.

Raczynski, S. 2006. *Modeling and Simulation: Computer Science of Illusion*, Chichester, England: John Wiley.

Rao, A. and M. P. Georgeff. 1991. Modeling rational agents within a BDI-architecture. In Proceedings of the Second International Conference on Principles of Knowledge Representation and Reasoning (KR'91) J. Allen, R. Fikes, and E. Sandewall, Eds. San Francisco: Morgan Kaufmann, p. 473–484. http://citeseer.ist.psu.edu/rao91modeling.html (Accessed January 11, 2009.)

Schall, D., H.L. Truong, and S. Dustdar. 2008. On unifying human and software services for ad-hoc and process-centric collaboration. *IEEE Internet Computing,* 12(3):62–68.

SDL. 2008. What is SDL? http://www.sdl-forum.org/SDL/index.htm. (Accessed January 11, 2009.)

Shapiro, S., Y. Lespérance, and H.J. Levesque. 2002. The Cognitive Agents Specification Language and Verification Environment for Multiagent Systems. *AAMAS'02,* 15–19 July, Bologna, Italy. 2002 ACM.

Siorpaes, K., M. Hepp, A. Klotz, M. Hackl, and the myOntology Consortium. 2008. myOntology in a Nutshell: Release Primer. http://www.myontology.org/docs/releaseprimer-myontology-ksiorpaes-20080617.pdf (Accessed January 11, 2009.)

Smith, B. 2004. Beyond Concepts: Ontology as Reality Representation. In *Proc. Third Int. Conf. on Formal Ontology in Information Systems* (FOIS 2004). Amsterdam: IOS Press, p. 73–84.

Smith, B. and C. Welty. 2001. Ontology: Towards a New Synthesis. Introduction to the *Second Int. Conf. Formal Ontology and Information Systems.* FOIS'0l, 17–19 October, Ogunquit, ME, p. iii–ix.

Smith, B., M. Ashburner, C. Rosse, J. Bard, W. Bug, W. Ceusters, L. Goldberg, K. Eilbeck, A. Ireland, C. Mungall, the OBI Consortium, N. Leontis, P. Rocca-Serra, A. Ruttenberg, S. Sansone, R. Scheuermann, N. Shah, P. Whetzel, and S. Lewis. 2007. The OBO Foundry: coordinated evolution of ontologies to support biomedical data integration. *Nature Biotechnol.*, 25(11)1251–1255. http://www.nature.com/nbt/journal/v25/n11/full/nbt1346.html. (Accessed January 11, 2009.)

SML. 2008. Service Modeling Language, Version 1.1. W3C Working Draft 3 March. http://www.w3.org/TR/sml/. (Accessed January 11, 2009.)

SUMO. 2008. Suggested Upper Merged Ontology. IEEE. http://www.ontologyportal.org/. (Accessed January 11, 2009.)

Sunik, B. 2005. The specification language T. *ACM SIGPLAN Notices*, 40(5):28–38.

Trencansky, I. and R. Cervenka. 2005. Agent Modeling Language (AML): a comprehensive approach to modeling MAS. *Informatica* 29:391–400.

van Linder, B., W. van der Hoek, and J. Meyer. 1998. Formalizing abilities and opportunities of agents. *Fundamenta Informaticae*, 34(12):53–101.

Vassev, E. and J. Paquet, 2007. ASSL—Autonomic System Specification Language. *Proc. 31st Annu. IEEE/NASA Software Engineering Workshop (SEW-31)*, Baltimore, MD, 6–8 March 2007, p. 300–309.

Voice. 2008. VoiceXML Overview. http://www.voicexml.org/overview.html. (Accessed January 11, 2009.)

W3C. 2005. Scalable Vector Graphics (SVG) Full 1.2 Specification: W3C Working Draft *13 April 2005. http://www.w3.org/TR/SVG12/.* (Accessed January 11, 2009.)

Wand, T. and R. Weber. 1993. On the ontological expressiveness of information systems analysis and design grammars. *J. Information Systems*, 3:217–237.

Wand, T. and R. Weber. 1995. On the deep structure of information systems. *J. Information Systems*, 5:203–223.

Weitzenfeld, A., M. Arbib, and A. Alexander. 2002. *The Neural Simulation Language: A System for Brain Modeling*. Cambridge, MA: MIT Press.

WordNet. 2008. WordNet: A Lexical Database for the English Language. Cognitive Science Laboratory, Princeton University, Princeton, NJ. http://wordnet.princeton.edu/. (Accessed January 11, 2009.)

Chapter 9

Time

The concept of time is a critical element in systems engineering. It leads directly to the subject of change, and in particular, raises the following question: How do we adequately address change in systems engineering? This chapter discusses concepts of time, types of change, and especially levels of change. A running theme is the application of change models from biology.

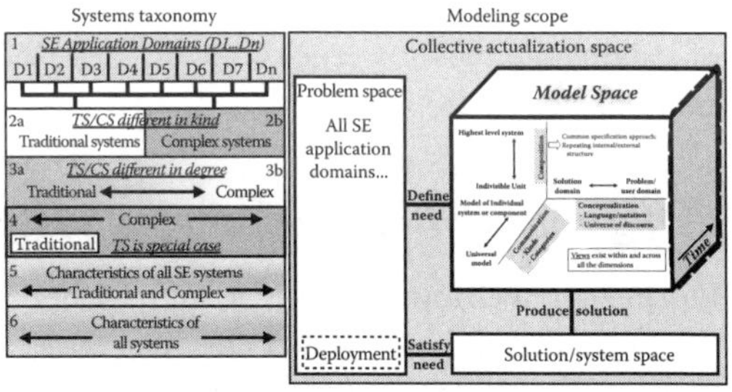

Time and Change: An Introduction to Basic Concepts

Previous chapters, especially Chapter 7, have discussed the concept of similarities and differences. In this chapter we look at the temporal aspects of these concepts. One way to think about this chapter is that the time dimension of the model space is a temporal specialization of similarities and differences. We think of similarity across time as *stability*, and differences or variability across time as *change*. A bridge between the two is that invariants represent what is common and stable, and variants represent what differs and changes. If everything were stable, then time could be abstracted away, and we would not need to include this chapter, or the time dimension in the model space. However, change is ubiquitous; that is, it is everywhere and it is continual — in the natural world and the engineering world. Therefore, change is the aspect of time that motivates and dominates this chapter.

We believe that traditional systems engineering (TSE) has not given sufficient attention to temporal change in systems engineering (SE) processes and in the assumptions about engineered systems and their environments. As Dörner (1989) points out, humans can more readily detect forms and patterns in space than over time. A key message of this chapter is that MOSES assumes continual change as the norm, not the exception. Paradoxically, that assumption behooves us to find stability at appropriate places in the model space; not to force stability or to pretend that there is stability, but to find natural stability where it exists and where it can provide a context for change. A major aim of this book is to articulate the stabilities of systems engineering. Major specific examples

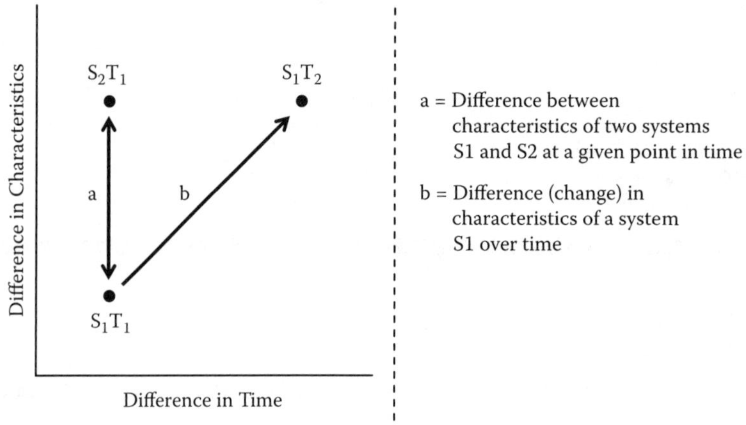

Figure 9.1 Differences within and across systems.

of stability are the SE science concepts described throughout this book and general models in the commonization dimension of the model space, described in Chapter 7.

Differences within and across Systems

Differences in characteristics of system S1 between time T1 and T2 are not inherently distinguishable from differences between characteristics of systems S1 and S2 at time T1. Figure 9.1 shows both types of differences. The arrow "a" represents the difference between S1 and S2 at a given time, and the arrow "b" represents the difference or change of system S1 across time. S1 at time T2 may look like S2 at time T1, so the differences "a" and "b" may be indistinguishable. Differences represented by "a" were the subject of Chapter 7. In this chapter we look primarily at differences or changes represented by "b" that occur across time. And to some degree we also look at the combination of both types of differences.

Whether Something Is Static or Dynamic

We might think that the question of whether a given system or model is static or dynamic is trivial or obvious, but this section discusses why it is sometimes neither trivial nor obvious. Keep in mind that in the context of MOSES, the pertinent question is not necessarily whether something *is* static or dynamic; the question is how do we choose to *model* the situation?

Consider an entity, a characteristic, and a specified time range. The entity is static with respect to the characteristic and the time range if the value of the characteristic does not change over the course of the time range. The entity is dynamic in that context if the value of the characteristic does change over the course of the time range.

Nann (2004) provides an example of scale in the interpretation of static versus dynamic:

> ... "dynamic" models are those that are frequently rebuilt or retrained. ... "Frequent" is defined as when enough historical data has been gathered to provide information that is statistically relevant to the model. In the context of this analysis, "frequent" was approximately every sixty days.

Here is a scale in which a model that changes every 60 days is considered dynamic. There are many scales in which a model that does not change for 60 days is considered a static model. This includes a defined world of any time period within those 60 days.

The distinction is related to cycles, as discussed in this chapter, and to views as discussed in Chapter 10.

A model can be static or dynamic, and its view of the system can be static or dynamic. A structure diagram or class diagram is an example of a static model with a static view of a system. Examples of static models with dynamic views of a system are a state machine model, Petri net model, differential equations of a system trajectory, software source code, a music score, or a food recipe. In each of these cases, the model is a static set of symbols; when driven or acted on by an appropriate agent, it depicts dynamic system behavior.

Measuring Change

How do we determine if a system changes and the nature of those changes? The way we measure change can significantly affect the answers we obtain and affect our perception of a process. We illustrate the measurement effect with an example.

Suppose a team of five extraterrestrials wants to conduct some research on certain characteristics of the "Earth" and the "Sun," and they have very limited knowledge. We spell out exactly what they know and some key things they do not know.

Their knowledge is based on an ontology that consists only of these terms and meanings:

- Earth: big rock floating in space (in our ontology, it is equivalent to our planet Earth, with its attendant characteristics).
- Fixed point on surface of Earth: place on the big rock where they will conduct their research.
- Sun: bright object overhead, visible from fixed point (in our ontology, it is equivalent to our Sun, with its role in the solar system).
- Orientation: north, south, east, west, with our usual meaning of these directions.
- Measurement: collection of information about location and orientation of Sun relative to fixed point.
- Time point and time interval: they have our usual meaning of these concepts, except that they do not use our units of time intervals (second, hour, day, week, month, year). They use their own intervals, which they call N1, N2, N3, N4, and N5. The relation of their intervals to ours is given below.
- Numbers, counting: They use our number system, and they can all count, at least up to six.

Their general knowledge does *not* include our concepts of solar systems, rotation of a planet, revolution of a planet around a sun, or the roles played by the sun in a solar system. To them, Sun is simply a word that refers to that bright object visible from the fixed point.

Purpose of research: Using fairly precise measurements, determine whether Sun changes position relative to fixed point and orientation, and if so, what those changes are. *Method:* Take periodic sample measurements of position of Sun from fixed point and fixed orientation. Each researcher will take six samples at a different rate (intervals corresponding respectively to N1 through N5) to make sure they get good coverage of change if any exists. They all select a common start time at a common fixed location, which turn out to be (in our terms, not theirs) noon local standard time on March 1 at a point that is not near either the North or South Pole. Our assumption for this research project: the Sun will not be hidden by clouds from the fixed point whenever they want to take a sample measurement.

The five researchers are identified as ET1 through ET5. Each of their experiments is defined as follows:

- ET1 samples at a rate of once every N1 time interval, which turns out to be, in our terms, once every 30 minutes, so the first sampling period is 2½ hours, from noon to 2:30 p.m. on March 1.
- ET2 samples at a rate of once every N2 time interval, which turns out to be once every 23½ hours, so the second sampling period is nearly 5 days (noon on March 1, 11:30 a.m. on March 2, etc.).
- ET3 samples at a rate of once every N3 time interval, which turns out to be once every 2 weeks, so the third sampling period is 10 weeks.
- ET4 samples at a rate of once every N4 time interval, which turns out to be once every 3 months, so the fourth sampling period is 15 months.
- ET5 samples at a rate of once every N5 time interval, which turns out to be once every year, so the fifth sampling period is 5 years, at noon on March 1 each year.

Each researcher writes up a report at the end of the project. Their executive summaries, each expressed in terms of whether and how Sun moves with respect to the fixed point and orientation, are:

- ET1: Sun moves acyclically from east to west.
- ET2: Sun moves acyclically from west to east.
- ET3: Sun moves acyclically from north to south.
- ET4: Sun moves cyclically, north to south to north to south.
- ET5: Sun does not move.

The committee that sponsored this research concludes that further research is needed.

Types of Change

The early Western concept of time was not an arrow, a one-directional linear flow with which we often conceptualize time; it was the concept of cycle: everything (day-night, lunar cycle, seasons, etc.) coming back to the same state periodically (Novikov 1998). In SE we make extensive use of both model types — acyclic and cyclic flow. This subsection discusses several aspects of cyclic and acyclic change models.

Cyclic Models

The cyclic model type has several variations, some of which are shown in Figure 9.2:

- Variant 1: repeating exactly the same cycle; can be modeled by a circle. Example: state machine implemented in software. Say it has repeating sequential states A-B-C-D-A-B-C-D-A-B-.... Every cycle repeats exactly the same states as the previous cycle.
- Variant 2: repeating cycles but differs slightly in a more or less uniform way. It can be modeled by a spiral. Example: gradual wearing out of physical machine (for a period of time, prior to when it experiences drastic drop-off or stops working).

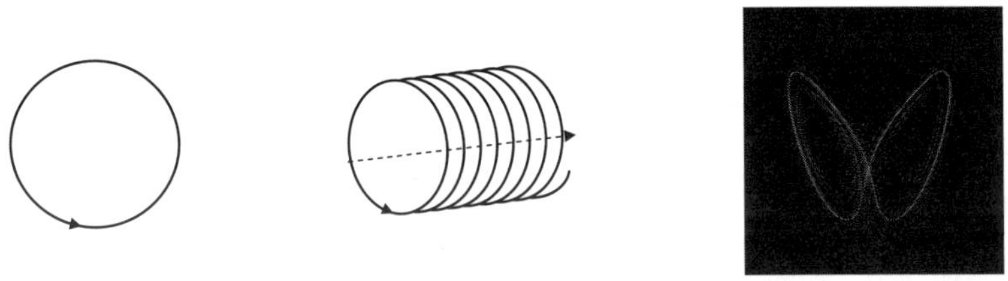

1. Same cycle: Model = Circle

2. Similar cycle: Model = Spiral

3. Non-repeating cycles: Model = Strange attractor

Figure 9.2 Cyclic model variations. (*Source*: Wikipedia 2008.)

- Variant 3: Repeating cycles but no cycle repeats exactly the path of any previous cycle, and may change back and forth between different cycles. It can be modeled by a strange attractor such as the Lorenz strange attractor (in the figure, strange attractor image is from Wikipedia, 2008). Example: Behavior of person acting in two alternating roles that map to the two different cycles, such as employee and parent.

There are also at least two views of any cyclic model: (1) a state-similarity view and (2) a timeline view. The three variants above were described and illustrated in the state-similarity view. Using Variant 1 from Figure 9.2, the two views are illustrated and contrasted in Figure 9.3. The state-similarity view (circle) only shows the repeating state, not the time when each occurs. The timeline view shows the time point when each state is revisited.

Scale Effects on Acyclic and Cyclic Models

Cycles exhibit a huge time-scale range: from billions of cycles per second for certain atoms, up to billions of years for galaxies. Time scales introduce interesting relations and views of acyclic and cyclic models.

Consider that any given system change vector can be described in terms of both variables — its cyclic change and its acyclic change — where one or both of these variables may be zero for a given time period. Suppose a system has a cycle of 1 hour. Now let's examine its change at a time scale of seconds (how it changes over the course of 1 second), at a scale of hours (how it changes over the course of 1 or a few hours), and at a scale of months (how it changes over the course of a month). Figure 9.4 is a notional view of all three scales, as well as the mapping between the scales.

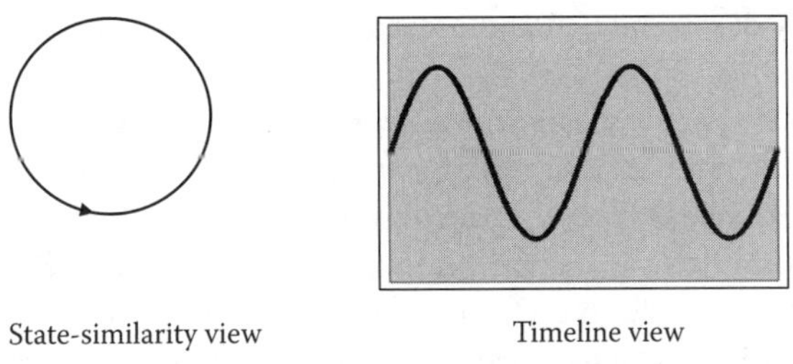

State-similarity view

Timeline view

Figure 9.3 State-similarity and timeline views of cycles.

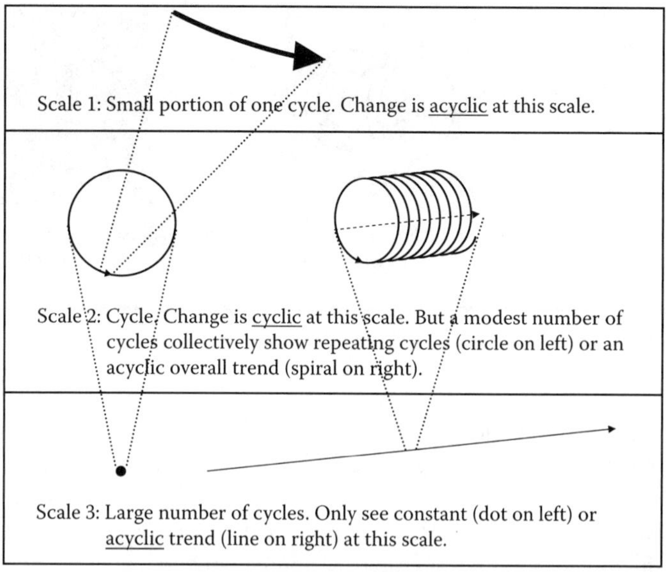

Figure 9.4 Scale effects on views of cyclic and acyclic change.

- Scale 1 (e.g., second): The change of the system appears to be acyclic at this scale. The cyclic change would be regarded as zero. The acyclic change is either zero or a small increment over the course of this second.
- Scale 2 (e.g., hour): The change of the system appears to be cyclic at this scale. The amplitude of the circle or spiral is visible and is the amount of cyclic change. If the system is seen as returning to exactly the same state (the circle on the left), the acyclic change is seen as zero.
- Scale 3 (e.g., month): The change of the system appears to be acyclic at this scale. The cyclic change would be invisible and regarded as zero. The acyclic change is either zero, if the system follows exactly the same circle at Scale 2, or an acyclic trend if the system follows the spiral at Scale 2.

This may appear to be an esoteric or unusual case but it represents a large number of situations — including the life of many systems, which go through the same routine cycles each day or some other periodicity, and over time also experience acyclic change, such as aging or growing, as illustrated in the three cyclic variants above and the levels of change that will be discussed later in this section. We now briefly discuss human and organization examples from the scale perspective.

> *Example 1: Human.* Each cycle in Scale 2 represents a day in the life of a human — correlating with the circadian rhythm. In this case, the Scale 2 model is the spiral on the right because a human is a bit different from one day to the next. Scale 3 represents change over the life span of a human. At Scale 1, say a few minutes, we cannot observe the daily cycle.
>
> *Example 2: Organization.* Each cycle in Scale 2 represents a year in the life of an organization with seasonal variations, so it has an annual cycle of fluctuating size. Scale 3 represents change over multiple years of the organization; for example, it grows in size 2% per year for 10 years. At Scale 1, say a week, we cannot observe the annual cycle.

Change of Interest

In previous chapters we discussed the concept of system of interest. In this chapter the concept of change of interest is useful. An observer is interested in some changes that occur and not in other changes that occur. The change of interest may be specified in terms of states, conditions, periods, observation points, or views. We discuss three of potentially many views of change of interest: invariant view, scale view, and hiding view.

Invariant View

An invariant is a model — a description or assertion — of a system characteristic or relation that is preserved in all observable states (see Meyer (1997) for a good description of invariants). What is of interest is the change, or in this case the lack of change, in the characteristic or relation in the observable states. What is not of interest is any change in the characteristic in non-observable states. One implication of this is that an invariant depends on what is observed, which generally is defined by the viewpoint from which observations are made. If a characteristic that is to be invariant is continuously observable from viewpoint 1, and is observable once a year from viewpoint 2, then the characteristic must hold continuously in viewpoint 1, but only needs to hold at each annual observation from viewpoint 2. In the latter case, the characteristic could change throughout the year; but if it had the same value at the observation point each year, it would be considered invariant by an observer at viewpoint 2.

Examples: In mathematics, an invariant is something that does not change under an operation or transformation. In software, an invariant is an expression that is the same before and after execution of a piece of observed software. The before and after viewpoint could range from a single line of code to a scripted sequence of multiple programs. In a car, the constraint that a tire must be mounted at all times while the car is in operation could be considered an invariant. The observable states for the mathematics and software examples are conveniently defined in terms of pre and post conditions. A quantity or expression must be the same at the end of transformation or execution as it was prior to transformation or execution. We do not care if it changed in the course of the operation, as long as its state at the end was the same as the state at the beginning. So the change of interest is defined at the start and end observation points, and an invariant defines that change to be zero. The situation with the car and tire, on the other hand, is continuously observed: The tire must be mounted on the car persistently across the span of time during which it is operating. Another example is a safety deposit box. The condition of the material when it is placed in the safety deposit box must be the same as the condition when it is removed from the box. It could be in any condition or kept anywhere during the interval between placing it in the box and retrieving it from the box, as long as it is in the same condition at both of those observation times.

Scale View

Changes of interest often vary by scale. Figure 9.4 illustrates one example of this. An observer who chooses scale 1 is interested in the modest acyclic change on a small part of what is a cycle at scale 2. An observer at scale 2 is interested in the cycle of change. An observer at scale 3 does not see and is not interested in cycles, but is interested in the longer-term change of the system, which is either no change (if scale 2 is the repeating circle) or a trend (if scale 2 is the moving spiral). The measurement example in the section entitled “Whether Something Is Static or Dynamic” illustrated that measurements can be misleading at different scales or from different views. The measurement must match what is of interest.

Another scale example is stepwise revolution, that is, a system such as an organization that transforms itself but does so via a series of gradual steps rather than a sudden big bang change. This distinction can be seen in Figure 9.1 vector b, which represents transformation of an organization S1 across the time period between T1 and T2. Suppose an external observer observed the organization S1 just prior to time T1 and again just after time T2. To this observer, the transformation would appear to be more sudden, while another observer who tracked the change during the T1 to T2 interval would view it as gradual.

A similar example is the use of the fossil record as an indicator of the history of biological evolution. In the example above, suppose the external observer is the fossil record. Because this is at the scale of geologic time, it compresses time spans into instantaneous points in time. Suppose a species S1 at time T1 evolves to a new species S2 at time T2 (along vector b in Figure 9.1). The time span between T1 and T2 may be hundreds of thousands or millions of years. The fossil record shows S1 at T1 and S2 at T2 but it does not capture the intervening period; T1 is followed immediately by T2. This view shows the species S2 appearing suddenly at T2. This is one of the difficulties of trying to obtain the historical details of biological evolution, and especially speciation.

Hiding View

The basic engineering principle of encapsulation or information hiding facilitates modular design. Parnas (1972) enunciated this as a criterion for designing a system, especially for decomposing a whole system into parts. The basic idea is that each part hides a design decision that is likely to change. The goal is for each part to expose information that is expected to be relatively stable, and to hide information that is expected to be relatively volatile from a design perspective.* The concept of change of interest relates to information hiding in the following way. Suppose we have two parts of a system called P1 and P2, and P2 depends on P1. This means that changes to P1 potentially have an impact on P2. Now suppose that we separate potential changes in P1 into two groups: external changes visible to P2 and internal changes hidden from P2. In this structure, clearly the P1 changes of interest to P2 are the external visible changes, and the P1 changes not of interest to P2 are the internal hidden changes. This step has made life easier for P2 from a change perspective. If in addition P1 arranges to hide internally the things *most* likely to change, and to expose externally the things *least* likely to change, this benefits P2 even further. But P2, and all other parts of the system that depend on P1, are not the only beneficiaries of P1 information hiding. P1 itself benefits because if it keeps most changes internal, it reduces its workload of overcoming resistance to change, and redefining and renegotiating interactions and agreements, with those parts that depend on it.

In the invariant view, the changes of interest are held to zero, while the changes not of interest are unconstrained except by the bounds of the larger invariant constraint. In the scale view, the changes of interest are not necessarily constrained, but they are a simplified subset of the total picture by virtue of the fact that they are the only ones visible at a given scale. In the hiding view, the changes of interest are minimized, and the changes not of interest are unconstrained except by the bounds of the external view. Described in this way, all three patterns or views are tools for

* There is another important perspective in which the hiding described here is not desirable: monitoring and tracking status of a system. The goal in this perspective, for everything from bridges to networks to human health, is to expose volatile information to monitor system status.

managing change by reducing the amount of change we must deal with in a given view. These are therefore part of the more general topic of dealing with change in the collective actualization process discussed in Chapter 13.

These three patterns or views — invariant, scale, and hiding — are not necessarily distinct or separate. For example, we could reasonably conceptualize an external model of a system as *hiding* changes exhibited by the internal model, or as an *invariant* that describes what is true of the externally visible behavior or data, while not being interested in changes in the internal behavior or data.

Other Concepts: Persistence and Duality

This section discusses examples of state and role persistence, followed by the concepts of duality and contrast in the time dimension.

State and Role Persistence

Some states and roles are perceived as persisting across multiple cycle levels. Examples of both are presented.

State: In a time-slicing computer operating system, consider a software program P1 that is loaded and executing, and is time-sliced with other programs. One cycle is the one-time process of being loaded at time T1, executing, and then finishing and being unloaded at time T2. The other cycle is the time slices: suppose the period T1 to T2 is divided into 1000 slices identified as t1, t2, …, t1000; and that P1 is allocated one of every 10 slices — say, t10, t20, …, t1000. Now we ask a question: When is P1 executing? Clearly it is executing during slice t10. But is it executing during slice t15? The answer might depend on the context and purpose of our question. If in the period between T1 and T2, we interrogate the system at the level of what programs are loaded and executing, the system will tell us that P1 is executing. On the other hand, if we are conducting an autopsy of the computer execution because the computer crashed between T1 and T2, and it turns out to have crashed during slice t15, then we might answer no, P1 was not executing at that moment and therefore may be a less likely suspect for causing the crash than the program that was executing during t15.

Role: Assume that Bob plays the roles of employee, husband, and Little League baseball coach. Further assume that he only plays one of these roles at a time. For example, he is in the role of employee during the day Monday through Friday, the role of baseball coach Saturday morning, and the role of husband all other times. Further, assume the baseball season, including preliminary meetings, practices, and games, goes from May through July, and that he is a coach from the 2001 through 2006 seasons. Now we ask a question: If someone asks Bob "Are you a Little League baseball coach?", when is the answer yes? Suppose the question is asked (1) on a Saturday morning in June 2002; (2) on a Tuesday evening in June 2002; (3) in September 2003; and (4) sometime in 2007. In general, the answer would probably be yes for all except (4). Bob would consider himself in the role of baseball coach even in the off-season until he stopped being a coach. This example illustrates three levels of cycle: weekly games, annual seasons, and the life span of his coaching career.

Furthermore, assuming Bob was an employee and husband from 2001 to 2006, then at any given moment during that period, he would say: I am a husband, I am an employee, and I am a baseball coach. But the question of when he is a coach could also depend to some degree on the context of the question. If two boys on his baseball team got into a fight and someone asked if he was their coach, there is a difference between whether the fight occurred on (1) a Saturday morning in June 2002 or (2) a Tuesday evening in June 2002. Bob had significantly more responsibility for their behavior in the first case than the second.

There are four related points to be made here regarding the question of whether a given state of affairs exists:

1. It may exist in different senses at different cycles.
2. Whether it exists at a given time in a given cycle may depend on perspective.
3. The state is often perceived to persist across longer cycles even when it does not hold at times in a shorter cycle.
4. Multiple different states of affairs of the same individual, especially roles, are often conceived as existing simultaneously.

Duality, Tension, and Paradox in Time Variables

We mentioned the concept of duality and tension in Chapter 1 as a characteristic of complex systems. The time dimension contains a few of these tensions. First, stability and change are in constant tension in systems and environments. A summary of evolution that reflects this is from Tang and Epstein (2007): "Evolution of biological complexity involves an environmentally regulated balance between genetic conservation and variation."

Second, cyclic and acyclic change reflect a balance and tension that shows up in different views, scales, and times, as illustrated in earlier discussions and Figure 9.4. Third, cyclic change itself is a tension or paradox of constancy and change: it continually changes, but at the same time it repeats itself in a constant repetition.

Another somewhat separate tension is planning. We need to set goals and define plans to reach those goals, but we rarely follow a plan. The goal may change; or even if the goal stays the same, the plan to get there usually changes. The balance and resolution of this tension is described in Chapter 13.

Modeling Levels of Process and Change

We define three levels of process and change.

- Level 1 = what a system does; that is, change accommodated or variation supported within the range of behavior of a given system.
- Level 2 = what a system is; that is, change involving modification of the system (parts).
- Level 3 = change in how we view a system; that is, change involving modification of system models — such as changing the design or requirements of a system.

In some circumstances, change can occur in any one of these levels without changing the other two levels. In other circumstances, if one level changes, one or both of the other levels also changes.

A car and a person can be used to illustrate the levels. For a car, Level 1 represents normal processes or changes in behavior, such as the engine being on or off, or the car changing direction or speed. Level 2 includes replacing or repairing engine parts, tires, etc. Level 3 could include changing internal or external models, such as changing the engine type, or a performance requirement. For a car, we usually consider Level 1 and Level 2 to be clearly separated: changing behavior seems clearly distinct from replacing a part, even though Level 2 change often results in Level 1 changes.

Now consider a person. Level 1 represents the changing behavior of the person, from one minute to the next, or from one role such as employee to another such as parent.

Level 2 is a change in the person. This could be at an extreme level of parts replacement, such as a heart transplant or replacing knee joints with artificial ones. It could be at a less extreme level of parts repair such as various types of surgery. But humans also change continually in at least three categories: First, they improve physically, mentally, and socially through development, growth, learning, and accumulated experience. Second, they deteriorate in various ways through aging. Third, they change in response to their interaction with the environment throughout their lifetime. This means that Level 1 and 2 changes occur continually and simultaneously, and in many cases directly influence each other. In humans, and by extension complex systems (CS) in general, the distinction between what a person does (Level 1) and what the person is (Level 2) is not so clear.

Based on this discussion, we can reconsider the car. It gradually deteriorates with use. From this perspective, what the car is seems more closely connected to what it does: as it behaves (Level 1), it gradually deteriorates (Level 2). Note that conceptual systems such as software programs do not in general deteriorate through use, so their Level 1 changes seem clearly distinct from their Level 2 changes. However, the overall software system performance may deteriorate, for example through accumulation of registry entries or disk fragmentation. Furthermore, self-modifying software closely connects Level 1 changes and Level 2 changes by modifying itself while it is operating, due to conditions of its operation.

Level 3 covers changes in models of humans. For example, our model of human anatomy may be fairly mature and stable at the current time, but our model of human genes has evolved significantly in recent years because of significant increases in knowledge that have led to improved fidelity of the model.

A fair conclusion of the foregoing discussion of Level 1 and Level 2 changes is that for systems in general, it might be useful to have two views: one in which Level 1 changes are clearly distinguished from level 2 changes, and one in which they are not clearly distinguished.*

Process or change models representing these three levels, and combinations of levels, are discussed in the sections that follow. Figure 9.5 compares the life span of processes at each level. The figure applies to all SE systems — TS, CS such as a human, and hybrid combinations of both. Formation of a traditional system is the initial development, which ends at initial operational capability. Formation of a human is reasonably modeled as gestation, which ends at birth. Formation of an organization may include incorporation and setup activities such as obtaining funds, facilities, equipment, and personnel. It ends at the point the organization begins operations or opens for business.

In all cases, system behavior begins at the end of formation. A maturation period occurs for both TS and CS, followed by full functioning. Changes continue to occur during full functioning. This is, of course, a simplification. For example, the line between maturation and full functioning is very fuzzy. In all three levels, processes and change stop at retirement.† In the case of the model (Level 3), the life span shown is the life span of the correspondence of the model with a system. This context focuses more on specific models and less on general models. In the extreme, a specific model may be defined for, and only correspond to, this system, in which case its life span is the extent of the line for L3 in Figure 9.5. But in the general decoupled case, a model, and especially a general model, has its own independent life span, which is not shown in the figure but is discussed

* Views are discussed in more detail in Chapter 10.

† This is by definition. Retire means "end of operational period of interest," which for a person could be retirement from the workforce, or an earlier point, or a later point, depending on what is being modeled.

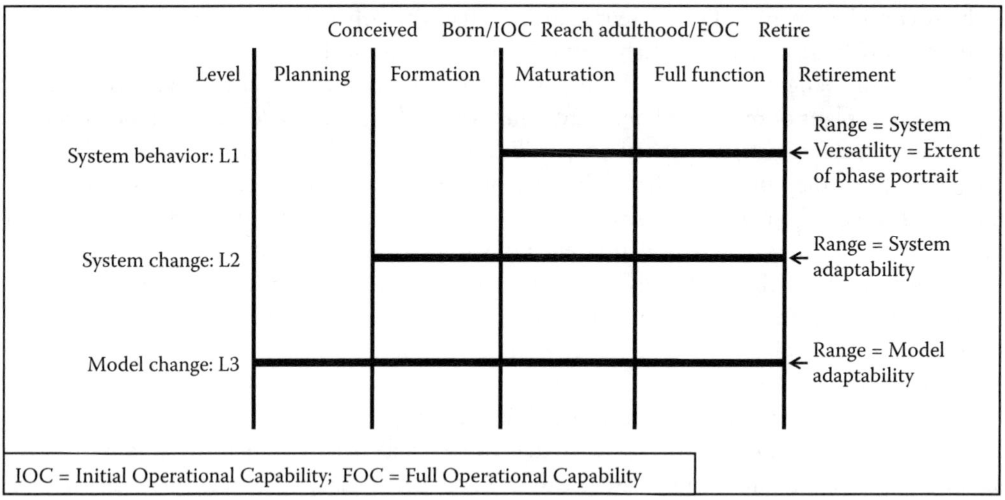

Figure 9.5 Process and change levels during system life span.

later in the chapter. A system can deal with environmental change at Level 1 by being versatile, that is, having a broad range of behavior in its repertoire without changing or being changed; and at Level 2 by being adaptable, that is, being able to change to better fit the environment. A model can also deal with change by being adaptable.

Models of Level 1 Change

Level 1 processes correspond to system behavior. A large number of models of these processes exist in both TSE and CS arenas. Many of these are traditionally called dynamic models. We list and briefly describe a few of them.

Unified Modeling Language (UML) (OMG 2007a, b): The general category is behavior diagrams (as opposed to structure diagrams). There are four types: activity diagram, interaction diagram, state machine diagram, and use case diagram. The interaction type, in turn, has four types: sequence diagram, interaction overview diagram, communication diagram, and timing diagram.

An activity diagram shows the coordination and flow of control, data, or objects between nodes. A sequence diagram shows interchange of messages between two or more entities (called lifelines). A communication diagram also shows message interchange but focuses on the architecture of the internal structure. A timing diagram shows interactions in a way that supports reasoning about time. It includes changing conditions along a time axis. A state machine diagram models states and transitions between states. A use case is a combination of static and dynamic model that is used to define the required behavior of a system and its interaction with an environment.

SysML uses a subset of the UML behavior diagrams: activity diagram, sequence diagram, state machine diagram, and use case diagram (OMG 2007c).

Another category in software engineering is algorithms, scripts, and programs. These similar model types all express processes of sequential (or sometimes parallel) actions to be taken by an agent. Usually the agent is a computer — especially in the case of scripts and programs.

All the above diagrams and artifacts map to internal models in MOSES. The use case diagram can also be used to express an external model in the sense of describing the required behavior of the system of interest.

Other traditional models include Petri nets. These are mathematically based graphical models of discrete behavior of systems. The key elements are places, arcs, and transitions, where an arc connects places with transitions. Extensions to the basic Petri-nets have included colored Petri nets and hierarchical Petri nets that allow sub-nets (Girault and Valk 2003).

Dynamical system models include the phase space orbits or trajectories of systems. Both a trajectory and an orbit are the sequence of system states over time. Typically the term "orbit" is used if the behavior sequence is cyclic, while "trajectory" is used if the behavior is acyclic. Differential equations are used to model system behavior on a continuous time scale, and difference equations or other tools, such as Petri nets, are used to model system behavior on a discrete time scale. These models tend to map more to MOSES external models because they characterize the behavior of the system as a whole, including for physical systems their position and velocity, for example. A subset of the dynamical system models characterizes behavior of chaotic systems, including orbits around strange attractors.

This context provides an additional view of the law of requisite variety (Chapter 2). Assume a system has six degrees of freedom, or six variables of behavior. Then its phase space has six dimensions. Assume that its possible behavior trajectories are captured in a six-dimensional phase portrait. Now assume a regulator needs to regulate three of these variables. Then the regulator needs to have a three-dimensional phase portrait that is the same as the three-dimensional subspace of the system's phase portrait, where the subspace is based on the three variables to be regulated. The point is that a more precise statement of the law is: to successfully regulate a system, the variety of the regulator must be at least as large as the variety *of interest* of the regulated system. The variety of interest to the regulator in this example is three out of the six variables. This is consistent with the "Good Regulator" theorem, discussed in Chapter 2, that posits a regulator of a system as a model of that system.

Models of Level 2 Change

Level 2 process models show how a system itself changes, as opposed to how it behaves. As shown in Figure 9.5, Level 2 changes occur during system formation, maturation, and full functioning. This section discusses change models for traditional and complex systems, including biological models of change, and then discusses Level 2 boundary crossing change.

Traditional and Complex System Change Models

Table 9.1 shows common terms for change in both traditional systems and complex systems, such as humans. Managing change is also included in the table.

Biology has multiple models appropriate for Level 2 processes. Table 9.2 summarizes a number of biological models of change. In each case, the information includes the change model and the unit of change. Primary sources of material in this table are Futuyma (2005) and Mayr (1982).

More general models that use some of the above processes are inheritance and reproduction.

Development occurs over much of the life of an individual. The other processes listed tend to occur more suddenly. Note that the designation of sudden change versus change over time is not inherent or absolute. In cases where the time scale is very fine grained, all the changes in Table 9.2 could be regarded as occurring over time.

Another source of biological processes is the open biological ontology (OBO), and specifically the biological process ontology (OBO 2008).

Table 9.1 Level 2 Change and Change Management

System Type	Change in Phase: Formation	Change in Phase: Maturation	Change in Phase: Full Functioning	Managing Change
TS	Development	Improvement, refactoring	Sustainment (External)	Configuration mgt
CS (humans)	Gestation	Growth, learning	Adaptation (Self)	Feedback, peer pressure, homeostasis

Boundary Crossing Change

Most of the change discussed in this section is change of an entity within itself; that is, characteristics change but the entity retains its identity or the categories it is in or the correspondence to models. Sometimes, however, change is so significant that an entity becomes another entity or moves to a different category, as illustrated in Figure 9.6. For example, a system changes to the point that it no longer satisfies its requirements model, so it crosses the boundary of the model region from inside to outside. Alternatively, a system could change or be changed so it now does satisfy its requirements — it moves from out of the region to in the region, from not corresponding to corresponding to the model. An even more drastic change, from the system's point of view, is a change to the point where it becomes a different system. In addition, the correspondence changes can also be achieved if the model (e.g., the requirements) changes and the system does not change. That is a Level 3 type of change but it is included here to present a more complete set of boundary crossing change types.

Comparison of Selected Engineering Models

We compare three engineering models: big bang model, incremental model, and evolution model. They are illustrated and contrasted in Figure 9.7. Assume a situation in which (1) users define their needs at a point in time, say T1; (2) a system is developed over time to satisfy the user needs; and (3) user needs change while the system is being developed. In the figure the full time period in which this occurs starts at time T1 and ends at T5. The relationship between changing user needs and the changing system varies according to the three engineering models. The big bang model develops

Table 9.2 Biological Models of Change

Change Model	Unit of Change
Development/growth	Individual organism
Mutation	DNA/RNA
Recombination	DNA
Division	Cell
Replacement	Cell
Birth	Individual organism
Death	Individual organism

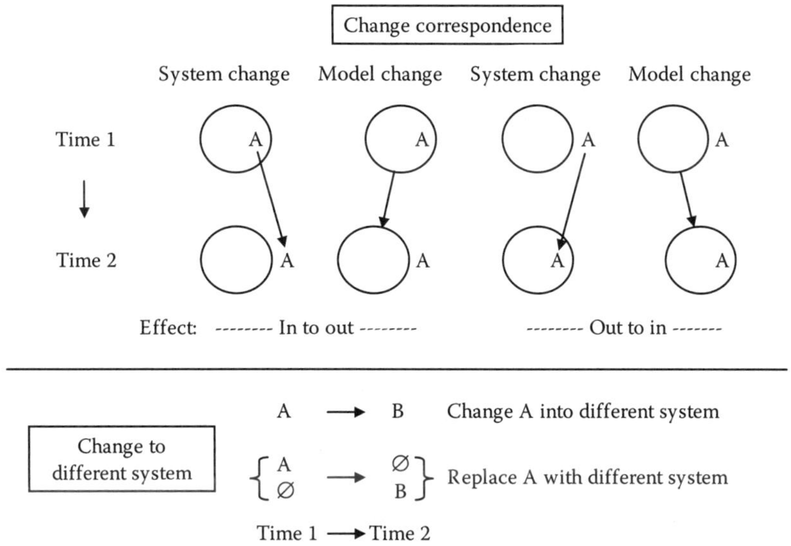

Figure 9.6 Boundary crossing change examples.

the entire system to satisfy the full set of requirements. The system is developed and ready to use at T5, and it satisfies the user needs defined at T1 (the system is designated as S_B5 in the upper left part of the figure). Because user needs have changed between T1 and T5, there is a significant gap between the system capability and what the users now need at T5.

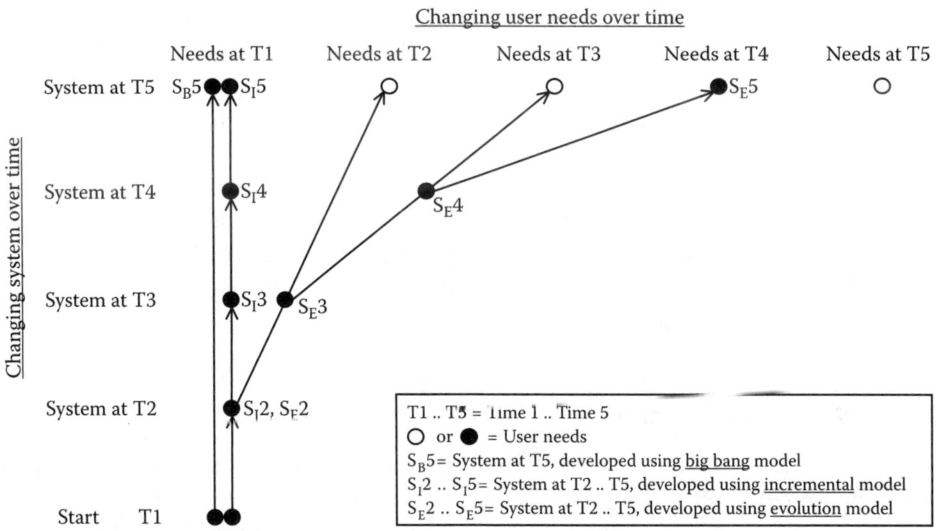

Figure 9.7 Comparison of Engineering Models

The incremental model develops the system in several increments, with all requirements defined and allocated to increments (S_I2, S_I3, S_I4, S_I5) at T1. The first increment has minimal capability, and each increment adds to it, until S_I5 represents the full system. However, like S_B5, the incrementally developed system S_I5 is still targeted to the user needs at T1, so there is a similar gap between the system and the user needs at T5.

The evolution model divides the development of the system into pieces, as in the incremental model, but the evolution model assumes change and does not commit to a final goal. At each iteration, it revisits user needs, and only defines detailed requirements for the next step (toward the updated user needs at that point). Thus, for example, at T2, the evolution model delivers S_E2, and reassesses user needs at T2. Based on that reassessment, it defines the next iteration as S_E3, which is the next step toward the now revised user needs.

The big bang and incremental models are based on a commitment to fixed needs and requirements that do not change over time. The evolution model, however, is designed to expect and adapt to change. As a result, the system S_E5 ends up closer to user needs at T5 than the systems developed by either of the other two engineering models.

Actually, what the figure does not show is that the evolution model continues to look further into the future at each iteration, beyond T5, to track the changing long term direction. This aspect of the SE evolution process is discussed in Chapter 13, especially in terms of the rolling wave model.

Models of Level 3 Change

This section concerns how models change. A distinction is made between change of specific models and change of general models.

Change of Specific Models

Specific models tend to exist more in the workspace portion of the model space than in the knowledge space portion. In this section we describe the change of a specific model associated only with a specific system; that is, it exists only in the context of a specific correspondence set. The life span of such a model is tied closely to the system life span.

Formation of a specific model occurs during the system planning phase. It is formed through creation, or through tailoring.

Maturation of a specific model occurs during system formation (and maturation) phase. For example, requirements often change and mature in the course of developing the system and are developed and refined further while the system is gaining its full functionality.

Full functioning occurs during system full function phase. Both system and model are fully functional in this phase, but both may change.

Change of General Models

General models tend to exist more in the knowledge space portion of the model space than in the workspace portion. A Level 3 change of a general model represents a category change that is decoupled from or does not rely on individual change. General model changes that are associated with individual (L2) change are called L2-3 changes and are discussed in the section entitled "Models of Change Involving Levels 2 and 3."

General models tend to be relatively stable, and their role is to provide a guide for other changes, such as L3 specific model change or L2 individual change. They typically represent technologies, theories, meta-models, reference models, and patterns. Changing a general model tends to be a community effort because it is part of the evolution of the knowledge space. An exception is the category of general models that represent the abstract vision or goal of a specific system. That category of models tends to be in the workspace.

Models of Change Involving Levels 1 and 2

This section discusses processes and changes that cover both Level 1 and Level 2.

Autopoiesis

Autopoiesis (Maturana and Varela 1980) is the process of self-creation or self-production. The key concept is maintaining its own organization, preserving the relational network that defines it as a systemic unity (Maturana 1975). The system that gave rise to this term was the biological cell. The term is contrasted with allopoiesis, which is a process in which a system of one type produces another type, such as a factory producing a car. By our definitions, autopoiesis is a Level 1 behavior that has a Level 2 effect, whereas allopoiesis is strictly a Level 1 behavior. Note, however, that the effect of the autopoietic behavior is in the direction of maintaining structure in the face of change rather than changing structure.

Phase Space Concepts

Phase Trajectories, Orbits, Portraits

We indicated earlier that phase trajectories or orbits represent dynamic behavior sequences of a system, which makes them Level 1 processes. However, we could also conceivably define a phase trajectory of what a system is, that is, how the system itself changes over time, as opposed to its behavior. This would be a Level 2 process.

There is also conceivably the notion of the phase space and phase trajectory of a model, which defines how the model changes over time. This would be a Level 3 process. In fact, the SE model space may perhaps be viewed as the phase space of models of SE systems.

Phase Transition

In thermodynamics, phase transition is the sudden change in a system from one phase to another, classically transitioning between solid, liquid, and gas forms. In change terms, phase transition is a Level 1 change (in state) that results in a Level 2 change (in form).

Self-Organizing Processes

Self-Organization

Self-organization is a process in which the internal state or structure of a system tends to become more organized without external direction or control. The concept has been applied to a variety of disciplines and systems, both natural and artificial. The issue of measuring the degree of organization is not entirely resolved. Polani (2003) argues that it cannot be measured as an intrinsic characteristic of a system and needs the addition of an observer — analogous to the case for emergence, which is often associated with self-organization.

Self-Organized Criticality

Self-organized criticality (SOC) (Bak, Tang, and Wiesenfeld 1987) is in one sense related to both phase transition and self-organization. The criticality corresponds to the critical point of a phase transition. If this critical point is an attractor, in the dynamical systems sense, then the system is drawn to this point in a self-organizing manner. However, there is some debate on the degree to which SOC represents self-organization (Vespignani and Zapperi 1997).

SOC is contrasted with highly optimized tolerance (HOT) as alternate views in Chapter 10.

Panarchy

Panarchy (Gunderson and Holling 2002) is a comprehensive process intended to scale up to resilience and sustainability at the level of interacting ecosystems, economies, and societies. The broad subject area is the interaction between people and nature. The word panarchy was picked to contrast with control hierarchies and to incorporate the concept of time and cycles. Three themes are brought together in this process:

1. Stability, change, and resilience
2. Cross-scale interactions across a wide range of scales, including the time scale and also most of the scale views we discussed in Chapter 1
3. Adaptive change and learning: slow cycles interspersed with rapid change cycles

This process embraces the tensions and paradoxes of exploitation and conservation, destruction and birth, release and reorganization, and stability and change. The integrated elements of this process are fully consistent with the MOSES approach to extension and unification of traditional and complex systems. The panarchy model could be a significant contributor to the SE processes of the future.

Models at Runtime

A movement addressing models at runtime has emerged. A primary theme is extending modeling into runtime execution of software and information systems. The motivation is the change occurring in runtime expectations of software. Software is more dynamic, and more decisions and changes are made at runtime. In MOSES terms, this is part of the blending of Level 1 and 2 changes where the system changes during operation.

Software needs to be adaptable, flexible, and reconfigurable not just in the design environment, but in the runtime environment. In addition, more software autonomy, as in self-managing or autonomic functions, is expected. Self-awareness and reflection are also important characteristics. Model-driven approaches have strengthened design and production of software, but modeling support during runtime is weak. That is what this movement is attempting to remedy.

Summaries of the first two workshops of this group are in Bencomo, Blair, and France (2006) and Bencomo, France, and Blair (2007).

Models of Change Involving Levels 2 and 3

Example

Level 2-3 change involves change in a model (L3) that is intimately associated with change in individual systems (L2). The model plays the role of category and the individual systems play the role of instances of that category. L2-3 change is the locus of biological evolution. An example will help reveal the simplicity of the L2-3 concept.

Let's consider the following scenario of models and populations in the domain of human anatomy. First, assume that essential anatomy of humans on Earth has not changed during the millennium between years 1009 and 2009. Assume also that humans will survive in some form for the next million years, and essential changes to human anatomy will occur over that period, that is, between years 2009 and 1002009. Further assume that the actual process of biological evolution of populations on Earth is susceptible to change, but that during the period 1009 to 1002009 it does not change.

Now define four models:

> A = the model of human anatomy defined by the community of human anatomists at any given point in time.
> FA = the high-fidelity model of human anatomy at any given point in time, that is, the model that reflects actual human anatomy at that time.
> C = the model of biological evolution defined by the community of evolutionary biologists (or the equivalent) at any given point in time.
> FC = the high-fidelity model of biological evolution at any given point in time.

Table 9.3 shows these models at several points in time.

Given the assumptions above, what can we say about the differences between these models and what levels they represent? The relevant points include:

1. FA1, FA2, and FA3 are all the same but FA4 is different because evolution of anatomy occurred in the human population from 2009 to 1002009. It is an L2-3 change because it is a model change (L3) based on change in the population of individuals (L2).
2. A1, A2, and A3 are all somewhat different; that is, although actual human anatomy did not change during this period, our understanding and model of it did. These changes therefore represent L3 — change in model but not in systems.
3. A4 differs from A3 for two reasons. First anatomists presumably learned more about anatomy, which is L3 change. Second, anatomists tracked the change in human anatomy over the course of the million years, so that change is an L2-3 change.

Table 9.3 Sample Model Changes across Time

		Year			
Model	*Change Level*	*1009*	*1909*	*2009*	*1002009*
FA	L2-3	FA1	FA2	FA3	FA4
A	L3 (L2-3)	A1	A2	A3	A4
FC	L2-3	FC1	FC2	FC3	FC4
C	L3	C1	C2	C3	C4

4. FC1, FC2, FC3, and FC4 are all the same. No change in the essential evolution process occurred across this time period.
5. C1, C2, C3, and C4 are all different. C1 was the model of change in human anatomy in 1009, which was something like this: human anatomy does not change over time; it is constant. C2 represents essentially Darwin's theory of evolution. C3 represents the current theory of evolution, which is more or less the modern synthesis that clarified the genetic role in variation (Futuyma 2005). C4 will reflect whatever additional learning takes place of the process of evolution over the next million years. All these changes are to the model only (L3), not to the evolution process itself, which we assume does not change during this period.
6. The change from C1 to C2 was probably greater than the change from A1 to A2 because, although our knowledge of anatomy increased significantly, the change probably was not as radical as that represented by Darwin's theory.

This example illustrates the distinction between an L3 model change, which is not associated with system change; and an L2-3 change, which is model change based on system change. In this case, L3 change was based on learning more about the systems being modeled, that is, about increasing fidelity of the model.

Applying Biological Evolution

Evolution has been a dominant and growing topic in SE in recent years, much more than other biological models of change. We explore in this section a definition of biological evolution and how the term "evolution" is used in SE.

Evolution: Definition and Introduction

Biological definition: Biological evolution, according to Futuyma (2005, 2 and 270), is:

> "change in the properties of groups of organisms over the course of generations. The development, or ontogeny, of an individual organism is not considered evolution. Individual organisms do not evolve. Groups of organisms, which we may call **populations**, undergo *descent with modification*. … The changes in populations that are considered evolutionary are those that are passed via the genetic material from one generation to the next. Biological evolution may be slight or substantial: it embraces everything from slight changes in the proportions of different forms of a gene within a population to the alterations that led from the earliest organism to dinosaurs, bees, oaks, and humans. … The 'genetical theory of natural selection' … is the keystone of contemporary evolutionary theory. … Natural selection is not the same as evolution. Evolution is a two step process: the origin of genetic variation by mutation or recombination, followed by changes in the frequencies of alleles and genotypes, caused chiefly by genetic drift or natural selection."

Engineering definitions: In contrast to biological evolution, in engineering a common meaning of evolution is change in a given individual system, rather than in a population. In MOSES terms, this type applies to the Level 2 change of an individual system. This meaning of evolution is reflected in phrases such as "This system evolves over time." Biology uses terms such as "adaptation," "growth," "development," and "maturation" for this type of individual change. Because this usage of evolution is established in engineering, the goal at this point is to promote recognition of both meanings of the term and making the distinctions explicit: (1) evolution is gradual change in an individual system or entity, and (2) evolution is gradual change in a *population* of individual systems or entities. Which definition is being used in a given situation should be made clear.

The remainder of this subsection discusses definition (2).

MOSES perspective on definition (2): In the context of SE and the issue of change, the MOSES position is that it makes sense to designate a population as a *category,* with its members as instances of the category. Each biological population has a defined set of characteristics that holds for all its instances. So, evolution of a population is change in a category, which is a change in a general model — a Level 3 change. And because it is intimately associated with the change of the population individuals, it is designated as L2-3.

Define the Unit of Selection

Evolution can be applied to any domain in which units of selection are defined and they exhibit variety on which selection operates. Baldwin and Clark (2000) discuss the application of evolution in this sense to the general area of design. Units of selection in design evolution can include artifacts, systems, designs, and task structures. Units of selection in industry evolution can include enterprises and organizations. In modular systems, the unit of selection does not have to be the whole system. It can be a subsystem or component anywhere in the composition hierarchy. This provides greater granularity in selecting fit elements over unfit.

In some applications, the unit of selection is a behavior of a system. The behaviors represent a variety, and certain behaviors are selected and others are not. If we consider the individual behaviors to change, or the frequency of behaviors to change in the population of behaviors, then we could call this an L2-3 change. We should note that while this is a reasonable application of evolution, it is not clear that this is more useful than the simpler biological models that exist at L2, such as adaptation or maturation of the system, or even L1 behavior of the system.

Applications

One clear and appropriate way to apply evolution to SE is to use general models that represent categories. A prominent example is types or generations of technologies. A perspective that helps explain this type of application is that of evolution in a changing environment. If the environment changes too rapidly, the population cannot evolve quickly enough and it dies out.

In technology and SE, the environment sometimes changes quickly. When the Internet emerged in the 1990s, it was a rather rapid and significant change for individuals, and a somewhat more gradual change for organizations. Many individuals quickly adapted and exploited the new environment, but some did not. After a period of time, if organizations did not adapt, they were probably at a disadvantage. Now there is another generation of technologies called Web 2.0, such as blogs and wikis. In the context of organizations, some call this Enterprise 2.0. This concept is not gaining traction in organizations as fast as expected (Church 2008). Based on anecdotal evidence, there is a generational difference in acceptability of Enterprise 2.0 technologies: the older generation of people is more content to stay with first-generation Internet technologies such as email, while the newer generation is using the new technologies more.

In SE, evolution in populations such as organizations can occur in several ways. One is by adaptation of the individual members to a new environment, which is what happened, to a large extent, with the Internet. Another is by replacement (e.g., through retirement and hiring) of members who have not adapted by new members who already fit the new environment, which apparently is happening to some degree with Enterprise 2.0. A defined process of applying evolution to organizations is presented in Axelrod and Cohen (2000) and discussed further in Chapter 13.

The same pattern exists with more traditional systems. If the environment of a system changes in a way that requires significant changes to the system, an analysis needs to be done to compare the options of changing (adapting) the system to the new environment, or replacing it with a new system that has a better fit.

The more general pattern is that generations of technology represent populations, and systems using or representing those technologies are individuals. Technologies evolve via selection of systems.

Scientists and scientific theories have exhibited the same tension, the same pattern. Max Planck (1936) expressed the pattern this way in 1936:

> An important scientific innovation rarely makes its way by gradually winning over and converting its opponents ... What does happen is that its opponents gradually die out and that the growing generation is familiarized with the idea from the beginning.

In fact, the pattern in science includes the phenomenon where even a scientist who proposes a new theory cannot accept it, while the newer generation in the field embraces it. A well-known example is Einstein, who made substantial contributions to quantum theory but could not accept many aspects of the theory, including some of his own implications (Isaacson 2008, 330; Stone 2005).

Distinctions, Differences

Some distinctions between the evolution process in biology and in SE are discussed here.

Inheritance and its role in evolution: In biology, there is a strong genetic link from the old to the new through the inheritance chain. No individual system (organism) is simply created with a whole different set of genes. Everything is linked through common descent. But in engineering, there is a much cleaner break in many cases from one technology generation to the next. There are many cases of inheritance in families of products, such as variants of Unix operating systems or Windows, or much of the technology for aircraft or cars. But many new technologies are more accurately modeled as a clean break, a whole new category that appears suddenly.*

An inheritance process does exist in SE, mostly in software engineering, information systems, and conceptual modeling, and it maps partially to biology. Both biology and software have two types: (1) individual and (2) category or taxonomy.

1. *Individual:* In biology, inheritance in much of the animal kingdom is the mixing of genes from two parents to produce a new individual similar to the parents. For many forms of life, such as bacteria and viruses, inheritance is through cloning of a single parent. In software, objects at runtime can be cloned from another object.
2. *Taxonomy:* In both biology and software, kinds in the generalization hierarchy inherit from parent kinds. For example, in biology, dog inherits from mammal; while in software, a subclass inherits from a class.

* We can model it as appearing suddenly even though innovations are often applications of old ideas to a new domain.

In biology, mutations occur when copying the genome, which is one source of the variation necessary for evolution. The counterpart in software is genetic programming.

Artificial evolution: There is a research community exploring artificial evolution, which is related to artificial life. This field defines and uses models of agents, fitness landscapes, selection criteria, etc. According to Ducoulombier and Sebag (1998), artificial evolution usually considers fixed environments or fitness landscapes. For application to the future of SE, this limitation would need to be revisited.

Conclusions: Needed Change Models

This section briefly characterizes the suite of change models needed by the future SE, building on the preceding sections. First, it is extremely important to include models of change in the arsenal of SE of the future. We believe this area is quite weak in traditional systems engineering. Biological evolution is one of a number of biological models of change. SE needs to examine others as well. The appropriate selection of biological models of change, and appropriate ways to apply them and tailor them to SE at various levels, is a subject for further research.

We can, however, describe certain characteristics of the needed suite of change models for SE. The suite includes models at all rates from very gradual change over time to sudden change. It includes models of self-change, such as self-adapting, self-modifying, self-organizing (e.g., Holland 1995; Kauffman 1993), growth, decline, and termination or death. It also includes models of change induced by an external agent, such as a designer or controller or upgrader or teacher. It includes purposeful change, accidental change, and influenced change.

It also includes desired change and undesired change, as shown in Table 9.4. In general, a given situation has both forces for change and forces for stability. Sometimes we desire change and sometimes we desire stability. For example, if we are engineering a new system, we — or the person requesting the new system — desire change of some type. If we desire change, we should exploit the forces for change and counter the forces for stability. In many cases of organizational change, some people desire the change, while others do not. Those desiring change will exploit the forces for change, such as the difficulties caused by the current situation, and those opposed to change will exploit the forces for stability, such as cost, difficulty of change, inertia.

In keeping with the modular nature of MOSES, it is important to emphasize that change is decoupled from the value or desirability of the change, as stated in Chapter 4. The analysis and modeling of change is one activity. A separate activity is the designation that the change represents progress or decline, or is desired or undesired. Table 9.4 represents a state in which such a designation has been made. Evolution is decoupled from valuation of the change that occurs.

Table 9.4 Desired and Undesired Change Actions

	Forces	
Desired	*Change*	*Stability*
Change	Exploit	Counter
Stability	Counter	Exploit

In selecting models for systems engineering, and especially in the context of MOSES, we emphasize a point made in earlier chapters — we want models that are useful; we are not so concerned with whether they are true. This should be kept in mind when we are examining biological models for their usefulness to SE. The concern of systems engineers is about the fidelity of biological theories or models to the socio-technical systems we build.

An important aspect of this argument is that biological models, and CS models in general, often have greater fidelity to many traditional systems than do mechanistic models, as argued in Chapter 4. That claim is an element of MOSES unification.

Time and Change Related to Other Dimensions

We have mentioned some connections of time with other dimensions in earlier sections in this chapter. Here are a few more connections.

Dual Hierarchy

The concept of dual hierarchy has been discussed in several places in the book. It occurs in the composition, commonization, and conceptualization dimensions of the model space. Is there an analogous duality in the time dimension? The answer is yes, and Stephen Hawking (1988, Chapter 2) has provided a physics version of it that we will use as the basis for our more general SE version.

Figure 9.8 shows the basic structure of the dual hierarchy. The present is the point in time that connects and separates the past and future spaces. More specifically, the present is the current state of a system, the past is the set of states that could lead to the current state, and the future is the set of states to which the current state could lead.* All other states lie in the "elsewhere" of the current system state — they cannot lead to the current state, and the current state cannot lead to them. The figure could have several interpretations. Two of them are:

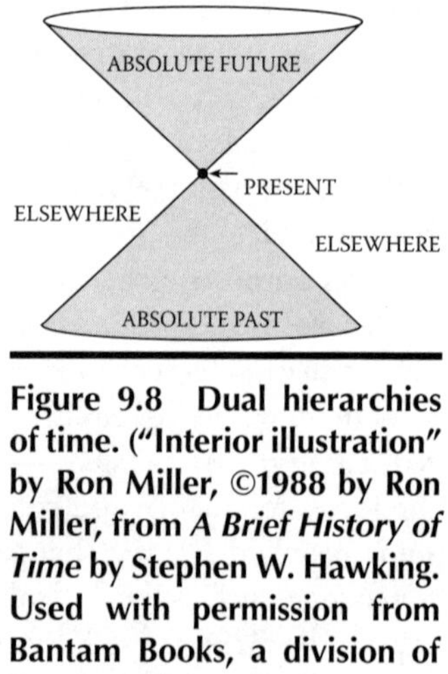

Figure 9.8 Dual hierarchies of time. ("Interior illustration" by Ron Miller, ©1988 by Ron Miller, from *A Brief History of Time* by Stephen W. Hawking. Used with permission from Bantam Books, a division of Random House, Inc.)

1. *Trajectory potential view:* The past, present, and future represent a snapshot of the constrained phase space of the system, where the constraint is the current state:
   - Past = all possible previous states and trajectories of system that lead to the present state.
   - Future = all possible future states and trajectories of system reachable from the present state.

   This interpretation is the view, along with the complete phase space, in which we define and analyze models of system behavior.
2. *Trajectory knowledge view:*
   - Past = actual previous states and trajectory of system.
   - Present = current state of system.

* Hawking's description (1988, 25): "The absolute future of [an event P] is the region inside the future light cone of *P*. It is the set of all events that can possibly be affected by what happens at *P*."

- Future = actual future states and trajectory of system.

  The constraint here is the amount of knowledge of past, present, and future: what do we know about the actual past states and trajectory that led to the present state; and about the present state; and about the actual states and trajectory that lead from the present state to the future?

  This is the view in which we define and analyze systems in terms of actual past states and predicted future states. It would be reasonable to suppose that we have knowledge of the past and certainly of the current state of a system, but in general we only have partial knowledge of both the past trajectory and the current state, as well as the future, especially of large or complex systems. A few examples:

  - *Past:* (1) Historians analyze information about the past, for large systems such as societies, to determine the past states and the trajectory that led to the current state. (2) Detectives and jury trials analyze past states to determine if a defendant actually committed a crime. If the defendant can provide convincing evidence that he was in the elsewhere of the crime event — for example, he was in a physical location different from the crime scene — then the jury concludes that he could not have committed the crime. (3) Systems engineers analyze a failure of a system to determine the cause or the circumstances that led to it.
  - *Present:* What is the current state of an organization, a power grid, an economy, or even a machine? We typically have varying degrees of knowledge of the current state. Many times we gain more knowledge of a current state at a later point. We can recognize speculative bubbles in a sector of the economy much more clearly after the bubble has burst than while it was building. We can recognize that a machine was wearing out after it stops working.
  - *Future:* All of us to some degree forecast what will happen in the future. Systems engineers predict budgets, costs, future states of a system, future states of the environment(s) in which a system exists or will exist, operating performance of a system being developed, etc. All predictions are based on partial knowledge.

Other Relations

All the dimensions have some variation of static views and dynamic views. In the *composition* dimension, a static view is that a contained part can only be in one whole; a dynamic view is that a contained part can be in multiple wholes consecutively, at different times. A shared part can be in multiple wholes either consecutively or simultaneously (in parallel).

- *Commonization:* individual can be an instance of multiple categories simultaneously or consecutively. A kind can be a subclass of multiple classes simultaneously or consecutively. Regions of models can be considered as space-time regions: they vary across space at a given point in time; and they vary across time at a given point in space.
- *Conceptualization:* languages, semantics, and meaning change over time. Ontologies typically include time-based elements such as continuants (streams) and occurrents (events).
- *View:* clearly there are static and temporal or dynamic views, and space-time views.

Temporal logic (Galton 2008) is important in specification (Chapter 11), especially verification, and in automation. A view that is important to SE and MOSES is that temporal logic is a subset of modal logic. Temporal logics and their uses include linear temporal logic for specifying and verifying serial processes, and computational tree logic for specifying and verifying parallel processes.

References

Axelrod, R. and M. Cohen. 2000. *Harnessing Complexity: Organizational Implications of a Scientific Frontier*. New York: Basic Books.

Bak, P., C. Tang, and K. Wiesenfeld. 1987. Self-organized criticality: an explanation of 1/f noise. *Phys. Rev. Lett.*, 59:381–384.

Baldwin, C. and K. Clark. 2000. *Design Rules. Volume 1: The Power of Modularity*. Cambridge, MA: MIT Press.

Bencomo, N., G. Blair, and R. France. 2006. Summary of the Workshop Models@run.time at MoDELS 2006. http://www.comp.lancs.ac.uk/~bencomo/MART06/modelsatruntimeSummary.pdf. (Accessed January 11, 2009)

Bencomo, N., R. France, and G. Blair. 2007. 2nd International Workshop on Models@run.time, in *Lecture Notes in Computer Science, Satellite Events at the MODELS 2007 Conference*, Springer-Verlag. http://www.comp.lancs.ac.uk/~bencomo/MRT07/modelsatruntime07Final.pdf (Accessed January 11, 2009)

Church, Z. 2008. Enterprise 2.0: On Its Way but Hardly There. http://searchcio-midmarket.techtarget.com/news/article/0,289142,sid183_gci1317448,00.html?track=NL-1014&ad=646456&asrc=EM_USC_3893588&uid=1104791#. (Accessed January 11, 2009)

Dörner, D. 1989. *The Logic of Failure: Recognizing and Avoiding Error in Complex Situations*. Germany: Rowohlt Verlag. (English translation: 1996. New York: Metropolitan Books.)

Ducoulombier, A. and M. Sebag. 1998. Continuous mimetic evolution. *Proc. Eur. Conf. Machine Learning (ECML)*, 1998:334–345.

Futuyma, D. 2005. *Evolution*. Sunderland, MA: Sinauer Associates.

Galton, A. 2008. Temporal Logic. *Stanford Encyclopedia of Philosophy*. http://plato.stanford.edu/entries/logic-temporal/. (Accessed January 11, 2009)

Girault C. and R. Valk. 2003. *Petri Nets for Systems Engineering: A Guide to Modeling, Verification, and Applications*. Berlin: Springer-Verlag.

Gunderson, L. and C. Holling. 2002. *Panarchy: Understanding Transformations in Human and Natural Systems*. Washington, D.C.: Island Press.

Hawking, S. 1988. *A Brief History of Time: From the Big Bang to Black Holes*. New York: Bantam Books.

Holland, J. 1995. *Hidden Order: How Adaptation Builds Complexity*. Boston, MA: Addison-Wesley.

Isaacson, W. 2008. *Einstein: His Life and Universe*. New York: Simon and Schuster Paperbacks.

Kauffman, S. 1993. *The Origins of Order: Self-Organization and Selection in Evolution*. New York: Oxford University Press.

Maturana, H. 1975. The organization of the living: a theory of the living organization, *Int. J. Man-Machine Studies*, 7:313–332.

Maturana, H. and F.J. Varela. 1980. (1st edition 1973). *Autopoiesis and Cognition: The Realization of the Living*. In Robert S. Cohen and Marx W. Wartofsky (Eds.), *Boston Studies in the Philosophy of Science* 42. Dordrecht: D. Reidel Publishing Co.

Mayr, E. 1982. *The Growth of Biological Thought: Diversity, Evolution, and Inheritance*. Cambridge, MA: Harvard University Press.

Meyer, B. 1997. *Object-Oriented Software Construction*. Englewood Cliffs, NJ: Prentice Hall.

Nann, B. 2004. Static vs. Dynamic Model Impact on Predictive Performance in Early Stage Collections. Fair Isaac White Paper.

Novikov, I. 1998. *The River of Time*. Cambridge, U.K.: Cambridge University Press.

OBO. 2008. Biological Process Ontology Guidelines. Open Biomedical Ontologies. http://www.geneontology.org/GO.process.guidelines.shtml?all. (Accessed January 11, 2009)

OMG (Object Management Group). 2007a. Unified Modeling Language (OMG UML), Superstructure, V2.1.2. OMG Document Number: formal/2007-11-02 Standard document URL: http://www.omg.org/spec/UML/2.1.2/Superstructure/PDF. (Accessed January 11, 2009)

OMG (Object Management Group). 2007b. Unified Modeling Language (OMG UML), Infrastructure, V2.1.2. OMG Document Number: formal/2007-11-04 Standard document URL: http://www.omg.org/spec/UML/2.1.2/Infrastructure/PDF. (Accessed January 11, 2009)

OMG (Object Management Group). 2007c. Systems Modeling Language (OMG SysML™), V1.0. OMG Document Number: formal/2007-09-01 Standard document URL: http://www.omg.org/spec/SysML/1.0/PDF. (Accessed January 11, 2009)

Parnas, D. 1972. On the criteria to be used in decomposing systems into modules. *Commun. ACM*, 15(12): 1053–1058.

Planck, M. 1936. *The Philosophy of Physics* (quoted in http://www.calresco.org/lucas/cas.htm). (Accessed January 11, 2009)

Polani, D. 2003. Measuring self-organization via observers, in W. Banzhaf et al. (Eds.), *Advances in Artificial Life, 7th European Conference (ECAL 2003), p. 667–675.*

Stone, A. D. 2005. Einstein's unknown insight and the problem of quantizing chaotic motion. *Physics Today*, 58(8):37–43. http://www.eng.yale.edu/stonegroup/publications/phys_today.pdf. (Accessed January 11, 2009)

Tang, C. and R. Epstein. 2007. A Structural Split in the Human Genome. PLoS ONE 2(7): e603. doi:10.1371/journal.pone.0000603. http://www.plosone.org/article/fetchArticle.action?articleURI=info%3Adoi%2F10.1371%2Fjournal.pone.0000603. (Accessed January 11, 2009)

Vespignani, A. and S. Zapperi. 1997. How self-organized criticality works: a unified mean-field picture. *Phys. Rev.* E, 57(6):6345–6362.

Wikipedia. 2008. Lorenz Attractor. http://en.wikipedia.org/wiki/Lorenz_attractor. (Accessed January 11, 2009)

Chapter 10

Views

> [Progress] has been made recently in finding "dualities" or correspondences between apparently different theories of physics. These correspondences are a strong indication that there is a complete unified theory of physics, but they also suggest that it may not be possible to express this theory in a single fundamental formulation. Instead, we may have to use different reflections of the underlying theory in different situations.... This would be a revolution in our view of the unification of the laws of science....
>
> **—Stephen Hawking**
> *Foreword, The Illustrated A Brief History of Time.*

The concept of view is a general and important tool in engineering in general and MOSES in particular. A view allows us to focus on an area of concern and ignore everything else. This chapter gives a definition, examples, and types of views, and describes some areas of further research.

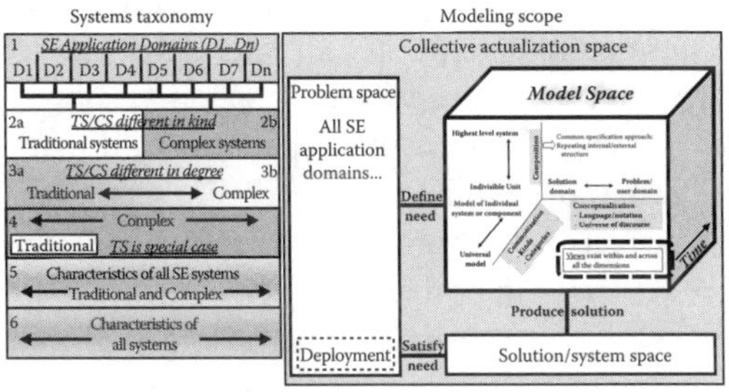

Introduction: Intuitive Concepts

In some important respects, the idea of a view as commonly understood is consistent with its use in MOSES. A view, according to the dictionary, is the extent or range of vision, a manner of looking at or regarding something. Informally, a view is defined in MOSES as a subset of information observed of one or more entities. The entities are collectively called the target or object of the view. The physical or conceptual point from which a view is observed is the viewpoint, which can be motivated by one or more observer concerns. Different views of the same target can be observed from different viewpoints.*

* These notions are defined more formally later in the chapter.

The concepts of view and model have much in common. So, if view is commonly understood, and similar to model, why do we need a chapter on views in this book? The answer is that we believe the potential benefit of the concept of view in SE goes well beyond what has been exploited to date, and we want to encourage more focus on this important concept.

We believe view is a basic, universal concept that is simple, powerful, and flexible. To some degree it is a systems engineering/science analog of the concept of set in mathematics, as a general foundational tool. The traditional architecture profession has long been using views in designing buildings, separating aspects such as floor plans, electrical systems, plumbing systems, heating/cooling systems, and the like. In the past few decades, the information systems community has embraced the concept of views, for example in CAD/CAM tools, software architecture, enterprise architecture frameworks, and software aspects. MOSES considers the concept of views to be applicable to the full scope of SE, including traditional and complex systems, natural and artificial systems, and supporting sciences and mathematics.

The definition indicates that a view does not exist in isolation. It exists with respect to what is potentially viewed, its target or object, and it is envisioned as one of potentially multiple views of the same target. The target can be more or less any ontological or conceptual element in SE, including another view, a region, a specification, or a collection of things, but most often in our discussions the target will be a model or a system, analogous to the notion that a model has a target system. A view is an abstraction of its target. A view can be either a model or a specification, and hence can be either explicit or (partially) implicit.

One important motivation for views is that, in general, we cannot see or understand everything of interest about a system in one visualization or representation. This is the sense in which Hawking (1996) discussed the constraints on how a unified theory of physics may eventually be expressed (see quote at the start of this chapter). Another motivation, which is serendipitous with the first, is that most stakeholders are not interested in everything about a system, but rather the things or features related to their concern. The first motivation introduces the need to separate the information about a system into views; the second motivation suggests how to separate, that is, the criteria for each view.

We should stress that separating information into views carries no constraint to partition the information into mutually exclusive sets. Different views of the same target typically overlap; the same piece of information may appear in multiple views. In fact, there are very few constraints on what views can be defined or the contents of any given view. The only theoretical constraint is that the view be an abstraction of the target. The only practical constraint is that the view includes the information related to the concern that motivated its definition.

Two important requirements for a view are separation and integration. These sound somewhat contradictory, but both must be true. It is self-evident that a view must be separable from the rest of the information about its target. That is, after all, the primary motivation for defining a view. However, a view must also be integrated with the rest of the information about its target. View integration requires a bit more discussion.

The concept of view integration is slightly different from what we typically think of as integration of a system or of multiple systems. That is, integrating the *views* of a system is different from integrating the *parts* of a system. One difference is that parts are distinct and interacting, while views are overlapping and largely non-interacting. In a minimal sense, a given piece of information must be consistent in all views that include it. (Not necessarily the same, but consistent. Sometimes a given piece of information appears in a different form in different views.)

In a broader sense, the information included in all views must form a consistent integrated coherent whole because the total set of information about the target/system must be consistent and coherent. Therefore, in a sense, the integration of views is more about the integration of the target information included in those views than it is about the direct interaction among the views. View integration is not an issue when the system information is defined first, and then views are defined based on that information; view integration in that case is automatic. A classic example of that case is a view defined in a database management system, in the domain of information systems engineering. But often a system is defined by a set of views, that is, the views are the vehicle for defining the system. That case is where view integration is an issue, and care must be taken to ensure the integration and coherence of the full set of information.

After discussing features of views, we return to the comparison of view and model. A view is like a model, in that both are abstractions. In fact, formally a view is a model. Further, a model can have multiple views, and a view can have multiple models. But by convention there are distinctions in how we use them, as we have been discussing. A view is potentially a member of a set, an acknowledgment that it represents only part of the story, not the whole story, and that it must be simultaneously separated from the rest of the story and integrated with the whole story. A model has more of a sense of a stand-alone entity.

The remainder of the chapter builds on these preliminary concepts with numerous examples, along with further concept development and definition. The scope includes a general context of SE along with both traditional and complex systems.

Variety: Engineering and Beyond

This section provides examples of views, first in a broader context and then within SE. The examples are used to develop and illustrate categories of views and characteristics of views.

Views in Broader Context

This section presents some views that go beyond traditional engineering. The reason they are included here is twofold. First, they illustrate the breadth and usefulness of the concept of view. Second, many of these examples in some way support the scope of MOSES, which goes beyond traditional engineering to include the characteristics of complex systems, systems of systems, organizations, and enterprises, as well as aspects of science.

A Universal Issue: Addressing the Hierarchy Problem

Hierarchy is an important and useful classification structure but efforts to define a hierarchy encounter a common problem: multiple competing hierarchies are possible. What is the right one?

Common proposed solutions include (1) pick one hierarchy and use it; or (2) do not use a hierarchical structure. In many situations, neither of these is a satisfactory solution. The problem with solution (1) is that no single hierarchy satisfies all situations. The problem with solution (2) is that hierarchy is an important modeling structure; depriving ourselves of this tool is not a good idea.

So what do we propose instead? The answer is views. That is, allow a set of hierarchies to be defined, where each one serves some purpose or addresses some concern. Treat each one as a different view. This solution addresses both problems mentioned above. One of the primary elements of the views solution is that it removes contention about which is right, and about issues of control. There are two variants of the views solution. The first variant is where all views are treated equally; none is dominant. The second variant is where one view is designated as the primary structure and the other views are considered secondary or supplemental.

Two examples are

1. *Philosopher ontologists* debate what occupies the second spot of the ontology hierarchy or the hierarchy of being. The top is "being" or "entity." "Some schemes take the division between universals and particulars as the next step in the hierarchy, others the division between abstract and concrete entities." (Honderich 1995, 634) The argument is based on the assumption that there is one right answer and we need to find out what it is. Views help resolve this. In some contexts, the universals versus particulars distinction is applicable, while in others, abstract versus concrete is more applicable. The two can be seen as alternative views.
2. *Organizational structures:* We typically think of organizations such as companies or government agencies as requiring a dominant view, usually called the line organization. Traditionally this has been hierarchical. Other views, such as project management structures, temporary teams, or informal interaction structures, are defined in contrast to the dominant line organization, and are in important respects subservient to it.

Scientific Disciplines

Each science offers a different view of the world. Examples are

- *Physics* view of the world: matter, energy, forces, ranging from the smallest elements (particle physics) to the largest systems (astrophysics).
- *Chemistry* view of the world: molecular scale of matter, compositions of atoms such as molecules, crystals, and metals, and their transformations and interactions.
- *Biology* view of the world: life, from cells to organisms to ecology.
- *Psychology* view of the world: human brain, individual human perceptions and behavior.
- *Sociology* view of the world: human interaction, groups of humans, societies.

Each of these disciplines may be considered a different view of the same defined world, which is more or less planet Earth as seen and understood by human beings on the planet over the past few thousand years. The world is an integrated whole; we have separated our understanding of it into different views or disciplines to get more traction in each view. A prominent motivation for systems science was the increasing specialization and separation of disciplines, and the lack of integration of knowledge into a unified whole.

Are Science and Mathematics Rational and Logical?

Over the past few decades there have been debates about the basis of mathematical proofs and scientific theories (including Kuhn's (1970) scientific revolutions). Argument 1: both follow a rational process based on logic (proofs) or empirical evidence (science). Argument 2: both follow a social process where the result is accepted or rejected by popular consensus, not rational thought; errors occur and if the examination and discussion is active, the errors are discovered and corrected over time.

The MOSES position is that the two arguments are not mutually exclusive, but rather are different views, both of which are correct and incomplete. They emphasize different aspects of the disciplines. Both science and mathematics are a human process, and therefore subject to psychological, social, and political considerations. Both are also the product of rational and logical thought. The complete picture is an interplay between the two views (and possibly additional views).

Courtroom Trial

At least three views are presented in a courtroom trial: an evidence view and two different views of the evidence view. The testimony and exhibits presented in a trial constitute an abstraction of an event — an evidence view of the event. The defense offers one view or interpretation of the evidence view-*innocent*. The prosecutor offers the opposite view of the same evidence view-*guilty*. The jury deliberates and decides whether to agree with the defense view or the prosecutor view.

Clearly, the two conflicting views are separated. The prosecutor view is generated by selecting and emphasizing the parts of the evidence that most strongly suggest guilt, and then adding interpretation to that evidence to reinforce the guilty view. The defense view is generated by selecting and emphasizing the weakest parts of the evidence, and then adding interpretation to that weakness to reinforce the innocent view. But are the two opposite views integrated? Yes, the integration is the evidence view on which both are based.

Universal Distinction between Discrete and Continuous Phenomena

To start with, the distinction is itself a view of the world, in that of all the properties and characteristics of things in the world, it selects only one to look at: what things are discrete, and what things are continuous? We believe the part of the world we are interested in is one or the other, and sometimes we debate which it is. After all, a given thing can only be discrete or continuous, not both, right?

In many cases, however, it is useful to ignore the question of whether the part of the world we are examining is discrete, or continuous, or both, and simply consider whether it is useful to view this part of the world as discrete, or to view it as continuous, or to allow either view, depending on the circumstances or our purpose. This is fully consistent with a basic tenet of modeling we discussed in earlier chapters: the criterion of a good model is not whether it reflects the world as it is, but whether the model works, that is, is useful for its purposes.

In physics, a classic example of including both discrete and continuous is wave–particle duality. Is the world continuous (waves) or discrete (particles)? It appears to be both, depending on the purpose or viewpoint. From Hawking (1988, 56):

> "The theory of quantum mechanics is based on an entirely new type of mathematics that no longer describes the real world in terms of particles and waves; it is only the observations of the world that may be described in those terms. There is thus a duality between waves and particles in quantum mechanics: for some purposes it is helpful to think of particles as waves and for other purposes it is better to think of waves as particles."

In other words, waves and particles are different models, or views, and both are important in understanding the overall picture. This example previews the duality concept discussed further in the section entitled "Duality, Definitions, Categories."

Basic SE Related

A number of views have been previously encountered in this book. In the model space, each of the dimensions of composition, commonization, conceptualization, and time can represent a view. For example, we could view a system as a multi-level composition structure of repeating whole–part relations. We could define a view of the system at a specific composition level. We could define a view of alternative compositions of a given whole, or a view of alternative environments of a given whole.

Static and dynamic categories of views of a system have a long tradition in engineering. Static views are achieved either by holding time constant or by selecting elements that do not change over the time period in which the view is defined. Dynamic views reflect change over time.

From a different perspective, two categories of views are associated, respectively, with internal and external models. An internal model represents an internal view of a system, or of a class of systems. Further, an internal model or view can be represented in the form of multiple views. Architecture views, discussed in the section entitled "Variety: Architecture Related," are examples. Likewise, an external model represents an external view of a system or class of systems. If that is all we considered in these categories, views would give us nothing in addition to models. But an external model can also be represented in the form of multiple views. For example, possible views of an external model, or possible external views of a system, include:

- The set of all stuff, such as goods, people, currency, or information, that goes into the system
- The set of all stuff that comes out of the system
- The relation between the stuff that goes in and the stuff that comes out
- Externally observable properties and attributes of the system
- Externally observable behavior of the system
- Services the system offers its environment (provided services)
- Dependencies of the system on its environment (required services)
- Characteristics (stuff, properties, provided and required services) of a specific interaction point of the system boundary (i.e., a port)
- Any of the above sets of information represented in one language or representation (versus a different language or representation, which could constitute a different view)
- Any of the above sets of information could be interpreted differently in different environments in which the system might exist; each interpretation could be a different external view of the system

On the subject of environments and interpretation, another example of views is the differing interpretations of an ontology in different contexts. See Obrst and Nichols (2005).

The language and interpretation examples are in the conceptualization dimension. Views also exist in the commonization dimension. For a given entity S, a view can be defined of possible generalizations of S; another view can be defined of possible specializations of S. The biological taxonomy example in Chapter 7 showed four roles of Dog: as a category of Lassie, as an instance of species, as a generalization of Collie, and as a specialization of mammal. Each of these roles is a different view of Dog.

Views are often defined across multiple dimensions. A common view defines the position of a given model in the model space, which, considering time as the fourth dimension of the model space, could be regarded as a four-dimensional view.

Table 10.1 Potential Targets of Nested Views

View	Potential Targets S	V1	V2	V3	V4	C1
V1: A(S)	√					
V2: static(V1)	√	√				
V3: structure(V2)	√	√	√			
V4: components(V3)	√	√	√	√		
V5: ext C1(V4)	√	√	√	√	√	√

Because the target of a view can be a view, views can be nested. Nested views are transitive if the relation on which the nesting is based is transitive. That is, for a transitive relation, if V1 is a view of V2, and V2 is a view of V3, then V1 is also a view of V3. In such cases, the target of a nested view can be any of the views or systems in the "transitive closure" of the nested view. An example will illustrate this. Suppose we have a system S and an internal model that we call the architecture A of S, A(S). Define these views:

- V1 = A(S)
- V2 = static aspects of V1
- V3 = logical component-connector structure of V2
- V4 = set of components in V3
- V5 = external view of component C1 in V4

Then the set of potential targets of each view is shown in Table 10.1. Each check mark for a view means that the view can be a view of the item in that column. Thus, V5 can be potentially be a view of component C1, or a view of system S, or a view of any of the views in between. System S is a potential target of all five views. Selecting a target from multiple potential targets is therefore a matter of choice, usually indicated by the purpose of the view. This possibility of multiple potential targets is one of the reasons MOSES advocates clearly identifying the target of every defined view.

Views of Humans

The example of the general user model ontology (GUMO) (Heckmann et al. 2005) was introduced in Chapter 8, and is now appropriate in this "Views" chapter. The Chapter 8 view was of GUMO as an ontology. The view of this chapter is that GUMO presents a different perspective on people: not just external users, but now an integral part of a system. As noted in previous chapters, the most common TSE view of a person was as a user external to a system. That view provided information on how a person would use the system but included little or no psychological or social interaction characteristics. The related arenas of man-machine interface, and human-computer interface (HCI), and ergonomics have continued to model humans as external to the system, but have further explored human norms and variances in physical comfort, visualization, and perception capabilities, to make machines more human friendly, simpler, clearer, and safer. So these arenas have taken an important step toward expanding models of humans. The extension of SE to include complex systems, enterprise engineering, and systems-of-systems brings humans inside the system and significantly expands our view of humans. We now model a fuller range of human psychological and social characteristics because they are factors in the behavior of humans viewed as complex systems in their own right and as components of larger enterprises and partially automated systems.

At first glance, the GUMO example seems to be a very general person model that goes beyond the needs of engineering. However, the foregoing discussion indicates how SE has broadened its scope to where the GUMO features are now in play. Expanded SE/CSE/Enterprise Engineering (that is, the scope of MOSES) now takes into account how organizations, social interaction, politics, and psychological states affect not only the use of systems, but also their development and acceptance. In that expanded context, GUMO makes more sense.

At the same time, we need to be careful not to swing the pendulum too far the other way; our goal is still to model only what is relevant to our purpose. We do not want to gratuitously add elements or features or properties to our model beyond what is needed. In TSE, when an engineer brings a preexisting component — such as COTS — into a system, he focuses on the external model of that component, not the internal model. When a systems engineer now brings a human into a system as a component, the pattern of focusing more on the external than the internal still has some validity.

The purpose of discussing this example at some length is (1) to illustrate how views of the same target (humans in this case) can change over time, even if the target does not change; and (2) to reinforce the substance and scope of how MOSES views and models humans. Of course this is not the only way MOSES views humans. In some views, an agent can represent either a human or a machine such as a software program; the distinction is deliberately abstracted away for the purposes of such a view.

Development versus Operation of a System

Chapter 9 discussed three levels of processes and change, and Figure 9.2 showed the life span of each level. From the standpoint of views, we are interested in level 1 (system behavior) and level 2 (system change). TSE and CS have different views of the relation between these two levels.

In TSE, development and operation are partitioned and usually treated as separate life-cycle phases. For example, the classical "big bang" model is full development followed by full function operation. In Figure 9.2, this represents L2 change during Formation, followed by L1 behavior during Full function, thus skipping over the Maturation phase. A model commonly used by at the U.S. Department of Defense (DoD) includes initial operational capability (IOC) and full operational capability (FOC), which adds a maturation phase during which the system operates but also changes. In both cases, L2 change (maintenance, sustainment) can occur in a phase when the system is operational, but the development of the change is still kept separate, and the system stops operation (L1) while the change is being installed (L2).

In CS, change during formation is still separate from operation because operation has not started. However, during maturation and full function, L1 and L2 (behavior and change) are often intertwined and in some respects are indistinguishable. The system — for example, a human — learns, grows, and adapts while it is operational. But there is still a useful distinction between L1 behavior and L2 change for all types of systems — mechanistic, organic, and hybrid systems. In MOSES, the distinction is best modeled as views. The L1 behavior view is concerned with what the system does at any point in time. The L2 change view is concerned with changes in the system having to do with growth or repair. Those changes can be discrete or continuous, and self-induced, such as growth or learning; or externally induced, such as an engineer replacing a part on a machine. Typically, self-induced changes are more continuous, while externally induced changes are discrete and periodic. The L1 and L2 views coexist.

Defined Worlds

A defined world is a view. Its target can be the set of all possible worlds, or some more constrained set. A few examples of defined worlds, some of which have appeared in earlier chapters, include:

- Universe of discourse (UD): a defined set of concepts or elements referred to by a language or set of terms; can be the basis of one or more ontologies.
- A system of interest (SOI): Chapter 6 defined five views/models in the context of this world.
- Namespace: a namespace is a defined world of names or identifiers. The requirement for uniqueness holds within the namespace. That is, a name does not have to be unique in all possible worlds, only within its defined namespace.
- Any defined spacetime region, such as a period of time in a given location, or a given organization.

Views of Interacting Systems

Suppose we have a set of interacting systems as shown in Figure 10.1. Multiple views of this situation are possible. We want to discuss and contrast two types of views.

- *View 1*: Input – process – output view of a given system, such as S2 in the figure. This view is common in a pipeline or service chain style where the inputs come from one or more provider systems and the outputs go to one or more consumer systems that differ from the provider systems. Examples: supply chain such as auto manufacturer, consumer goods retailer such as Wal-Mart, the biological food chain, a data processing pipeline, etc.

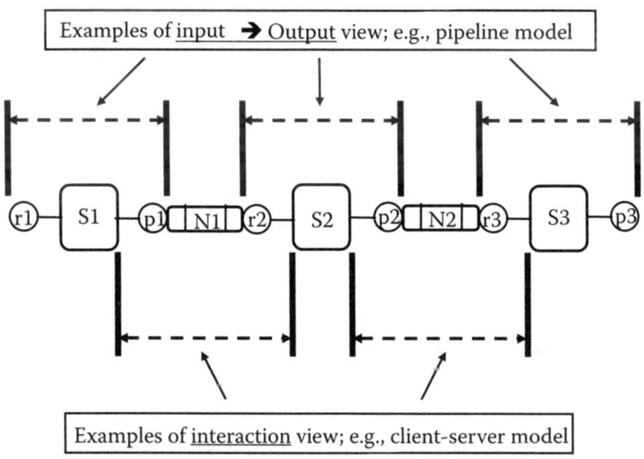

Figure 10.1 Contrasting views of components in interaction.

- *View 2:* Interaction between two or more systems via a connector, such as between S2 and S3 in Figure 10.1. This view is common in a request-reply style or protocol where the service provider gets the inputs it needs from the same system to which it provides its output. Examples: client-server interaction, tax preparer-customer interaction, and software module that calls another module. Chapter 6 presented a number of patterns or categories of this view.

View 1 focuses on a system, and looks at outputs as a function of inputs. View 2 focuses on an interaction, and looks at how a system responds to a request from another system.

SOI and Relative Holarchy

Recall two concepts discussed in previous chapters. First, we said that a holarchy is a repeating structure of wholes and parts. Second, we discussed the idea of SOI, and five models or views associated with a SOI. Now we combine these concepts into the notion of a bounded holarchy as a view, based on relative bottom and top. For this we use the concept of atomic from the foundations part of the Reference Model of Open Distributed Processing (RM-ODP 1996): an entity is atomic at a given level of abstraction if it cannot be subdivided at that level of abstraction.

Suppose we have a system of interest S that exists as a holon in a holarchy, an encompassing system E of which S is a part, and S has parts S1 and S2. Also assume encapsulation of all holons in the hierarchy. Then from the viewpoint of S, the following views exist:

- From the external viewpoint of S, S views E as a closed system, in the sense that S only sees the inside of E; E is the top of the holarchy that S can see.
- From the internal viewpoint of S, S views S1 and S2 as atomic units, in the sense that S only sees the outside of S1 and S2; S 1 and S2 are the bottom of the holarchy that S can see.

In this way, the view of a holarchy from a holon SOI in the hierarchy is a relative and bounded holarchy that extends only one level up and one level down from the SOI.

Example of Perspective: Alexander's Architecture Patterns

Christopher Alexander (1977 and 1979) defined a set of architecture patterns in the 1970s for everything from a room in a house to a town to a region, as discussed in Chapter 5. An enthusiastic audience of Alexander's approach to patterns has been the software engineering community, which has applied his concepts in the form of software design patterns. There are several reasonable views of Alexander's patterns work, including:

- *View 1: model for software design.* The patterns are a model for software design patterns, each of which is a problem-solving and generic design template for software designers to use in tailoring a specific design.
- *View 2: engineering complex (social) systems.* Using knowledge of human and social patterns and proclivities to arrange and influence homes, neighborhoods, and towns to encourage natural organization, interaction, adaptation, and development of a healthy living "organism," something that is alive. This is clearly in the arena of complex adaptive self-organizing systems, which is an area of increasing interest and research in the SE

community. A more subtle aspect of this view is that these patterns are abstractions that capture, or at least attempt to capture, stability in the face of continual change. The rules of the patterns are stable; they foster the evolution and growth of systems (towns, etc.) that support natural human preferences and interactions.

- *View 3: unification of engineering.* Alexander's approach works for both CS and TS. View 1 is more or less an application to TS. View 2, Alexander's application, was a CSE approach to architecture and planning. It also represents multiscale or multilevel engineering that we think works for both TS and CS: The same pattern template and approach is used at all scales (room, house, street, neighborhood, town, region), but the contents are different at each level or scale. This multiscale aspect has not been exploited by the software community (View 1).

Miscellaneous

The set of possible views is virtually unlimited. Here are a few other miscellaneous potential SE views:

- An intensional definition of a function represents a different view from an extensional definition of the same function.
- Different scales can represent different views.
- Internal component types, such as logical versus physical, can represent views.
- A structure of layers can be a view of an architecture; or each layer can be a view.
- Similarities and differences can be views; in some views we want to focus on what is common while in other views on what is different.

Variety: Architecture Related

The engineering community has defined various sets of views in the context of design and architecture. The collection of views that have been defined are of some importance to SE because they illustrate a variety that can serve as a starter set of common views for the future of SE. Therefore, we include a fair number of examples in the survey that follows. At the same time we note that the stance of most of the examples is that, to varying degrees, the definers and proponents regard their set of views as "complete." MOSES has a much more expansive concept of view, which fosters a more open-ended and flexible approach to views and a significantly increased role for views in the future of SE. Therefore, our position on the examples that follow is that they provide an important start on filling the space of potential views, but the SE community has barely begun to explore this space and its huge potential.

Composition Views

TSE tends to focus on differences between composition levels. This is reflected in different specific names for each level. Examples are:

- IEEE 1220-2005 (IEEE 2005, 3): defines the hierarchy of system – product – subsystem – assembly – component – (subassembly) – subcomponent – part.

- Kossiakoff and Sweet (2003, 31–33): Acknowledges that system can apply to any level, but then defines a model of complex systems with these levels: (system of systems) – system – subsystem – component – subcomponent – part
- INCOSE Handbook (INCOSE 2007): System – Element – Subsystem – Assembly – Subassembly – Component – Part.

MOSES, on the other hand, describes in Chapter 6 a repeating pattern of {whole–part} or {system – component} as many levels as necessary. The emphasis is on what is common at all levels, while acknowledging that there are also differences. It scales up or down to any number of levels. In Chapter 7, general mechanisms are defined for describing differences.

Now in the context of this chapter, the full MOSES position on this contrast is described. SE needs to allow and support both similarities and differences at different composition levels. We need to exploit what is common by not duplicating it at every level, and we need to account for what is different by capturing those differences at each level. The mechanism, of course, is view. All composition hierarchies are a combination of two views. In one view reside all characteristics that are common to or the same at all levels of a given composition hierarchy. In the other view reside all the characteristic differences between levels. The view with the common information is analogous to a holarchy or a fractal. The two views could perhaps be called the autopoietic view of composition (similarities) and the allopoietic view of composition (differences). Alternatively, they could be called the similarities view of composition and the differences view of composition.

The reason for including this section is not that it is difficult to model the similarities–differences distinction or to capture the information. The reason is to emphasize the importance of considering, and making explicit, the existence of both similarities and differences in composition hierarchies. It lends clarity and efficiency to composition modeling.

This approach of similarity and difference views also is useful in addressing a specific topic that has received attention over the past decade and that we have mentioned in the book, namely, the concept of system of systems (SoS). Sage and Cuppan (2001) emphasized the differences between SoS and system, stating that an SoS exists when the majority of the following characteristics are present: operational and managerial independence, geographic distribution, emergent behavior, and evolutionary development. Valerdi, Ross, and Rhodes (2007) note that there are three schools of thought on the concept of SoS. The pro-difference school claims fundamental differences between systems and SoS; the second school believes an SoS is simply a system and there is no need for the term (we might call this the pro-similarity school); and the third school is agnostic, not committing to a position until a clearer answer emerges.

The MOSES position is that the level of SoS, like all levels of composition, potentially exhibits both similarities and differences, and the approach of capturing both in views as described above addresses the issue. The pro-difference and pro-similarity schools are both partially correct but the whole picture needs both aspects. There may well be more differences when the Sage-Cuppan characteristics are present but the views approach to this issue is flexible enough to handle those differences.

Design Views

An early example, originally published in 1987, is the *IEEE Recommended Practice for Software Design Descriptions* (IEEE 1998, 6) shown in Table 10.2. In MOSES terms, the first three

Table 10.2 Recommended Design Views

Design View	*Scope*	*Entity Attributes*	*Example Representations*
Decomposition	Partition of the system into design entities	Identification, type, purpose, function, subordinates	Hierarchical decomposition diagram, natural language
Dependency	Description of the relationships among entities and system resources	Identification, type, purpose, dependencies, resources	Structure charts, data flow diagrams, transaction diagrams
Interface description	List of everything a designer, programmer, or tester needs to know to use the design entities that make up the system	Identification, function, interfaces	Interface files, parameter tables
Detail description	Description of the internal design details of an entity	Identification, processing, data	Flowcharts, N-S charts, PDL

Source: IEEE. *IEEE Recommended Practice for Software Design Descriptions.* IEEE 1016-1998. New York: IEEE Computer Society, 1998. Used with permission.

views (decomposition, dependency, and interface) are different views of an internal model of a system — including external models of each component — while the fourth (detail) is an internal model of each system component.

Enterprise Architecture Frameworks

Arguably the most important category of extant views for SE is that of enterprise architecture (EA) frameworks. The purpose of the survey that follows is to focus on the views defined by or illustrated in these frameworks, not a comprehensive description or assessment of the frameworks. The ones surveyed include Zachman, RM-ODP, the U.S. Department of Defense Architecture Framework (DoDAF)/UK Ministry of Defense Architecture Framework (MODAF)/NATO Architecture Framework (NAF) family, The Open Group Architecture Framework (TOGAF), the Federal Enterprise Architecture (FEA), and the Service-Oriented Modeling Framework (SOMF).

Zachman EA Framework

The Zachman Enterprise Framework (Zachman 1987; Sowa and Zachman 1992; Zachman International 2008) is summarized in Figure 10.2. It is a conceptual high-level framework captured in one matrix of six rows and six columns. The rows are categories of models that have traditionally been associated with phases of a system in an enterprise while the columns are cross-cutting categories of models that correspond respectively to What, How, Where, Who, When, and Why. Each cell is a model or set of models that corresponds to both its row and column. For example, row 2/column 1 cell is a business entity model.

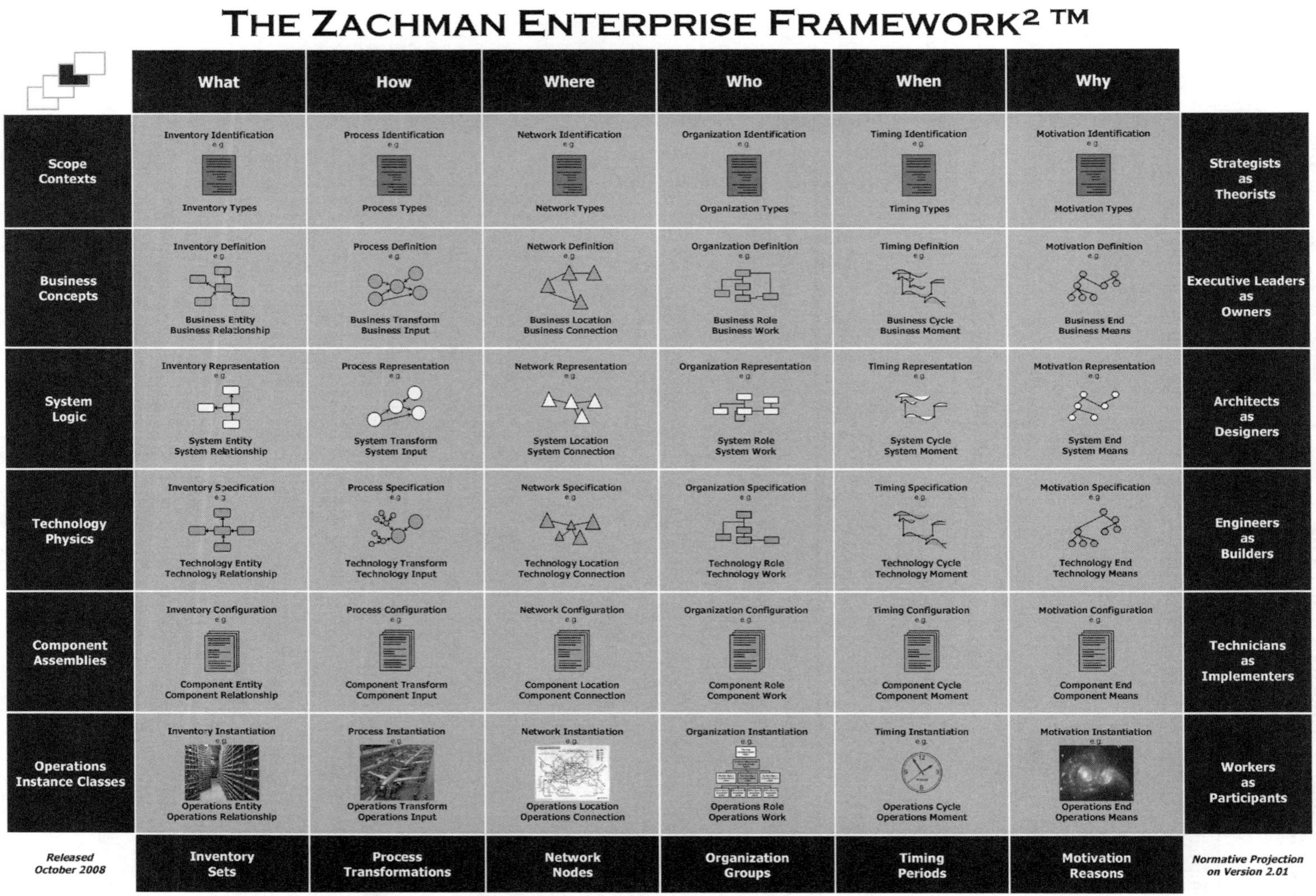

Figure 10.2 Zachman enterprise framework. (*Source:* Published by Zachman International, 2008. http://www.zachmaninternational.com/index.php/the-zachman-framework.)

Zachman Correspondence to MOSES Model Space

The Zachman rows have some correspondence with composition levels. Rows also correspond to the conceptualization spectrum, ranging from the problem domain in the top two rows to the implementation domain in the bottom row. In MOSES terms, the columns are cross-cutting views. From the MOSES viewpoint, the Zachman framework includes multiple views:

- Each row is a view. For example, the top two rows are views of an enterprise; the bottom three rows are views of a system or system part.
- Each column is a view, as the framework identifies.
- Each cell is a view (in our terms, not Zachman's).

If we look more closely at the Zachman framework, we see the following:

- Each cell, especially in rows 2 through 5, represents one or more internal models. In common parlance, Zachman cells represent architecture information, not requirements information. That is consistent with its original name — Framework for EA. But external models are implicitly intertwined throughout the structure, most notably in the fact that all internal models include the external models of their parts.
- We could argue there is a missing row between row 2, the business models, and row 3, the system models. Row 2 is an internal business view of a given enterprise. Row 3 is an internal view of a given system in the given enterprise. What is missing between these two is a system view of the enterprise, in which the enterprise components are systems, and the internal model shows an external view of the systems within the enterprise and how they interact. That view would connect row 2 with row 3, and provide a more explicit link between the two.

There is no intended correspondence of Zachman to the MOSES commonization dimension. Zachman portrays models whose intended target is a given enterprise or a given system, as opposed to reference architectures or patterns targeted to a whole class or category of systems.

RM-ODP Framework

The RM-ODP is an ISO/IEC standard for open distributed processing systems. It defines a framework consisting of foundations,* architecture characteristics, and architectural semantics (RM-ODP 1996). The architecture part defines a set of viewpoints, which we would call views or view types: enterprise, information, computational, engineering, and technology. Each viewpoint has a language that defines concepts and rules for specifying ODP systems from that viewpoint.

The *enterprise* viewpoint associated with an ODP system is a set of rules about the system and its interactions with its larger environment. The focus is on the purpose, scope, and policies for that system, and the policies and roles of the interactions. A role is defined using deontic logic, that is, in terms of permissions, obligations, prohibitions, and behavior of an entity fulfilling the role. The enterprise viewpoint corresponds in MOSES to an internal model of the interaction of a system with the other parts of a larger system — that is, view V5 in Figure 6.1 in Chapter 6.

---

* The foundations part of the RM-ODP standard is a well-crafted, general ontology for system models, targeted to distributed information systems but in many respects applicable to systems in general.

The *information* viewpoint defines the information that is exchanged between a system and its environment, and the semantics of the information and its processing. The information viewpoint corresponds in MOSES to a model of the stuff exchanged in the interaction of a system with the other parts of a larger system, where in this case stuff is limited to information.

The *computational* viewpoint is an internal model of a system that defines the parts or objects, their external interfaces, and their interactions. It is a logical view in that it hides the distribution of the objects and focuses only on their interfaces and interactions. This viewpoint corresponds in MOSES to an internal view of a system (view V3 in Figure 6.1) where the components and connectors are logical entities that hide their distribution.

The *engineering* viewpoint is an internal model of a system that defines the parts or objects, their external interfaces, and their interactions. It is a physical distributed view in that it focuses on the distribution of the objects at defined nodes. This viewpoint corresponds in MOSES to an internal view of a system (view V3 in Figure 6.1) where the components and connectors are entities that exhibit their distribution.

The *technology* viewpoint is an internal model of a system that defines the parts or objects, their external interfaces, and their interactions. It is a physical implementation view in that it focuses on the technologies of the objects and provides information for implementers. This viewpoint corresponds in MOSES to an internal view of a system (view V3 in Figure 6.1) where the components and connectors are entities that specify the implementation — that is, a product to buy or build.

Because RM-ODP is targeted to distributed systems, it defines a number of aspects of distribution that can be hidden. These are called distribution transparencies. Examples include migration transparency, which hides the change of location of an object from other objects; and failure transparency, which hides from an object the failure and possible recovery of other objects, to enable fault tolerance. In this way, even the physically implemented system can hide certain information to promote various desirable properties: facilitate flexibility, enable fault tolerance, and limit the impact of change. Each of the transparencies defines, in MOSES terms, another view of a distributed system.

DoDAF, MODAF, NAF Frameworks and UPDM

The DoDAF (2007) is an evolving architecture framework designed to apply to systems and organizations involved in DoD acquisitions. It has been the basis for at least two other frameworks: The U.K. Ministry of Defense Architecture Framework (MODAF 2008) and the NATO Architecture Framework (NAF) (Hagenbo 2008; NAF 2007). The latter two have added some concepts such as support for service-oriented architecture (SOA), and are now influencing the further evolution of DoDAF. The Unified Profile for DoDAF and MoDAF (UPDM) group was formed to define a common modeling standard to support these frameworks, especially DoDAF and MODAF (UPDM 2008). The UPDM profile is intended to be submitted to the Object Management Group (OMG) for standardization.

In the discussion that follows, we use the terms of each framework. Where appropriate, we indicate mapping of their terms to MOSES.

We now summarize the views of these frameworks. DoDAF defines three basic architecture views that we might call families of views or view types: operational, systems and services, and technical standards, plus an 'All-View' that describes the overall scope and vocabulary of an architecture. Each view has a number of products that could be described as views or models. Many of the products are consistent with the types of models in the Zachman Framework.

The *operational* view describes the operational nodes, activities, and information exchanges required. The *systems and services* view describes systems, services, and interconnections that support the activities in the operational view. The *technical standards* view is a set of technical standards, implementation conventions, and rules governing the structure and interaction of system parts or elements.

The DoD mandates compliance with this framework on acquisition programs to achieve a common approach to architecture in DoD. The goal is to facilitate comparison of architectures and communication about architectures.

MODAF started with the same set of views as DoDAF but calls them viewpoints, which are collections of views. MODAF has added two additional viewpoints: *strategic* (to support capabilities management) and *acquisition* (to describe programmatic details). In addition, MODAF allocates each view or product within each viewpoint to one of seven categories: tabular, structural, behavioral, mapping, ontology, pictorial, and timeline.

NATO started from the MODAF viewpoints and added a service concept that includes services provided not just by software but by any type of system, including people. They added a set of SOA views into their architecture framework:

- NSOAV-1 is service taxonomy: hierarchy of services.
- NSOAV-2 is service specification: interfaces, parameters.
- NSOAV-3 is service composition: services composed of other services.
- NSOAV-4 is service orchestration: how services support operations.
- NSOAV-5 is service behavior: functions, state machines, interactions.
- NSV-13 is service provision: which systems/people provide services.

TOGAF

TOGAF (2006) defines four architectures, which we would call architecture views:

1. *Business architecture:* addresses user concerns, describes business information flow.
2. *Data architecture:* focuses on database components of system and the concerns of database designers and systems engineers.
3. *Applications architecture:* focuses on application software components of system and the concerns of systems/software engineers.
4. *Technology architecture:* focuses on the hardware and software and the concerns of the acquisition and procurement personnel.

A differentiator of TOGAF is that it also defines a developmental approach, whereas most of the other frameworks are developmentally agnostic. This difference is significant as the views/products defined by other frameworks can be constructed using the TOGAF methodology, leveraging both government and industry standards.

FEA

The FEA program is directed by the U.S. Office of Management and Budget (OMB) (FEA 2008). It focuses on a business-driven approach intended to facilitate government-wide improvement in several areas such as budget allocation and information sharing. It defines the government lines of business and services in a set of reference models: Performance Reference Model, Business

Reference Model, Service Component Reference Model, Technical Reference Model, and Data Reference Model. We would call each of these a view of the FEA but the flavor and scope of the views and of FEA as a whole differ from the other frameworks. FEA is more a reference architecture than a framework; it is a general model or category whose region includes the architectures of all the agencies across the federal government.

The Business, Service, and Technical Reference Models are taxonomies, each decomposed into hierarchies. For example, the Business Reference Model is divided into business areas (e.g., services for citizens), each of which has lines of business (e.g., law enforcement, education), and each of those has subfunctions (e.g., criminal investigation, higher education). The Service Reference Model has service domains (e.g., business management services), each of which has service types (e.g., supply chain management), and each of those has components (e.g., procurement). Agency architectures and acquisitions are mapped to the reference models.

A discussion of many of these and other enterprise architecture frameworks is provided in Schekkerman (2004) and in Lankhorst (2005).

SOA and Other Treatments

SOA Frameworks

SOA has received much attention in the past few years. In general, there are two sets of views that exist in most approaches to SOA. In the enterprise context, the business view is the driver for what services are needed and how they are used, while the technology or IT view is where the services are implemented. In the service interaction context are the two complementary views of service provider and service consumer. Standards have been defined for service orientation, notably the OASIS SOA reference model (OASIS 2006) and a significant number of Web services standards, some of which were identified in Chapter 8.

Three approaches that could be called frameworks are based on service-oriented modeling.

1. IBM's Service-Oriented Modeling and Architecture (SOMA) uses a modeling approach to align the business process model with SOA (Arsanjani 2004).
2. The Service-Oriented Modeling Framework (SOMF) defined in (Bell 2008), emphasizes modeling across all environments, from the world in which the services will be used to the services themselves. It includes a substantial number of business architecture perspectives or views in two categories — conceptual and structural. It also includes service-oriented conceptual, logical, and physical architectures.
3. CBDI has defined a meta model or framework called the CBDI-SAE Meta Model for SOA. Version 2 of this meta model (CBDI 2007) is defined as a set of related packages (as in UML package). We judge each package to be a view, and therefore in the list below we use the term view to describe each package.
   - The *Service* view defines a notional service, which includes a name, an existence, and possibly a few attributes. The Service view corresponds to a general model of a service in this book.
   - The *Business* Modeling view defines business model elements called Business Service Subjects, which in Version 2 include business processes, capabilities, types, events, rules, and policies.
   - The *Specification* view defines, for each notional service, at least one service interface, which specifies the operations of the service. The operations can be defined

declaratively, via pre- and post-conditions, in which case they are consistent with the interface specification approach described in Chapter 11 of this book.

- The *Implementation* view defines implementation in terms of Automation Units, each of which implements or automates one or more services, or even non-service software such as legacy systems.
- The *Deployment* and *Runtime* view models service deployment and runtime behavior. The primary concepts are a deployment, which represents an installed automation unit, and an endpoint, which has an associated protocol and recognizes a collection of operations.
- The *Solution Modeling* view defines links between services and solutions with modified use cases; each use case step can be associated with a service operation.
- The *Organization* view is based on the general concepts of party, which it specializes into organization units, persons, and posts; and role. Each party can play multiple roles.
- The *Policy* view defines policies, where a policy is a rule or constraint, which might range from simple to complex. A policy governs an entity or set of entities in a defined scope, such as an entire enterprise or a specific organizational unit, and may apply to business elements or services.

Additional Treatments of Architecture Views

Kruchten 4 + 1 views: Kruchten (1995) defined five views of software architecture and then described their use in the Rational Unified Process (Kruchten 2004). They are:

- *Logical view:* addresses functional requirements, includes major design packages, subsystems, and classes.
- *Implementation view:* organization of static software modules, such as source code and data files, in the development environment.
- *Process view:* runtime processes of systems and their interactions.
- *Deployment view:* map of runtime components to underlying platforms.
- *Use case view:* scenarios or use cases; illustrates how the other views work.

This set is called 4+1 views, rather than five views, because the use case view is positioned as linking the other four together.

Maier and Rechtin views: Maier and Rechtin (2000, 146) describe these as major systems or architectural views:

- Purpose/objective: what the client wants.
- Form: what the system is.
- Behavioral or functional: what the system does.
- Performance objectives or requirements: how effectively the system does it.
- Data: the information retained in the system and its interrelationships.
- Managerial: the process by which the system is constructed and managed.

SEI views and view types: A team at the Software Engineering Institute (SEI) (Clements et al. 2002) discuss views in their book on software architecture. They recognize the separation of views;

but instead of integration, they discuss combining or collapsing views into one view. They define three view types,* and several styles associated with each view type. They are:

- Module view type has decomposition, uses, generalization, and layered styles.
- Component-and-connector view type has pipe-and-filter, shared-data, publish-subscribe, client-server, peer-to-peer, and communicating processes styles.
- Allocation view type has deployment, implementation, and work assignment styles.

Many of these SEI concepts are in MOSES, but factored in different ways. MOSES considers the component-and-connector styles to be different interaction protocols in the composition dimension, and the "uses" relation is the dependency relation in the composition dimension. Decomposition and generalization are different MOSES model space dimensions. Deployment is a physical view in MOSES. Implementation and work assignment are considered management views in MOSES.

The notion of view type, however, is an important element in MOSES. It is expanded here from the Clements notion of broad categories of views of styles to the approach that every view is considered an instance of a view type. There are an unlimited number of potential view types, including types for all the "instance" views we have cited in this section. For example, an information view, defining all the information exchanged in a system, may be useful for many projects and systems, so it can be defined as an information view type (which RM-ODP has done, except they called it an information viewpoint). Then if a systems engineer describes the information view of a system, that information view is an instance of the information view type.

IEEE 1471 Architecture description: The IEEE recommended practice for architecture description (IEEE 2000, 3–4) defines the concept of view and viewpoint but does not define a set of views. The definitions are:

- *View:* a representation of a whole system from the perspective of a related set of concerns.
- *Viewpoint:* a specification of the conventions for constructing and using a view. A pattern or template from which to develop individual views by establishing the purposes and audience for a view and the techniques for its creation and analysis.

The IEEE definition of view is largely consistent with the MOSES concept of view, although MOSES allows the target of a view to be an entity other than a system — for example, a model region. The definition of viewpoint is consistent with the MOSES concept of view type, as discussed above. As discussed at the start of this chapter, MOSES defines viewpoint more in the traditional sense, as a point or locus from where a view is observed. The IEEE 1471 standard is the starting point for an ISO/IEC architecture description standard (42010) that is being developed. A framework that is influencing developing ISO standards 15704 and 42010 is called the Generalized Enterprise Reference Architecture and Methodology (GERAM 1999). Many GERAM concepts are consistent with MOSES, including the areas of enterprise engineering and management, modeling, and views.

Comments on the framework and related examples: The comments here apply to the examples in the sections entitled "Enterprise Architecture Frameworks" and "SOA and Other Treatments." The dominant mapping of design and architecture views is to MOSES internal models, which is appropriate. However, *the intertwining nature of external and internal models* in the composition

* They use the single term "viewtype"; we split it into "view type."

hierarchy is not captured in these views. In most cases, the place of external views is not articulated. The external view is invisible between two composition levels of internal views in many of the frameworks. That is a general weakness that needs to be remedied.

Some views, especially DoDAF views and products, can span multiple levels of composition without making the levels clear and distinct. This can result in monolithic models rather than modular models.

Nevertheless, many important and useful views are available, as seen from the examples. Over time, some view types may achieve sufficient consensus to where it is appropriate to include them in a standard such as IEEE 1471 or ISO/IEC 42010. But in any case, the set of defined views and view types should always be open-ended.

Layered Architecture

This section discusses the venerable concept of layered architecture from three perspectives that differ somewhat from the norm: the infinite loop, leverage, and dual hierarchies.

The Infinite Loop

A key driver of layered architecture is the virtual machine concept. But a primary benefit of virtual machines, portability, is predicated on a rather subtle constraint or assumption. That assumption is that there is only one virtual machine and multiple real machines, or at least that the number of virtual machines on which the higher level desires to run is significantly smaller than the number of real machines on which the higher level desires to run. Suppose a program P wants to run on three real machines and there is one virtual machine. In this case, writing P to use (depend on) the services of the one virtual machine increases its portability to real machines 1, 2, and 3. However, suppose there are three different virtual machines, and P needs to run on all three. Now the virtual machines have not made P more portable. Instead of writing three versions to run respectively on real machines RM1, RM2, and RM3, the P programmer must write three versions to run respectively on virtual machines VM1, VM2, and VM3. If there are five virtual machines that P needs to run on, then P is now less portable than simply writing to real machines RM1, RM2, and RM3. So now we need to define a new virtual machine on top of VMs 1–3 or 1–5 to make P portable. This is what has happened over the past few decades. We wrote operating systems as virtual machines that hid hardware differences, and tried to standardize (e.g., POSIX). But we had too many operating systems. So we came up with middleware. But then we could not standardize on middleware. So that is what prompted OMG in their Model Driven Architecture (MDA) to come up with platform-independent models (PIMs), which were to be a virtual machine on top of middleware.

This succession of events reflects the history of platform independence in the form of a repeating pattern of platform hiding. Based on the history of these efforts, one could roughly define a "platform hiding infinite loop" algorithm that works something like this:

1. *Problem:* Software should be available to all but multiple specific platforms exist. This makes it too expensive to build and maintain software because of the multiple variations needed.
2. *Solution:* Raise the level. Standardize or hide the platform variants behind one common virtual machine at this new level.
3. *Goal:* Get everyone to use this virtual machine.

4. *Problem:* The world does not behave as we want. Multiple competing interfaces are defined at this virtual machine level. These competing interfaces now look a lot like specific platforms. The models that access these competing virtual platforms are no longer portable because they have to be ported to each one.
5. Go to 1.

This loop would presumably stop (at Step 3) if everyone agreed on one virtual machine at some level.

However, even with this situation over recent decades, we have continued to use layered architectures, so a virtual machine must not be the dominant driver for layered architectures. We suggest instead that the primary benefit, and therefore the primary driver (although tacit to this point), is leverage.

Leverage

We argue that the primary benefit of a layered architecture is that it achieves significant leverage, as defined in Chapter 7, by combining components ranging from very specific to very general. This combines the types of leverage so that the leverage of both a specific solution and a general solution, along with various layers in between, are exploited. The effects are to some degree cumulative, and the result is that most of the leverage space below the trade-off limit line is used or exploited. For example, suppose the three-layer architecture in Figure 10.3 is application layer 3, middleware layer 2, and platform layer 1. Then the total leverage available from this architecture is the union of the areas on the left in the B1-B2, C1-C2, and D1-D2 boxes.

Dual Hierarchies

Figure 10.4 shows that what layers are general and what are specific or unique depends on the viewpoint. From the top layer problem viewpoint: top layer is most specific (i.e., a specific application domain) and the bottom layer is most general (i.e., it can run on any platform). From the

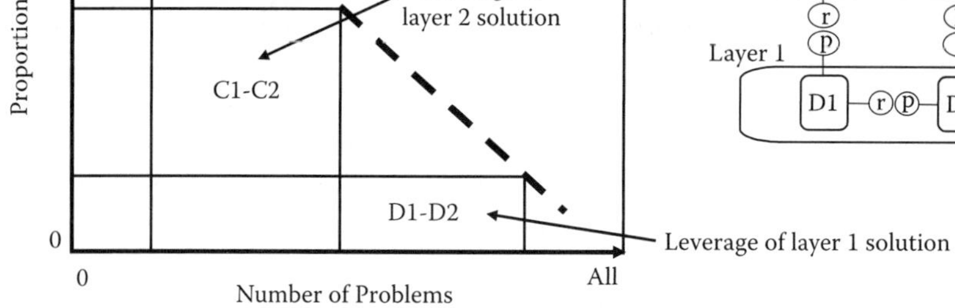

Figure 10.3 Combined leverage of layered architecture.

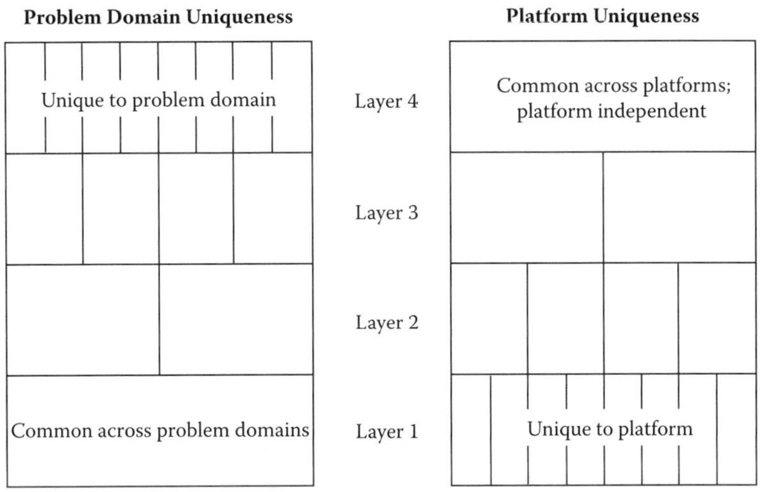

Figure 10.4 Basis of dual hierarchies in layered architecture.

bottom layer platform viewpoint: bottom layer is most specific (i.e., it is a specific platform) and the top layer is the most general (i.e., it can support any application domain).

This is another variation of the dual hierarchies shown in Figure 6.13. We already have the first two pieces — the hierarchies emanating from top and bottom, respectively. The third piece is the dual hierarchies emanating from the middle. From the viewpoint of the middle layer of the architecture (e.g., some type of middleware), it is specific. It supports multiple application domains and can run on multiple platforms, so from this viewpoint both the top (application domain) and bottom (platform) layers are more general.

Aspect-Oriented Development

The aspect-oriented software development (AOD) community has explored in depth the topic of aspects, which in MOSES terms are a type of view, namely, the cross-cutting software type. Views in MOSES can be cross-cutting or not, and can target any type of system, not just software.

Aspect orientation started in the programming arena, driven by inefficiencies of duplicating cross-cutting issues in many code modules (Kiczales et al. 1997). Functions such as logging, display, and audits often are repeated in many or all code modules of a software system. The goal of aspect-oriented programming was to factor out each of these common cross-cutting concerns as an aspect, code it in a separate module, and then use tools to integrate or "weave" the aspect code into all the necessary places in the base program. The scope of this approach has grown over the past decade in several ways. The concerns being investigated have extended to design and requirements, and to more nonfunctional issues, including performance, security, and other qualities. The Web site http://www.early-aspects.net/ is devoted to early aspects, which focus on crosscutting properties in requirements engineering and architecture design. The Web site http://www.aspect-modeling.org/index.html is devoted to aspect-oriented modeling, a concept that has brought together AOD and software modeling communities in a series of workshops.

In addition, a symmetric approach has emerged. Most AOD approaches have been asymmetric, where aspects are composed or woven into components of a base model or system. Aspects and components are different. In a symmetric approach such as multidimensional separation of concerns (MDSOC) (Tarr et al. 1999), no distinction exists between aspects and components; in fact, there is no base model. All concerns, including functionality, are factored into separate modules, which are then composed or woven into a system. See Harrison et al. (2002) for further discussion of the asymmetric and symmetric approaches. The AOD community often talks about the tyranny of the dominant hierarchy. The asymmetric approach alleviates the tyranny; the symmetric approach presumably removes it.

The movement of AOD has been toward increasing modularity. The belief is that every concern can be separated into its own module or component, and thus that every component will represent only one concern, in categories such as functionality, performance, correctness, or security.

Assessment

The AOD community has done important work in examining how to separate concerns and minimize duplication; and extending the scope across the development cycle is good. But there seem to be two constraints on achievable separation. First is separation: not every concern can be allocated to code that only addresses that concern. Second is integration: the challenge of integration is increased in relation to increases in separation. We discuss these constraints in turn.

Limits of separation. The AOD goal is to divide modules until each primitive resulting code module represents just one aspect or concern. This works for some concerns but not others. Often, two or more different concerns inseparably permeate a given entity or module. Consider the example of software code. The code to compute the basic function is not separate from the code that achieves performance or reliability or other qualities. The same code computes the function and achieves some level of performance. If we want to improve performance in this function, we do not write additional code; rather, we modify the functional code. We can, however, define and distinguish between a functional view and a performance view of the same code.

Extending the concept beyond software, consider the example of a human being playing different roles in different contexts, such as parent, spouse, employee, baseball player, etc. We might say that each role represents a different aspect or concern of the human. Clearly, we cannot divide a human into modules: this part of the person is a parent, that part of the person is an employee, etc. The clear modeling approach is that the whole person is a parent, and the whole person is an employee, etc.; and each role is a different view of the whole person. Whatever approach the community develops over time to deal with cross-cutting or overlapping aspects and views, it simply cannot depend on subdividing systems to the point where each part, component, or module only represents one view. We will have to come up with something that recognizes and embraces the fact that aspects, and more generally views, coexist in a way that can only be separated conceptually, not separated by modular decomposition. The point is not that extreme modular decomposition of systems never works; the point is that enough situations exist where it does not work that it cannot be a general solution.

Challenge of integration. Even within the limits of how much we can separate, the AOD community has taken separation to the point of having small component fragments, while assuming that a system can be composed in a straightforward manner. The AOD terminology of weaving, joining, and merging is a different concept from considerations of interaction and integration. Given the historical difficulties of integration, why should composing fragments into a whole system be easier or more straightforward than composing traditional components? We would expect it instead to be more difficult because there are many more pieces to bring together. Interactions and side effects and sequencing effects can have a significant impact on integration.

Furthermore, the whole area of emergent properties and behavior would seem to be made more problematic by the extensive separation. The expansion of SE to include more complex systems increases the need to understand emergent behavior, especially from the interaction of large numbers of parts. AOD seems to emphasize the piecemeal view over the holistic view, the separated view over the integrated view. Both separation and integration are needed to succeed in engineering systems of any complexity.

Because of these constraints, our minimal recommendation is to augment the AOD approach with the views approach described in this chapter. Even better, AOD and the idea of aspects should be brought into the views framework as a type of view. Reasons for this include

- Views support both separation and integration simultaneously.
- Views do not require physical separation. They facilitate conceptual and visual separation of all concerns, and support physical separation where appropriate.
- Views work in cross-cutting cases where AOD fails, namely, where physical separation is not possible.
- Views do not result in fragmentation. The use of views allows integration to be an invariant — that is, the system can remain integrated at all times even when we are conceptually and visually separating concerns. The MOSES term for a system in which all concerns coexist is an integrated system; AOD calls such a situation "tangled concerns."
- Views go beyond the scope of cross-cutting concerns to address all concerns with the same seamless and holistic approach; AOD addresses only cross-cutting concerns, so an entirely different method is needed to address other concerns.

Dominant view. There is, however, an important question that AOD has raised with the symmetric–asymmetric issue, a question that applies to views as well. The question is: Should any view be considered as the dominant or base view? This is an area that needs further research, and is discussed in the section entitled "Need for Dominant Structure or View." In the meantime, MOSES does not take a position on it, except a utilitarian one: if a modeling situation works better with a dominant view, define one; if it works better without a dominant view, do not define any view as dominant.

Did Parnas Anticipate Features of Aspects?

Parnas, in his famous decomposition criteria paper (1972, 1057), may have presaged or foreshadowed concepts that later became key elements of aspect-oriented programming. In his discussion of performance issues with his KWIC information hiding design (because it would cause many calls to the same Line Storage module), he said:

"To save the procedure call overhead, yet gain the advantages that we have seen above, we must implement these modules in an unusual way. In many cases the routines will be best inserted into the code by an assembler; in other cases, highly specialized and efficient transfers would be inserted. To successfully and efficiently make use of the second type of decomposition will require a tool by means of which programs may be written as if the functions were subroutines, but assembled by whatever implementation is appropriate. If such a technique is used, the separation between modules may not be clear in the final code. For that reason additional program modification features would also be useful. In other words, the several representations of the program (which were mentioned earlier) must be maintained in the machine together with a program performing mapping between them."

We suggest that this description contains concepts that have some resemblance to weaving and other processes that were defined 25 years later as features of aspect-oriented programming.

Duality, Definitions, Categories

Yin and Yang: Duality, Contrast, Harmony

Chapter 1 introduced the principle of reconciled tensions, dualities, contrasts, and paradoxes, where opposing and seemingly mutually exclusive concepts harmoniously and even synergistically coexist in the same entity. The contrasting concepts retain their separate identities and characteristics in the entity, but both contribute to unity and balance. This situation is ubiquitous in the world, including in the world of science and engineering. Traditional engineering has encountered and dealt with such contrasts, but more often implicitly than explicitly. The need for science and math to deal with complex systems over the past century has led to unavoidable recognition of many contrasts and contradictions, and they have been made explicit — but not so much in engineering.

A tool for reconciling contrasts, tensions, and paradoxes is the concept of views. This has not been recognized as a tool for dealing with contradictions in either science or engineering, but we propose that it is a natural tool for both. Complex systems are a locus not only for many contrasts and paradoxes, but also for their resolution. This section discusses, with a few examples, how views are an important, even crucial, tool in the harmonizing of contrasts and dualities.

In each contrast in the following list, the contrasting items can be represented as views of a common whole:

- Top down and bottom up
- Discrete and continuous (discussed in this chapter in the section entitled "Views in Broader Context")
- Simplicity and complexity (much has been written about this contrast, which is a great illustration of the yin and yang principles; for example, see Cohen and Stewart [1994])
- Order and disorder (chaos is the reconciler here)
- Deterministic and probabilistic
- Random and predictable
- Controlled and autonomous or self-organizing
- Homeostatic and adaptive
- Stability and change (see discussion of panarchy in Chapter 9)
- Unity and variety
- One and many, whole and part: a system is one, and is many parts
- Relative and absolute
- General and specific
- Instance and category
- Abstract and concrete

- Abstract and interpreted
- Context sensitive and context generalized
- Pure and impure: nature (and artificial systems?) needs impurities. From Merkle (1993): Making computers highlights another problem. It's not enough to make a pure crystal, it must also have an extremely precise and complex pattern of impurities.
- Robust and fragile (Carlson and Doyle 1999): "This 'robust-yet-fragile' feature is characteristic of complex systems throughout engineering and biology."
- Practice and theory
- Observation and theory
- Essence and accident; on purpose and accidental or random; damaging accidents or errors and serendipitous accidents or errors (e.g., mutation in adaptation, evolution)
- Stand-alone and interactive
- Never-ending change and cycle — birth/death, start/end; evolution embraces both at various scales

The classic prototypical example of the concept of simultaneously embracing contrasts is yin and yang. Although the contrasting concepts, yin and yang, are opposites, they coexist in harmony, are interdependent, and continually change from one into the other in a way that maintains overall balance. The principles of yin and yang appear to be a good model of the contrasts that exist in complex systems.

Harmonizing is an approach that on one hand acknowledges and models the opposing characteristics in an entity, and on the other hand models their mutuality and role in the unity of the entity. Table 10.3 shows a few examples where complex systems are sometimes described as a simple contrast with traditional systems, but the more complete answer is a harmony and balance of the two opposites.

An example that reflects several contrasts in one package is the comparison of self-organized criticality (SOC) with highly optimized tolerance (HOT) given in Carlson and Doyle (2002). Some of the contrasting variables in the comparison profiles are robustness (the critical point is stable in SOC, but systems in the HOT model are "robust, yet fragile," a major theme in HOT); the number of design degrees of freedom (small in SOC, very large in HOT); and the mechanism for power laws (in SOC, power laws are a signature of an internal self-sustaining critical state; in HOT, power laws are a symptom of the robust fragile condition resulting from tuning multiple variables). The article advocates HOT over SOC and claims that real systems match the HOT profile rather than SOC.

Our purpose is not to weigh in on the debate directly but rather to conjecture that SOC and HOT represent two different and complementary modeling views, at least for some of the variables used for comparison. We discuss this point using one variable: the similarity of parts in the internal configuration. The SOC view is that parts are homogeneous and self-similar at different scales. The HOT view is that parts are extremely heterogeneous and self-dissimilar, especially across scales. We have pointed out in the book that both views are needed in the total picture. SE seeks to exploit similarities within and across scales, while recognizing and accounting for differences.

Kosko (1993) advocates fuzzy thinking as a tool for dealing with contradictions and paradoxes. We will illustrate with a form of the liar's paradox. Statement A = "This statement is false." According to the law of the excluded middle, either A is true or it is false. Which is it? If we assume it to be true, then it turns out to be false. If we assume it to be false, then it turns out to be true. How do we resolve this?

Table 10.3 Examples of Harmonizing Traditional and Complex Dualities

Traditional	*Complex*	*Harmony/Balance*
Order	Disorder	Both coexist
Simple	Complex	Both coexist
Top down	Bottom up	Both: back and forth, opportunistic
Control	Autonomy, *laissez faire*	Both: arrangement, influence, higher-order patterns
Avoid change	Encourage change	Both: find and exploit stability, find and exploit change, use stability to guide change
Reductionism: Analyze parts	Emergence: Analyze whole	Analyze both whole and parts
Pure dictatorship/ one over all	Pure democracy/ everyone is equal	Everyone has a voice, but leadership arises, emerges, sanctioned by the group; a few are leaders, many are followers
Avoid risk, even if it arises from significant opportunity	Embrace opportunity, even if it entails significant risk	Risk and opportunity; evaluate appropriate balance
Do what we know	Try something new	Explicitly trade off the two; or sometimes do a combination of both; or do both in parallel and see which works out better. In situations where a binary choice must be made, having the perspective, the sense, the awareness, the gestalt, to know which to pick is an important kind of knowledge or intuition. This is a universal trade-off.

- The fuzzy answer: Remove the law of the excluded middle, and allow A and not A. Take this further and say A = not A, which results in the answer: A is half true and half false.
- The yin yang answer: Remove the law of the excluded middle, and allow A and not A. A is both true and false.
- The views answer — an augmented yin yang answer: The question is not harmonized in an integrated sense; it remains a paradox. But it is resolved in a separated sense; A is completely true in one view, and completely false in another view.

The MOSES position is that fuzzy thinking and fuzzy logic have an important place in systems engineering, especially in specifying continuous ranges instead of discrete distinctions. That means that fuzzy measures may have a role, in addition to views, in addressing many of the contrasts listed above because few of them are paradoxes. But resolving paradoxes is not an appropriate role for fuzzy thinking. To say that a paradoxical statement is half true and half false is not a satisfying response in the way that legitimate applications of fuzzy measures are (for example, saying that someone is half tall and half short, or a glass is half full and half empty).

Contradictions and dualities exist not only in CS, but also in TS. Here is an example of a TS duality. Assume that the wheel has not yet been invented, and we want to write a requirements spec for the wheel. It could reasonably look something like this (assuming we defined "road" suitably):

1. Wheel shall move relative to road.
2. Wheel shall not move relative to road.

If the engineers of the day had looked at this, they probably would have said: "This is obviously impossible." This spec could have delayed the invention of the wheel by centuries. Fortunately, those who did invent the wheel did not realize they were doing something that was impossible. Now as we look back on the wheel and its design, we can explain the impossible requirements in a way that resolves the contradiction:

1. Wheel shall move relative to road: the wheel rotates, moves as a whole. The inner hub of the wheel satisfies this requirement, moving in a line parallel with the ground, which is where the axle fits and connects to the wheel on the other side, and this supports the vehicle, which moves along the road.
2. Wheel shall not move relative to road: The wheel should not slide on the road; there must be friction, grip. The outer edge of the wheel satisfies this requirement; each portion of the edge, when in contact with the road, grips the road and does not move.

Definitions

We have informally defined view and related concepts earlier in the chapter. Now we present the full set of definitions. We propose that the following elements exist in connection with every view: {viewpoint, object or target, observer or agent, selector or interpreter or abstractor, abstraction or view}. We now explain each element.

- *Viewpoint:* a spacetime region (not just a point; can be physical or conceptual) that represents a position from which the target is observed. Example: external view of 3D block is observed from viewpoint = {inside a 3D sphere surrounding the block in a time range or time point}.
- *Target:* system or other entity that is being observed or viewed. The target may take the form of a defined world. There can be multiple views of a target. Two common targets are (1) a specific system that exists or that we plan to build, and (2) all possible systems that correspond to the model (i.e., the model region).
- *Observer:* person or other agent that observes the target, possibly with the aid of supporting tools or agents such as (for physical targets) telescope, microscope, thermometer, sensor, probe, camera.
  - View observers/agents do more than just see the view. They define the view, the viewpoint, the object, and serve as selector, interpreter, or abstractor. So view observers go beyond, and are more active than, system observers/definers. They are as active as model definers/observers.
  - It is sometimes useful to distinguish an active agent such as a human from a passive observer such as a sensor (e.g., a thermometer).
  - Software can be an agent, and can range from passive to active.

- *Abstractor:* a function that selects info from the target to include in the view, or defines how target information is interpreted in the view.
- *Abstraction* or *view:* what the observer "sees" by applying abstractor to target from viewpoint.

Other elements exist in the context of a view. An important one is *view type*. In the MOSES recommended approach, the view in the set of elements above is an instance of a view type. However, we did not include it in the definition of view because it is not part of any individual view. View type is part of the supporting context.

Categories of Views

We have discussed a significant number of examples of views in this chapter. Is there a way we can classify them? A preliminary set of categories is shown in Figure 10.5.

The separate component category is one in which we focus on one element of a group of elements, such as a system of interest. The component type view is where the same space, such as an internal model of a system, can consist of different components and connectors or different levels of abstraction, such as physical and logical views, or the RM-ODP computational versus engineering viewpoints, or alternative decompositions (designs) of a system.

The cross-cutting/part category is a dominant type and motivator of the need for views. It represents situations such as where common elements existing in different components or systems are brought together into a view. A classic example of this type is an aspect-in-aspect oriented

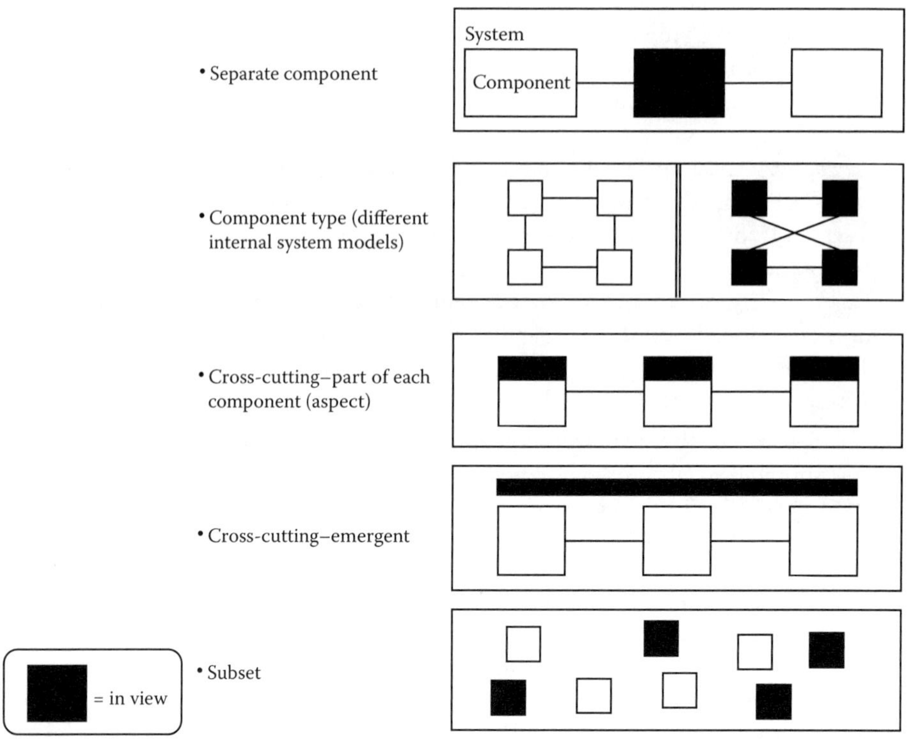

Figure 10.5 Categories or types of views.

development. The cross-cutting/emergent category is where a view observes an emergent property or other characteristic of a whole, separate from the parts from which it came. The subset view is a very simple category exemplified by a database view. More research is needed on view categories, as discussed in the subsequent section.

Areas for Additional Research

Integration

The needed research is on mathematical representations and manipulations of views that support integration of views. One possible approach is to use the mathematical model of unlimited dimensions called a polytope. Some work has been done to capture different views as part of this data structure (Baker 2006). This could be a vehicle for integrating views.

Need for Dominant Structure or View

The question was raised earlier in the context of aspects and hierarchies whether any view is to be considered the dominant or base view. We might suppose that any use of the term cross-cutting assumes a base or dominant view, which is the thing being cut across. On the other hand, a cross-cutting view in the absence of a base could simply be a view that cuts across any other view. Example of a dominant view: in an organization, a matrix management structure cuts across the line organization, which is considered the base or dominant structure of the organization. But is a dominant structure needed in an organization or in any complex system, or in any system at all? If so, is it always needed or only under certain circumstances or for certain types of systems? What circumstances or system types?

Example: Do organizations of the future really need a dominant structure? Can a healthy organization exist with a set of individuals who form multiple different structures or views, and none of the views is dominant? If so, is anyone in charge? Does anyone need to be in charge?

View Categories

We identified some possible categories of views. But a large number of possible variables could be used in deciding common categories. There is a need for research that explores what categories are good candidates for common view types the community would find useful, and the ways in which view types could be extended. Some other variables and aspects to be considered in differentiating views include levels, scales, focus, resolution, model space dimensions, interpretation, representation, field of view, role of observer/agent in shaping what is seen, the possibility of adding information, not just subtracting (abstracting) it, and meta-information in a view.

Summary

The following features of views were discussed in this chapter:

- The concept of view is highly flexible and has significant potential for future SE.
- Views can exist within a model space dimension or cut across dimensions.
- In a composition hierarchy, views can focus on similarities of levels, or on differences between levels.
- Views can be nested.

- System behavior and system change can be regarded as disjoint activities, as in TSE; or as two views of a system, as in CS.
- Defined worlds are views.
- An SOI is a view and has several commonly associated views.
- Different areas of science consider different views of the same overall world.
- Views can help resolve the competing hierarchies problem.
- Architecture frameworks have put forward a variety of views. As SE matures, conventional sets of views will likely emerge.
- The concept of layered architecture can be illuminated by views and the idea of leverage.
- Views can help address the limits of modularity facing the aspect-oriented development community.
- Views are a natural way of dealing with dualities and contrasts that are a prominent characteristic of CS and to some extent TS.
- A preliminary set of categories or types of views is offered but a more mature set will emerge from a combination of research and practice.

References

Alexander, C. 1979. *The Timeless Way of Building*. New York: Oxford University Press.

Alexander, C., S. Ishikawa, M. Silverstein, M. Jacobson, I. Fiksdahl-King, and S. Angel. 1977. *A Pattern Language: Towns – Buildings – Construction*. New York: Oxford University Press.

Arsanjani, A. 2004. Service-Oriented Modeling and Architecture: How to Identify, Specify, and Realize Services for your SOA. IBM. 9 November. http://www.ibm.com/developerworks/webservices/library/ws-soa-design1/. (Accessed January 11, 2009.)

Baker, S. 2006. Math Will Rock Your World. *Business Week*. January 23. http://www.businessweek.com/pdf/Matheducationaleprint.pdf. (Accessed January 11, 2009.)

Bell, M. 2008. *Service-Oriented Modeling: Service Analysis, Design, and Architecture*. Hoboken, NJ: John Wiley & Sons.

Carlson, J. and J. Doyle. 1999. Highly optimized tolerance: a mechanism for power laws in designed systems. *Phys. Rev.*, E 60:1412–1427.

CBDI. 2007. CBDI Service Architecture & Engineering Meta Model for SOA. http://www.cbdiforum.com/public/meta_model_v2.php. (Accessed January 11, 2009.)

Clements, P. et al. 2002. *Documenting Software Architectures: Views and Beyond*. Boston, MA: Addison-Wesley.

Cohen, J. and I. Stewart. 1994. *The Collapse of Chaos: Discovering Simplicity in a Complex World*. New York: Penguin Press.

DoDAF. 2007. Department of Defense, *DoD Architecture Framework*, Version 1.5. 3 volumes. 23 April 2007.

FEA. 2008. Federal Enterprise Architecture. OMB. http://www.army.mil/armyBTKC/enablers/pm/fea.htm (Accessed January 11, 2009.)

GERAM. 1999. GERAM: Generalized Enterprise Reference Architecture and Methodology, Version 1.6.3 © IFIP-IFAC Task Force on Architectures for Enterprise Integration, March 1999.

Hagenbo, M. 2008. The NATO architecture framework SOA (Service Oriented Architecture) views. 17 April. http://www.afei.org/brochure/8a05/documents/Hagenbo_.ppt (Accessed January 11, 2009.)

Harrison, W., H. Ossher, and P. Tarr. 2002. Asymmetrically vs. Symmetrically Organized Paradigms for Software Composition. IBM Research Report RC22685, IBM Research Division, Yorktown Heights, NY, December 20, 2002. http://domino.watson.ibm.com/library/cyberdig.nsf/0/2a4097e93456d0cf85256ca9006dac29?OpenDocument. (Accessed January 11, 2009.)

Hawking, S. 1988. *A Brief History of Time: From the Big Bang to Black Holes*. New York: Bantam Books.

Hawking, S. 1996. *The Illustrated A Brief History of Time*. Updated and expanded ed. New York: Bantam Books.

Heckmann, D., T. Schwartz, B. Brandherm, M. Schmitz, and M. von Wilamowitz-Moellendorff. 2005. GUMO — The General User Model Ontology, in *Proceedings of UM* 2005: Int. Conf. on User Modeling, July 24–30, Edinburgh, UK.

Honderich, T. (ed) 1995. *The Oxford Companion to Philosophy*. New York: Oxford University Press.
IEEE. 1998. *IEEE Recommended Practice for Software Design Descriptions*. IEEE 1016-1998. New York: IEEE Computer Society.
IEEE. 2000. *IEEE Recommended Practice for Architectural Description of Software-Intensive Systems*, IEEE Std 1471-2000, 21 Sept., New York: IEEE Computer Society.
IEEE. 2005. IEEE Standard for Application and Management of the Systems Engineering Process, IEEE Std 1220-2005, New York: IEEE Computer Society.
INCOSE. 2007. INCOSE Systems Engineering Handbook, v3.1. Seattle, WA: INCOSE (International Council on Systems Engineering).
Kiczales, G. et al. 1997. "Aspect-Oriented Programming," Proc. of Eur. Conf. Object-Oriented Programming (ECOOP 97), LNCS 1242, Springer, 1997, p. 220–242.
Kosko, B. 1993. *Fuzzy Thinking: The New Science of Fuzzy Logic*. New York: Hyperion.
Kossiakoff, A., and W. N. Sweet. 2003. *Systems Engineering: Principles and Practice*. Hoboken, NJ: John Wiley & Sons.
Kruchten, P. 1995. The 4+1 View Model of Architecture, IEEE Software, 12(6):42–50.
Kruchten, P. 2004. *The Rational Unified Process, 3rd ed.* Boston, MA: Pearson Education.
Kuhn, T. 1970. *The Structure of Scientific Revolutions, 2nd ed.* enlarged. Chicago, IL: University of Chicago Press.
Lankhorst, M. 2005. *Enterprise Architecture At Work: Modeling Communications and Analysis*. New York: Springer.
Merkle, R. 1993. Molecular Manufacturing: Adding Positional Control to Chemical Synthesis. Xerox Corp. http://www.zyvex.com/nanotech/CDAarticle.html. (Accessed January 11, 2009.)
MODAF. 2008. MODAF Meta Model. V1.2. 2008-07-11. http://www.modaf.org.uk/m3/. (Accessed January 11, 2009.)
NAF. 2007. NATO Architecture Framework Version 3. NATO/EAPC. July.
OASIS. 2006. Reference Model for Service Oriented Architecture 1.0. OASIS Standard. 12 Oct. http://docs.oasis-open.org/soa-rm/v1.0/ (Accessed January 11, 2009.)
Obrst, L. and D. Nichols. 2005. Context and Ontologies: Contextual Indexing of Ontological Expressions. MITRE Technical Paper, July 2005. http://www.mitre.org/work/tech_papers/tech_papers_05/05_0903/index.html. (Accessed January 11, 2009.)
Parnas, D. 1972. On the criteria to be used in decomposing systems into modules. *Commun. ACM*, 15(12): 1053–1058.
RM-ODP. 1996. Open Distributed Processing — Reference Model Part 2: Foundations, International Standard 107462, ITUT Recommendation X.902; Part 3: Architecture, International Standard 107463, ITUT Recommendation X.903, 15 Sept 1996.
Sage, A. and C. Cuppan. 2001. On the systems engineering and management of systems of systems and federations of systems. *Information, Knowledge, and Systems Management* 2:325–345.
Schekkerman, J. 2004. *How to Survive in the Jungle of Enterprise Architecture Frameworks: Creating or Choosing an Enterprise Architecture Framework*. Victoria, BC: Trafford Publishing.
Sowa, J. and J. Zachman. 1992. Extending and formalizing the framework for information systems architecture. *IBM Systems J.*, 31(3):590–615.
Tarr, P., H. Ossher, W. Harrison, and S. Sutton, Jr. 1999. N Degrees of Separation: Multi-Dimensional Separation of Concerns. In *Proc. 21st Int. Conf. on Software Engineering*, p. 107–119, May.
TOGAF. 2006. The Open Group Architecture Framework Version 8.1.1, Enterprise Edition. http://www.opengroup.org/architecture/togaf8-doc/arch/. (Accessed January 11, 2009.)
UPDM. 2008. UPDM Group. http://updmgroup.org/. (Accessed January 11, 2009.)
Valerdi, R., A. Ross, and D. Rhodes. 2007. A framework for evolving system of systems engineering, *CrossTalk: The Journal of Defense Software Engineering*, October. p. 28–30. http://www.stsc.hill.af.mil/Crosstalk/2007/10/0710ValerdiRossRhodes.html. (Accessed January 11, 2009.)
Zachman, J. 1987. A framework for information systems architecture. *IBM Systems J.*, 26(3):276–292.
Zachman International. 2008. The Zachman Enterprise Framework. http://www.zachmaninternational.com/index.php/the-zachman-framework. (Accessed January 11, 2009.)

Chapter 11

Specification Information

The approach to specification is a significant element of MOSES, and is the subject of this chapter. Key topics include the complex systems expansion of traditional categories of external specifications; the list of information types for external and internal specifications; and the use of probability in specifying uncertainty.

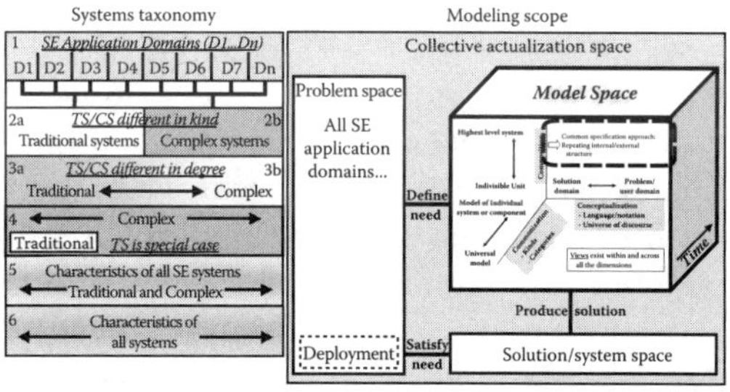

Introduction: Specification Types and Structure

In previous chapters, especially Chapter 6, we discussed external and internal models, and their intertwining structure that supports scale-up and scale-down through a repeating self-similar pattern. We have also, especially in Chapters 3 and 4, discussed the concepts of model versus specification, decoupling, and correspondence set. In this chapter we build on these concepts and drill down into the specification types to define the information content of each type.

This chapter is entitled "Specification Information" instead of "Model Information" for the following reason: it attempts to define more details of the types of information needed for shared understanding between parties but it does not make any statement or constraint on how much of this information is explicitly stated in a model.* Much of the information spelled out in this chapter will be implicit and will not show up in models, especially as SE becomes more mature.

Before proceeding further on this path, we mention an alternative definition of specification: a shared understanding of a correspondence set, which is a model M, a modality MP, a defined world W, and a system S, as defined in Chapter 4. This is a useful definition in cases where a correspondence set is defined, and the modality, the defined world, or the system is subject to differing

* The explicit information must be greater than zero, however. If it is zero, there is no model with which to associate the specification. The practical minimum is an explicit model name.

interpretations. In the discussions in this chapter and this book, we use the first definition as the default.

A brief recap of the concepts from previous chapters that we use in this chapter: specification types, and their intertwining relation, are derived predominantly from the composition dimension. The two general specification types are internal and external. An internal specification of a system is a set of rules about the parts (components and connectors) that are within the system, their structure and interaction, and the stuff they exchange, such as information or goods. An external specification of a system or component is a set of rules about the externally observable view of that system or component. This includes observable stuff, behavior, ports or interfaces, and services, as well as their observable properties or qualities. In this chapter we expose the decoupling of external specifications from the interfaces that are associated with them, as described in Chapter 4. Therefore, we describe three sets of specification content: internal specifications, external specifications, and interface specifications.

As shown in Figure 11.1, the relation between the two major specification types is that an internal specification of a system includes the external specifications of its components and connectors. The relation between external specifications and interface specifications is many to many, which is why they are decoupled: an external specification has multiple interfaces, and an interface specification can be used by or offered by multiple external specifications. Figure 11.1 also shows another convention that we have used throughout most of the book and which we now make more explicit. In the connection ontology in Chapter 3, we discussed the exchange of stuff and the possible performance of a service, which also involves an exchange of stuff. The terminology we have used, and continue to use in this chapter, is that of the more complicated case of a service being performed and stuff being exchanged. The shorthand terms (from Chapter 3) "provides a service"

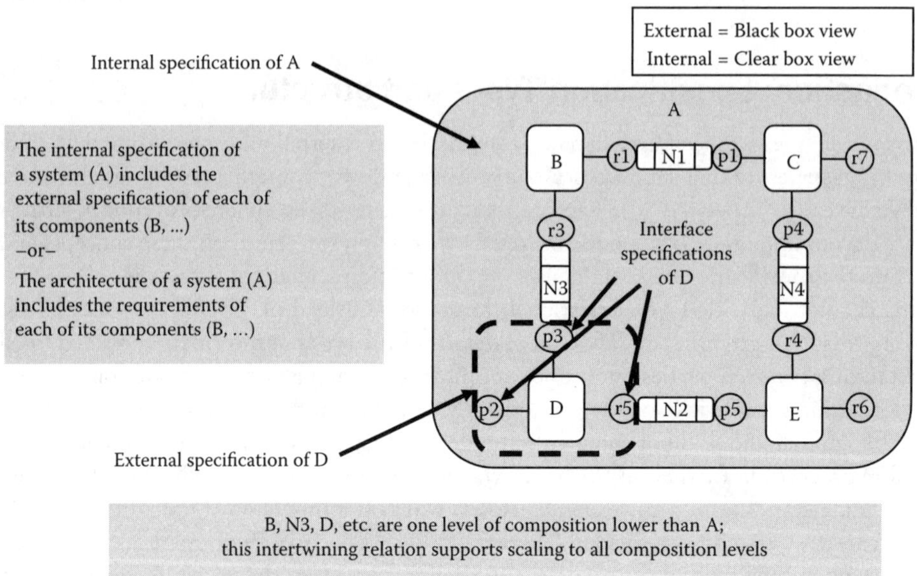

Figure 11.1 Internal, external, and interface specifications. (*Source*: Hybertson, D. Using a specification approach to facilitate component testing. In *Testing Commercial-off-the-Shelf Components and Systems*, S. Beydeda and V. Gruhn, Eds. New York: Springer-Verlag, p. 213–237. 2005. Modified with permission of Springer Science + Business Media.)

and "receives a service" refer to these actions and exchanges. In cases where no service is being performed, the meaning defaults simply to the stuff being exchanged. The services terminology is not meant to imply a restriction in this chapter or book to systems that adhere to the service oriented architecture (SOA) style. It is, instead, meant to apply to the models and specifications of all SE systems.

Elements and Background of Approach

A general source for the specification approach described here is Szyperski (1998), as well as the general SOA approach to service specifications. An important element of achieving precision in specification is a clear understanding of dependencies. Traditional systems specifications define what services or stuff are provided but not what are required. Luckham, Vera, and Meldal (1995) describe an "object connection architecture" as an architecture of a system in which component interfaces define services provided but not services used or required. They suggest, instead, "interface connection architectures" that specify both services provided and services required. An approach that supports this concept is Design by Contract (DBC), defined by Meyer (1997). DBC originally emphasized contract provisions in the code itself, and the focus was primarily on behavioral or functional — as opposed to non-functional or performance — issues. Other inputs to the approach defined here include UML (OMG 2007a, b), SysML (OMG 2007c), and Real-Time Object-Oriented Modeling (Selic, Gullekson, and Ward 1994).

The component-based development community has embraced the DBC concept and extended it for component use. First, the concept of trusted components has come into play. Meyer (2003) defines a trusted component as a reusable software element possessing specified and guaranteed property qualities. In the same paper, Meyer defines a high road and a low road toward establishing high-quality components. The low road is to start with today's COTS or open-source components and move to certification based on contracts. The high road is to produce new components with correctness proofs by defining correctness properties based on contracts.

Second, DBC has been extended to include nonfunctional information (performance, QoS) as well as cost. QoS in general has been defined and researched extensively over a period of time. Common metrics include performance, such as message throughput and reliability, as well as security. See, for example, Wang et al. (2004). Aagedal and Ecklund (2002) describe an approach to QoS modeling that features the notion of QoS profile. A profile defines a QoS region in which a component operates. Multiple different QoS profiles may be defined for a given component — for example, one profile defined for a PDA platform and another for a desktop platform. De Miguel, Ruiz, and Garcia (2002) adopt a similar idea in defining a QoS model using UML extensions and applying it to component frameworks. Urting et al. (2002) also use contracts to model nonfunctional constraints for embedded software.

A metric that has emerged more recently is quality of experience (QoE), which is the ultimate effect experienced by a consumer or user; for example, see Kilkki (2008) and Gallo, Siller, and Woods (2007). An earlier variation of QoE is work done on task-technology fit (TTF). Goodhue and Thompson (1995) proposed that for information technology to have a positive impact on individual performance, the technology must be utilized and it must be a good fit with the task that it supports. They measured the fit based on user perception and evaluation, which is what makes it a form of QoE.

QoE is positioned at the user end of the interaction, in contrast to quality of direct output of system, and QoS of the connection or medium between system and user. Although the QoE value of a service or transaction is defined by the user, it can be included in a system specification in terms of a QoE response level that is received from the user or consumer.

The cost might be the initial purchase of the component, the purchase of ongoing maintenance updates, or the cost of "renting" the component as a service on a pay-per-use basis. Cost may also be specified at a more granular level for each service.

Brucker and Wolff (2001) and Paige and Ostroff (2001) discuss different approaches to increasing formal support for contracts in UML and OCL. The specification approach described here incorporates these extended contract concepts. It does not require formal specification or automation but it is fully consistent with these ideas.

The well-developed area of reliability, with various metrics such as dependability, MTBF, etc., addresses much of the concept of probability of normal and abnormal characteristics. This provides a sound foundation for extending the approach of probabilistic specifications to autonomous agents and complex systems (CS).

One basic theme: MOSES does not require any particular language or method of specification; many can be used. On the other hand, most current specification languages and methods have some deficiencies as viewed from the MOSES defined approach. For example, adequate semantics, and making explicit the dependencies in an external specification are often lacking, although in recent years more attention has been given to semantics.

Expand External Behavior Categories

Traditionally we have two categories of system behavior: normal and abnormal. We say this system behaves normally .999 of the time and abnormally .001 of the time, or behaves properly .7 of the time and improperly .3 of the time. We also want to be able to specify and predict more than that. Because a system, in general, is capable of exhibiting many behaviors, both proper and improper, we want to know what proper behavior it will exhibit, and especially what improper behavior it will exhibit. On the improper side, what are the types of things that can go wrong? System faults and failures can vary significantly, from causing a minor temporary inconvenience to causing death and destruction. We want to know the likelihood of it causing death and destruction, or any lesser outcome of improper behavior, and if possible what the conditions are that will result in this behavior. Also, we want to know if we (the user) will even know if a system behaved improperly.

What all this means is that we need to map both the *expected* (or normal) behavior space, and the *unexpected* (or abnormal) behavior space (David Parnas called the latter space "undesired events"). We need to lay out the various behaviors as much as we can, the conditions that will lead to each of them, and the probability that each of them will occur. We have done this in the past in software engineering and to some degree in systems engineering. The desired space is the basic specification, and the undesired space is the delineated exception.

However, normal and abnormal behavior only cover part of the external specification space when we include complex systems, such as people and organizations in our scope of systems.

First consider normal behavior. Normal expectations are different for a machine and a CS such as a person, organization, or economy. For a machine, an expectation stays static over time. For a person, the expectation increases over time. When we hire a person for a job, we expect them to meet a minimum competency at the start, but then over time to learn, grow, improve, and increase their level of competency in that job. So for a CS, normal expectations are to some degree an increasing function of time rather than static.

Now consider behavior that is not normal. Remember that a human or an autonomous intelligent agent can do things that a machine, at least as reflected in traditional mechanistic models, cannot do. An intelligent agent can learn, adapt itself, grow, improve, be fault tolerant — of various kinds of faults and errors and abnormal behavior, both internal and external; be self-healing, change goals, and make decisions based on incomplete or uncertain information. Intelligent agents are capable of coming up with solutions that are not just better than expected, but of a completely different type, out of range of what was envisioned. For example, if a person is supposed to produce ten widgets a day, they might improve to where they can produce fifteen widgets a day. But they might also come up with the idea that the factory should be making different widgets. This is not something your typical machine is capable of. On the other hand, a person is capable of out-of-range negative behavior as well, causing significant damage and destruction.

The result is that the normal behavior category needs to have a broader definition to cover both machines and agents, and the categories of possible behavior of a system need to be expanded beyond the two traditional categories of normal and abnormal. Agents add three additional categories: better than expected; out of range good; out of range bad.*

If we address the decoupled situation discussed in Chapter 4, where valuation is decoupled from behavior, then this expanded set consists of three categories: Normal, not normal, out of range.† If we apply valuation to this, we obtain these five categories:

1. Out of range good
2. In range, better than expected
3. What is expected, normal
4. In range, less than expected; abnormal
5. Out of range bad

Only categories 2, 3, and 4 are addressed in an external model or specification. Categories 1 and 5 cannot be anticipated; they are out of range, outside the field of vision of the external specification. It is important to note, however, that internal models and specifications do not have this restriction. Systems design in the era of the future expanded SE needs to emphasize robustness at least as

* To some extent TSE arguably addresses the category "out of range bad" through methods such as failure analysis and failure modes, including catastrophic failure. But this category is intended to emphasize different types of willful or purposeful acts by humans.

† If we really get strict about removing valuation, then we cannot say something is normal and something else is not normal; we just identify neutral categories of behavior A, B, C, D.

much as correctness. Jen (2005) addresses robust design in systems across the range of biology, ecology, and engineering, and specifically mentions (p. 4) engineers explicitly designing for robustness beyond the range of the (external) specifications, beyond what is anticipated. The general topic of engineering or actualization processes for the expanded SE is discussed in Chapter 13.

The various categories are not inherent. They are relative in some cases to the systems (machines, agents) involved because expectations change not only across time, but also for different roles or circumstances. Examples include

- Three employees of differing capability: one is currently capable of producing 20 widgets per day, the second 30, the third 40. The employer sets different levels as adequate behavior for the employees (perhaps increasing across time), and pays them accordingly.
- A company produces widgets of type Y. The company challenges its employees to think of new company goals and ways the company can improve and grow. It also hires a consultant to come in and help the company strategize, reengineer, and think outside the box. What is expected behavior for the consultant would be out-of-range thinking for a junior employee. Normal behavior for the junior employee might be: The company should hire 10% more people and we could produce 10% more widgets. Normal behavior for the consultant might be: The company should switch from making widgets of type Y to making widgets of type Z.

The discussion thus far has associated valuation with each behavior type. Chapter 4 discussed the decoupling of behavior and valuation because a given behavior may be given different values (good or bad) by different observers or at different times. Sometimes it makes sense to couple them; other times not.

A different perspective on valuation and the relation between normal and abnormal characteristics is given in Dayan and Williams (2006). In their discussion of computational and mathematical models in neuroscience, they emphasize the point that an adequate understanding of disease requires not just models of disease, but also a context of models of normal functioning and how it is disturbed in disease. In general terms, we need to specify both normal and abnormal characteristics to understand and adequately address abnormal characteristics.

Specification Content Types

This section informally spells out the types of information that form the content of specifications. We reiterate that the information must be mutually understood to be a specification, but not all of it must be made explicit in a model. The first subsection provides a list segmented in three tables: internal, external, and interface specifications.* The remaining subsections add explanation and discussion of the tables. Some of the items identified here and their rationale and origin were described in the sidebar earlier in the chapter.

Content List

Tables 11.1, 11.2, and 11.3 show the types of information in an internal system specification. They reflect the relations shown in Figure 11.1 — namely, that an internal specification includes the external specifications of its components and connectors, and an external specification includes the specifications of its interfaces. Table 11.1 presents rules that describe the required information for

* An earlier version of these tables was presented in Hybertson (2005).

Table 11.1 Internal Specification Content

Rules	*UML/OCL*
Identifier of the model: unique in a defined namespace	Interaction name
Version of the model, if appropriate	Interaction stereotype or attribute
Mapping between views (if more than one view defined)	Associations, OCL, stereotypes, annotations
For each view in the model (if only one, model = view):	
Identifier of view	Interaction or class name
Identification of external models of system parts (components and connectors) in this view (may be general or specific models)	Interaction, component, deployment diagrams (or simple list)
Roles that may be assumed by parts in this view	Role in association end or interaction
Rules about stuff visible in this view: information, goods, definitions, invariants, initialization, states, relations, etc.	Class/object definition; stereotype
Process or behavior rules: flow of stuff, synchronization of parallel processes, interactions, state transitions, etc.	Interaction diagrams; state charts
Types of components allowed	Annotations, stereotypes, OCL
Types of connectors allowed (e.g., Internet-based; distributed types: signal, operation, flow (RM-ODP 1996); software types: procedure call, data access, linkage, stream, event, arbitrator, adaptor, distributor (Mehta, Medvidovic, and Phadke 2000)	Associations specialized via stereotypes
Attachment of connector interfaces to component interfaces	Interaction diagrams
Protocol specification: format, sequence, error checking, properties	Associations, OCL or stereotype or annotation
System-wide constraints on supporting environment (e.g., physical or information technology infrastructure)	Annotations, stereotypes, OCL
Implementation rules (e.g., use certain suppliers of material, or certain vendor products, or follow certain standards)	Annotations, stereotypes, OCL
Creation, use, and deletion of resources in system	Annotations, stereotypes
Other constraints the system places on each component, or on each connector (e.g., security constraints or policies)	Annotations, stereotypes, OCL
For each component and connector: whether shared or contained	Annotations, stereotypes, OCL
For each component and connector: If inclusion in system is uncertain, probability of inclusion	Annotations, stereotypes, OCL
For each component and connector: if inclusion in system is partial, fuzzy measure of inclusion	Annotations, stereotypes, OCL
External specification of each part (by inclusion or reference or hyperlink)—See Table 11.2	Class or object name
Optional information—e.g., architecture style, view type, explanatory notes, correspondence information	Annotations, stereotypes, OCL

Source: Hybertson, D. Using a specification approach to facilitate component testing. In *Testing Commercial-off-the-Shelf Components and Systems*, S. Beydeda and V. Gruhn, Eds. New York: Springer-Verlag, p. 213–237. 2005. Modified with permission of Springer Science + Business Media.

Table 11.2 External Specification Content

Rules	*UML/OCL*
Identifier of the model: unique in a defined namespace	Class or object name
Version of the model, if appropriate	Class stereotype or attribute
Mapping between views (if more than one view defined)	Associations, OCL, stereotypes, annotations
Cost of system (e.g., purchase, maintenance upgrades, cost-per-use)	Interface annotation or stereotype[1]
For each view in the model (if only one, model = view):	
Identifier of view	Package or class name
Identifiers of interfaces that provide a service	Interface name
Version of provider interface	Interface stereotype or attribute
Identifiers of interfaces that require a service	Interface[2]
Version of receiver interface	Interface stereotype or attribute
System invariant over all provided services and all required services (e.g., policies or other constraints)	Interface invariant in OCL or annotation
System externally observable properties	Interface attributes in OCL or annotation
Service or feature interactions among the interfaces of this system (sequence constraints, how one affects others, ...)	Interface invariant in OCL or annotation
Specification of each interface (by inclusion or reference or hyperlink)—See Table 11.3	Interface
Optional information—e.g., view type, explanatory notes, correspondence information	Annotations, stereotypes, OCL

[1] This assumes a cost is associated with the system; if cost is instead associated only with each interface, then this item will not apply.

[2] For all cases where Interface appears in the table, an additional mapping to UML for systems in a connector role is Association or Association End.

Source: Hybertson, D. Using a specification approach to facilitate component testing. In *Testing Commercial-off-the-Shelf Components and Systems*, S. Beydeda and V. Gruhn, Eds. New York: Springer-Verlag, p. 213–237. 2005. Modified with permission of Springer Science + Business Media.

an internal specification. Table 11.2 shows the rules for an external specification, and Table 11.3 shows the rules for an interface specification.

A given specification has a defined scope and purpose, so in many cases only a select subset of the information called for in these tables will be needed in the specification. More specifically, an agent that generates a model typically intends to leave certain specification items unconstrained, so those items would not have rules defined. This means that three nested sets of items exist. From most to fewest items, they are

1. The full set of constrainable specification items: Tables 11.1 through 11.3.
2. The subset of (1) constrained in a given specification: the set of defined rules, mutually understood.
3. The subset of (2) made explicit in the model associated with the specification: the set of rules expressed in one or more given languages and notations.

Table 11.3 Interface Specification Content

Rules	*UML/OCL*
Identifier of the interface: unique in a defined namespace	Class or object name
Version of the interface, if appropriate	Class stereotype or attribute
Service if provided through this interface:	
Access method	Interface operation
Semantics:	Interface operation in OCL or stereotype or annotation
What it requires of its users: precondition	
What it provides its users	
Part of invariant associated with service	
Provided post-condition	
Provided properties, QoS (performance, precision, reliability, throughput, capacity…)	
Policies or other constraints under which it operates	
Provided QoE (quality of experience) assessed by user	
Probability of meeting expectations (post-conditions, QoS, QoE)	Interface operation in OCL or stereotype or annotation
Ways of exceeding expectations and probability of each	Interface operation in OCL or stereotype or annotation
Ways of failing to meet expectations and probability of each	Interface operation in OC or stereotype or annotation
Cost of service (CoS)	Interface operation in OCL or stereotype or annotation
Service if received through this interface:	
What it offers its providers: precondition	Interface operation in OCL or stereotype or annotation
What it requires of its providers: usage dependencies	Interface operation in OCL or stereotype or annotation
Part of invariant associated with service	
Required post-condition	
Required properties, QoS (performance, precision, reliability, throughput, capacity…)	
Required QoE (quality of experience)	
Probability of meeting expectations (post-conditions, QoS, QoE)	Interface operation in OCL or stereotype or annotation
Cost of service (CoS) constraints	Interface operation in OCL or stereotype or annotation
Optional information—e.g., explanatory notes, correspondence information	Annotations, stereotypes, OCL

Source: Hybertson, D. Using a specification approach to facilitate component testing. In *Testing Commercial-off-the-Shelf Components and Systems*, S. Beydeda and V. Gruhn, Eds. New York: Springer-Verlag, p. 213–237. 2005. Modified with permission of Springer Science + Business Media.

All three tables also associate each set of rules with corresponding UML or OCL representations. It is important to point out that a wide range of languages and methods are available for representing the rules and information in these tables, as discussed in Chapter 8 — including the full spectrum of elements in Figure 8.3.

A common item in all three tables is optional information, including correspondence information. In general, correspondence set information should not be included in any model. As part of decoupling, correspondence sets should be spelled out separately, in part to enable the use of models and systems in multiple correspondences. This item is included primarily because the alternative definition of specification allows it to be applied to correspondence sets in some cases. When it is so applied, the correspondence can show up in the specification but still should not appear explicitly in the model unless the intent is to hardwire the model to a given system or category of systems.

Internal Specification Content Discussion

In general, an internal specification defines the rules for system parts, their structure and interactions, and the stuff they exchange. It covers concepts such as protocols, orchestration, coordination, sequence constraints, competition, state transitions, business processes, service interaction, agent interaction, and group/social dynamics, as well as patterns or styles such as service provide/consume style, peer to peer, client-server, controller-slave, publish-subscribe, and push-pull.

Much of the basis for and details of internal models and interactions were presented in Chapter 6. This section discusses some of the elements in Table 11.1 that need further elaboration.

Internal Specification Views

Sample views of an internal specification are summarized in Figure 11.2. The types of views are different sets or types of parts, language, and model specificity. Additional examples of internal views, such as architecture views, were discussed in Chapter 10.

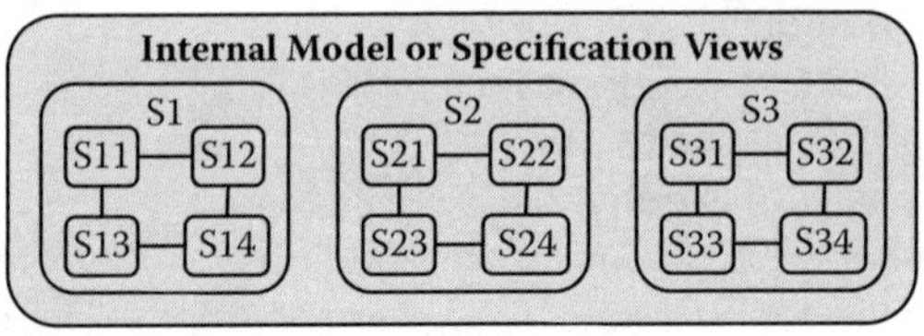

- Type of parts: Different internal views often have different types of parts. Examples: functional vs physical architecture view; or an enterprise architecture model that has an organizational view, an information exchange view, a budget view, and an IT system view.
- Language: An internal model can be written in different languages or domains of discourse. Example: Using RM-ODP, S1 might be computational language, S2 engineering language, S3 technology language.
- Model specificity: An internal model can be written at different levels of specificity. Example: S1 might be written at a general level where the parts are general categories, and S2 at a specific level where the parts are specific systems.

Figure 11.2 Sample internal specification view types.

Internal Interaction Involving Agents

More languages and methods are being developed to support the modeling and engineering of interactions involving humans and other agents. Some examples include:

- Agrawal et al. (2007a) have developed a Web services human task specification that defines human tasks in a way that can be part of a business process model or a service interaction. The same group (Agrawal et al. 2007b) has incorporated the human task specification concepts as a proposed extension of the Web Services Business Process Execution Language (WS-BPEL), called BPEL4People.
- Holmes et al. (2008) extend the BPEL approach to a more general view-based, model-driven approach that defines human business processes at different levels of abstraction.
- Schall, Truong, and Dustdar (2008) have defined a human-provided services (HPS) framework that enables the integration of software and human services in Web-scale collaborations.
- Some of the agent specification languages (Chapter 8) are quite relevant to this domain. For example, the Cognitive Agents Specification Language (CASL) supports specification of what an agent knows or believes, and what their goals are (Shapiro, Lespérance, and Levesque 2002). Ciobanu and Lucanu (2006) describe a specification language for coordination of autonomous objects.

External Specification Content Discussion

External Specification Views

Common views of external specifications are illustrated and explained in Figure 11.3. The variants are context specificity, language, and model specificity. We now briefly explore the distinction in the figure between the two engineering roles because they illustrate the matching of context specificity views. Suppose a system architect looks for a component to fulfill a need in his system. This could apply to component-based development, or to a system being integrated as a component into a larger system of systems. Simplistically, one could assert that there is one external component specification of interest to a component user — namely, the one defined in the system architecture. The architect/user simply finds a component whose external specification matches the architecture component specification, that is, selects an available component if and only if specification of available component *equals* specification of desired component. But several factors motivate a more realistic criterion. First, two specifications are rarely, if ever, equal. We need to think of "satisfies" rather than "equals," and we need to define what it means for one specification to satisfy another specification. Zaremski and Wing (1997) and Hemer and Lindsay (2002) discuss issues in specification matching from a formal perspective. Second, components evolve over time as COTS vendors produce new versions. The specification approach needs to support a component user in determining whether a new component version still meets the system requirements. Third, systems also evolve, and the system requirements for a component may change. The architect needs to assess whether the existing component meets the revised system requirements.

Interface Specification Content Discussion

All the elements of provided and required services in an interface specification can be defined statically or as a function of time. For traditional machines, the elements are typically static

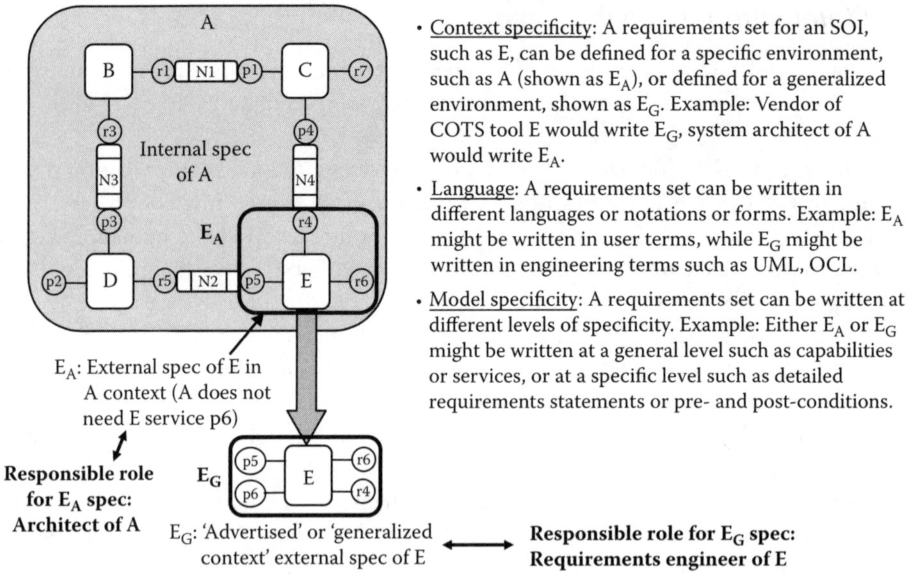

Figure 11.3 Common external specification view types.

definitions, but for organic systems such as people, organizations, economies, etc., the elements may be defined as a function of time. There are many ways characteristics, including behavior, can change over time. Two prominent ones are adaptation and growth. The normal expectation is that organic systems will grow and learn and improve, as well as adapt, over time, and that expectation can be reflected in the specification. The models used to express learning and growth can range from simple linear (or nonlinear) functions of time to more sophisticated human or agent learning models. Models of adaptation can range from traditional feedback models to more sophisticated models of agent adaptation. An example of the latter is Zhang and Cheng (2006), who took a temporal logic language called LTL (linear temporal logic) and added an adaptation operator to it to formally model adaptation.

The required service item is intended to include all dependencies, even if the interface is virtual or indirect. An example is the dependency of a software system on a computer processor. The software does not have an interface that calls or invokes a computer processor service the way it calls or invokes or sends messages to other software modules. But this dependency needs to be identified in an interface in a MOSES specification. (Remember that this does not mean it needs to be explicit in the model but it does need to be understood by all parties to the specification.)

Uncertainty in the specification of characteristics, whether traditional system or agent, is accounted for in the probabilistic items. That topic is the subject of the next section.

Uncertainty: Probabilistic Specifications

To introduce our discussion of probabilistic specifications, we will ask this question: How do the notions of deterministic, probabilistic, certainty, knowledge, predictability, and randomness relate to each other and to specifications of system behavior? Is there a binary distinction between deterministic, certainty, knowledge, and predictability on one hand, and probabilistic and random on the other? If certainty implies a probability of 1.0, then what does uncertainty or randomness imply? A probability of 0.0? Or 0.5? To answer all these questions requires some additional discussion but we can answer the binary question now. We take deterministic to be binary. System behavior cannot be "sort of" deterministic; it either is or is not deterministic. And if it is not deterministic, it is probabilistic and includes the element of randomness. But the other concepts are fuzzy. We can be "sort of" certain and have partial knowledge with some degree of predictability.

There are several ways of characterizing uncertainty. In this book, the range of certainty and uncertainty is reflected with probability. Probability values range from 0 to 1. If we are assessing the probability P that a certain condition holds, then P = 0 means that the condition certainly does not hold, P = 1 means that it certainly does hold, and P = .5 means there is a 50-50 chance it holds. P = 0 and P = 1 both represent certainty; the range in between 0 and 1 represents varying degrees of uncertainty.

A note on randomness in the context of probabilistic specifications and uncertainty: our assumption is that some uncertainty exists in our knowledge of the behavior of all systems. The uncertainty may be very small, for example, .0000001, but it is something greater than zero. Based on the uncertainty assumption, we claim that the appropriate way to model, and therefore specify, the behavior of all systems is that they reflect some degree of uncertainty, and that they behave randomly. That might sound odd — to model a system that exhibits near-certain behavior as behaving randomly — but after we explain, it will sound quite reasonable.

The subsections that follow discuss several aspects related to probabilistic specifications. First is a discussion of the probability mass and probability density functions that are used to characterize random variability and uncertainty. This is followed by an engineering example using conditional probabilities. The last two subsections briefly describe the issue of estimating probability distribution/density functions (PDFs) of human behavior, and list other uncertainty methods.

Probability Distribution/Density Function

To arrive at an understanding of the above issues and distinctions requires a brief but reasonably intuitive foray into the concept of probability density and probability mass functions. A random variable X can be characterized in terms of the probability of the values it can take. In the case of a discrete variable, the probability mass function assigns to each discrete value a probability that X will take that value. If X can take any of n values x1, x2, …, xn, then the probability mass function assigns a probability for each value: P(x1), P(x2), … P(xn).

The special case where all probabilities are the same — that is, P(x1) = P(x2) = … = P(xn) — is called a uniform probability distribution, in which case we also know that the probability of each value = 1/n. In many cases, we use uniform probability mass (or density) when we have no evidence upon which to judge that one outcome is more likely than any other. Perhaps the two most common examples are flipping a coin and rolling a die. Assuming the coin is not imbalanced, the probability of a coin flip resulting in heads is .5 and the probability of resulting in tails is also .5. Similarly, the probability of a roll of one die is 1/6 or approximately .167 for each of the six possible results (one through six dots).

But in the general case, the distribution function is not uniform. Consider the case of rolling two dice, in which the result is the total of the two, and therefore the set of 11 possible results is {2,3,4,5,6,7,8,9,10,11,12}. The probability of the 11 results varies because the number of ways

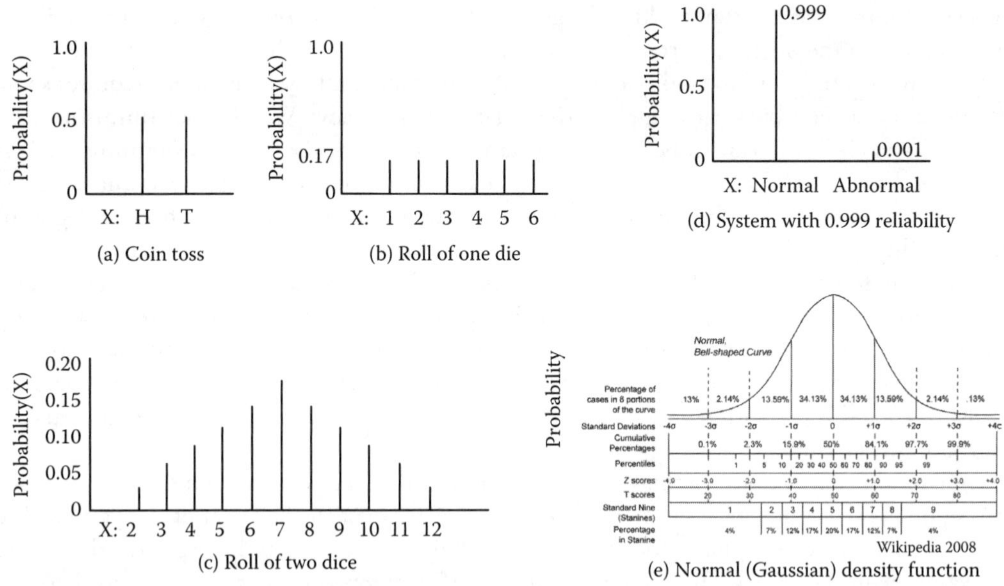

Figure 11.4 Examples of probability mass/density functions.

of obtaining the various results differs. There is only one way to obtain a result of 2 (both dice show 1) or 12 (both dice show 6), but there are six ways to obtain a result of 7 (the first shows 1, the second shows 6; or the first shows 2 and the second shows 5, etc.). Therefore, the probability of obtaining a result of 7 from rolling two dice is six times the probability of obtaining a result of 2. The probabilities are shown visually in Figure 11.4 but the important point of this example is that it shows a nonuniform probability mass function. The reason this is important is that it sets the stage for specifying the behavior categories of a system. If we have a system that exhibits normal specified behavior with probability .999 and unsatisfactory or abnormal behavior with probability .001, then we can characterize it as a highly nonuniform probability distribution function.

In the case of a continuous variable, the probability density function assigns to an interval [a, b] the probability that the value of X is in the inclusive range of a to b, or $P(a \le X \le b)$. The term "distribution function" refers to $P(X \le x)$, which is the definite integral (or sum) of the density (mass) function over all values of X less than x. The most prominent example of a continuous density function is the normal or Gaussian distribution, often called the bell curve because of its shape. It is characterized by a mean μ and a variance σ^2, and is therefore a family of functions (actually an infinite set of functions because the mean can be any real number and the variance can be any positive number). The density function f(x) of a normal variable x is

$$\frac{1}{\sigma\sqrt{2\pi}}\exp\left(-\frac{(x-\mu)^2}{2\sigma^2}\right)$$

where μ is the mean, σ^2 is the variance, and σ is the standard deviation.

Figure 11.4 shows the mass or density function for the cases just discussed: (a) one coin toss; (b) roll of one die; (c) roll of two dice, where the outcome is the total of the two; (d) system with reliability .999, that is, system with normal behavior 99.9% of the time and unsatisfactory behavior 0.1% of the time, and it is not known when the unsatisfactory 0.1% will occur; (e) normal or Gaussian distribution, taken from (Wikipedia 2008). These distribution functions are characterized as follows: (a) and (b) are both uniform and discrete; (c) and (d) are both discrete but not uniform; and (e) is neither uniform nor discrete.

If a system has an analog (or other continuous) input stream and an analog (or other continuous) output stream, we can characterize or specify the behavior of the system in terms of a continuous probability density function. But many of the system behaviors we specify in systems engineering, even complex systems, are discrete, and therefore we characterize the behavior in terms of a discrete probability distribution function.

If we have more information, we can specify the probabilities more specifically. An example is a light bulb. If the reliability of a light bulb is .999, we do not assume that the failure has equal probability of occurring at any time. First, the light bulb only burns out when it is on, so during periods when it is off, the probability of failure during those times is zero — assuming there is no other destructive action (such as the light bulb being physically smashed). Second, the probability of failure is significantly higher for a short interval of time (a second or less) after the light is turned on each time. After that, the probability decreases considerably to a low level until the light is turned off. Third, over time, the probability of failure increases slowly as the bulb approaches its expected life span. Because of these factors, we can specify the probability distribution in a way that has higher fidelity to the system than the assumption that at all times it has a probability of .001 of failure.

We are now in a position to answer the questions posed above. Given the assumption that deterministic is a binary concept that means complete certainty and a probability of 1.0, we characterize the behavior of all "real" systems this way:

- None is deterministic.
- All behave in a probabilistic manner.
- All behave randomly, in the sense that we do not know what their next behavior will be.
- None is completely predictable. It is necessary to frame the concept of "predictability" with a range that we can claim to include the characteristic in question with a specific probability (e.g., the variable X will fall in the interval (a, b) with probability of .95); such expressions are called "confidence intervals." The size of the range will be determined by both the form of the distribution used and its parameters, reflecting our state of knowledge. Consequently, systems with the least predictable behavior are described by uniform distributions, with equal probability over all possible behaviors; more predictable systems are described by probability distributions that are consistent with our detailed knowledge of the variability of system behavior.
- Because we rarely know the actual probability distribution, we say that systems exhibit an estimated range of predictability based on an estimated distribution. System behavior reflects the actual distribution over time, not the estimated distribution; but the good news is that the estimated distribution can be updated by recording or sampling actual behavior, and that should lead to the estimate moving closer to the actual — assuming the actual remains stable.
- More knowledge of a situation or context leads to better estimates of probability distributions for the situation or context.

Probability, PDF, and Knowledge

We discuss two small examples. The first is a case where we have a good estimate of the PDF but do not know the outcome of an event in the past. In the second case, we do not know about the occurrence of an event and we also do not have a solid basis for a PDF.

The discussion in Chapter 9 pointed out that uncertainty exists about past and present events as well as future events. Based on that, we can define probabilities about any uncertainties — past, present, or future.

Suppose a friend of mine is on the phone at time T1 and says he is about to flip a coin, and asks what the probability is that the result will be heads. I have a good estimate of the PDF of the possible outcomes: Both heads and tails have a probability of .5. Therefore, my estimate of the probability that the outcome of this specific toss is heads is .5; that is what I tell him, and he agrees. Now he tosses the coin, and at time T2 he observes that it comes up heads but does not tell me the result. This observation changes his knowledge of the event, but not mine. Now he asks me again: "What is the probability that the result was heads?" I repeat my answer: "The probability is .5 that it was heads." For him at this time T2, the probability that it was heads is 1.0, and the probability of it being tails is 0. Then at time T3 he tells me the result was heads. Now we both agree that the probability that it was heads is 1.0.* The estimated probabilities were based on knowledge of the PDF and presence or absence of knowledge of the event outcome. Probabilities do not have an independent or inherent existence in a situation; they depend on the knowledge of the observer.

The second topic is PDF: Assume that today is April 17 and we want to estimate the probability that it will it rain a month from now, on May 17, in Hong Kong; and we also assume that we know nothing about the climate or rain patterns in Hong Kong. How do we estimate the probability? Because we know nothing about the PDF, we could assume a uniform PDF of rain or not rain, and estimate the probability of rain on May 17 (or any other day for that matter) to be .5. That is a very weak basis for estimating. Now suppose we obtain a better basis for the PDF. We find out the following two items of information:

1. Rainy days as a proportion of all the days of the past century in Hong Kong is .37
2. Rainy days as a proportion of all the days in May of the past century in Hong Kong is .47

We interpret (1) as an estimate for the PDF of rain on a randomly selected day of the year in Hong Kong; and (2) as an estimate for the PDF of rain on a random day in May in Hong Kong. Which of these two PDF estimates gives a better basis for estimating the probability that it will rain on May 17? The first is

* Intentionally ignored are certain other probabilities that would reduce it from 1.0, such as: He did not flip a coin at all; or he might have not seen the outcome correctly; or he might not be telling me the correct outcome.

a larger sample, which generally means a better basis; note that in this case, the data are from two distinctly different distributions. However, the second indicates that there are cyclic seasons and May is a rainy month, so we would select (2) over (1) as our estimate of the PDF because it is the more specific population of which our event is a member. That is, our estimate of the PDF of the variable M = rain on a day in May is: P(rain) = .47; P(not rain) = .53. Based on all this, we would estimate that the probability that it will rain in Hong Kong on May 17 is .47. If the only PDF estimate we had was (1), and we had no knowledge of Hong Kong rain cycles, we would estimate that the probability is .37.

The collective point of the two examples is that our estimates of probabilities in system specifications are themselves relative and uncertain. That is, we make uncertain estimates of the uncertain behavior of systems. Does that mean the method is weak? No, it is still a useful method in specifying the uncertainty of behavior in systems.

Example Using Conditional Probabilities

The example shown in Figure 11.5 shows the role that conditional probabilities can play in modeling and specification. Introductory information includes:

- System C requires a service (r1) that system B provides (p1). To provide p1, system B depends on two services, r2 and r3, that must be provided (p2 and p3) by some service provider.
- Connectors N1, N2, and N3 will be used to deliver the services as shown in Figure 11.5.
- Situation 1 is prearranged service provision: system B has identified service providers S2 for p2 and S3 for p3; that is, S2 and S3 agree to the r2 and r3 respective contract requirements for p2 and p3.
- Situation 2 is dynamic just-in-time service provision: system B has not identified service providers for p2 or p3, but estimates that providers will be found, when needed, with probability P(find(p2,p3)). Implied in "finding" is that the providers will agree to the r2 and r3 respective contract requirements for p2 and p3.

In both situations, satisfaction of both ends of all contracts and the presence and proper functioning of all connectors are uncertain. We now develop an example that assumes Situation 1 and ignores the connectors — that is, makes the (false) assumption that the connectors will be present and functioning properly with probability 1.0. It would be a straightforward

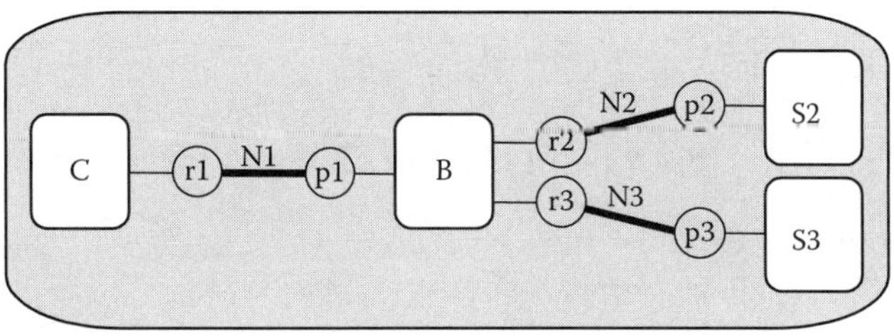

Figure 11.5 Example structure for conditional probability.

extension to add connector uncertainty to this example. In proceeding with Situation 1: What is the degree of certainty that C's required service r1 will be obtained or satisfied? And how does it relate to all the uncertainties in the service chain? We use probability as a measure of certainty and the concept of conditional probability to help answer these questions. First some notation:

- P(x) = Probability that x is satisfied
- P(x | d) = Conditional probability that x is satisfied given that *all* x dependencies are satisfied
- P(d) = Probability that *all* x dependencies are satisfied

Observations:

- P(x) = P(x | d) · P(d) + P(x | ¬d) · P(¬d), where "¬d" denotes "not d", and "·" denotes the multiply operator
- P(x) ≤ P(x | d); P(x) = P(x|d) only if (1) x has no dependencies, or (2) all x dependencies are satisfied, that is, P(d) = 1.0

The first observation is true because P(x) = P(x | d) · P(d) + P(x | ¬d) · P(¬d) by Bayes' theorem; the events "d" and "¬d" being collectively exhaustive and mutually exclusive. The first part of the second observation is true because all probabilities must be greater than or equal to zero; therefore P(x) ≤ P(x | d). The second part is true because (1) P(x) = P(x | d) is the definition of independence; and (2) if P(d) = 1.0, then P(¬d) = 0 and P(x) = P(x|d) · P(d) + P(x | ¬d) · P(¬d) = P(x | d) · (1) + P(x | ¬d) (0) = P(x | d).

In situation (1), the service chain dependencies are clear: ignoring the connectors N1, etc., satisfying r1 service depends on the contract ends r1, p1, r2, r3, p2, and p3 being satisfied.

In general, conditional probabilities (P(x | d)) are isolated, while nonconditional probabilities (P(x), P(d)) are aggregated and encapsulating. We will develop the example and use it to explain these probabilities and show how they are useful.

The equations that follow reflect an assumption that all identified probabilities are independent:

1. P(r1) = P(r1 | d) · P(dr1) + P(r1 | ¬dr1) · P(¬dr1); but P(r1 | ¬dr1) = 0 (because we ***assume*** that r1 cannot be satisfied if ¬dr1); so P(r1) = P(r1 | d) · P(dr1)
2. P(dr1) = P(p1) (because this example ignores connector uncertainty)
3. P(p1) = P(p1|d) · P(dp1) (same reason as (1))
4. P(dp1) = P(r2) · P(r3)
5. P(r2) = P(r2 | d) · P(dr2)
6. P(dr2) = P(p2)
7. P(r3) = P(r3 | d) · P(dr3)
8. P(dr3) = P(p3)
9. Therefore: P(r1) = P(r1 | d) · P(p1 | d) · P(r2 | d) · P(r3 | d) · P(p2) · P(p3)

Who has control? ---C--- --------------B--------------

C has complete control of P(r1 | d), and B has complete control of the three conditional probabilities shown; these are isolated from the effect of dependencies by their definitions.

Now let us proceed with a numerical illustration of the example:

- Given: P(r1|d) = .9999; P(p1|d) = .9995; P(r2|d) = .9999; P(r3|d) = .9999; P(p2) = .9995; P(p3) = .999. Determine P(r1) and P(p1)

The equations are listed in the order they are computed:

1. 8) P(dr3) = .999 = P(p3)
2. 7) P(r3) = .9989 = P(r3 | d) · P(dr3)
3. 6) P(dr2) = .9995 = P(p2)
4. 5) P(r2) = .9994 = P(r2 | d) · P(dr2)
5. 4) P(dp1) = .9983 = P(r2) · P(r3)
6. 3) **P(p1)** = **.9978** = P(p1 | d) · P(dp1)
7. 2) P(dr1) = .9978 = P(p1)
8. 1) **P(r1)** = **.9977** = P(r1 | d) · P(dr1)

How are probabilities and equations like this used? A couple of applications are briefly described.

Application 1: Assume a *service-oriented architecture,* in which B provides service p1 to the service consumer C who requires service r1. In this case, B assumes responsibility for satisfying its dependencies r2 and r3. Therefore, B publishes its service with a specified QoS of P(p1) = .9978.

Application 2: Assume a *component-based system* in which the architect is assembling components C, B, S2, and S3. In this case, the architect assumes responsibility for satisfying B's dependencies r2 and r3 in this environment. Therefore, the vendor of component B publishes its component specification with a QoS of conditional probability P(p1|d) = .9995.

Why the difference? An SOA service provider must provide the service, which means it must account for the uncertainty in its runtime environment. A component vendor does not provide a service, but rather provides a component that can be installed in multiple environments to provide a service. The component vendor cannot specify the probability that its dependencies will be satisfied; that varies by environment and must be specified by each architect that uses the component.

It seems that probabilities, and the conditional/unconditional distinction, provide a rather elegant and precise way of making this useful distinction.

Now assume that A can partially compensate for B not providing the promised service; for example, B is degraded, or is not performed at all. By "partially compensate" we mean that $0.0 < P(A|\neg B) < P(A|B)$ [where $P(A|\neg B)$ means probability of A given "not B", i.e., given less than expected service from B]. There are two different statements made in this expression:

- $0.0 < P(A|\neg B)$ says that A is not dead in the water because of the degraded or lost B service. In classical engineering, A would not be able to perform its function if its dependency was not satisfied, so the statement would be $P(A|\neg B) = 0.0$.
- $P(A|\neg B) < P(A|B)$ says that A cannot fully compensate for the degraded or lost B service. If A could fully compensate for it, A would not depend on B, and the statement would be $P(A|\neg B) = P(A|B)$, which is another way of saying that A is independent of B.

In effect, the combined statement says that if B does not do its job, A can still function, but not as well. Note that our statement about classical engineering assumes no fault tolerance or error recovery. Clearly, much work has been done in fault tolerance and error recovery for many years in most areas of engineering and systems engineering. We are not trying to ignore that work. On the contrary, we are trying to recognize it more explicitly in the MOSES approach to specification, and build on it as appropriate in the areas of self-healing and self-adaptation that the larger scope of autonomous agents and complex systems bring into this arena. The above formulation of the probabilities of an agent compensating for abnormal behavior outside its control is a first step toward that goal.

Benefits of this approach to specification of uncertainty include:

- Unifies specification of both traditional and complex systems; the same approach applies to both
- Supports both separation and aggregation of elements of uncertainty throughout the service chain
  - *Separation:* Helps a given component or view separate what they control from what they depend on; this helps facilitate accountability, by holding people or elements accountable for what they control; the separation into individual elements of uncertainty also provides a stronger basis for the aggregate probabilities
  - *Aggregation:* Supports encapsulation, hiding details from a given person, role, or view and revealing only what they need to see; also, the aggregate does not ignore the hidden dependencies — they are accounted for in the aggregate but are hidden when appropriate
- Scales well; approach works for any length of service chain, at any level of composition
- Is flexible; can support a large number and variety of views and stakeholders, and all the views will be consistent because they are based on the same framework
- Provides a way to incorporate uncertainties of fault tolerance, self-healing, and self-adaptation that characterize the larger scope of autonomous agents and complex systems

Other Uncertainty Methods

A list of uncertainty methods was presented by Rajabally et al. (2005). They include the following formalisms:

1. Probability theory/Bayesian belief networks
2. Interval analysis
3. Rule-based systems/certainty factors
4. Evidential reasoning/Dempster Shafer theory
5. Possibility theory/fuzzy logic

Another approach to uncertainty is modal logic, or the various logics of possibility. The Stanford Encyclopedia of Philosophy identifies some of the modal-related logics (SEP 2003):

Logic	*Symbols*	*Expressions Symbolized*
Modal Logic	□	It is necessary that ...
	◊	It is possible that ...
Deontic Logic	O	It is obligatory that ...
	P	It is permitted that ...
	F	It is forbidden that ...

Temporal Logic	G	It will always be the case that ...
	F	It will be the case that ...
	H	It has always been the case that ...
	P	It was the case that ...
Doxastic Logic	Bx	x believes that ...

The MOSES approach uses some of these in different ways. First, as this chapter shows, it uses probability theory extensively, as the primary tool for specifying uncertainty. Second, it uses modal logic primarily in expressing relations between model and system in a correspondence set, but also uses it in models and specifications themselves. Third, it considers fuzzy logic to be about partialness, not uncertainty. This is addressed in the next section.

Partialness: Fuzzy Specifications

Fuzzy specifications make use of fuzzy logic and fuzzy set theory. The basic concept is that concepts that we normally think of as being discrete — such as whether an element is a member of a set, whether something is true or false, or whether a person is in a room — are instead conceived as being a matter of degree, with a range from 0 to 1. So an element can be halfway in a set (0.5) or a proposition can be half true (0.5).

In MOSES, fuzzy specifications can be used for a number of purposes. Examples include the degree to which a component is in a system; or the degree to which a system is in a model region — that is, the degree to which a system satisfies a specification; or perhaps the degree to which a system is encapsulated.

Fuzzy set theory is sometimes included along with probability in the set of techniques for representing uncertainty, as in the previous section. As shown in Figure 11.6, indicators of fuzziness and of uncertainty both have a range from 0 to 1. However, the meaning is different. Probability is an indicator of uncertainty about whether something is true, such as whether the next coin flip will be heads or tails. Fuzziness is an indicator of partialness or gradualism. If I am standing in the doorway and am half in the room and half out, the 0.5 fuzziness measure does not reflect uncertainty about whether I am in the room; rather, it reflects the partialness of me being in the room: I am certainly halfway in the room. Therefore, fuzziness in MOSES is not used as an uncertainty mechanism.*

Examples

Here we give some very brief sketches that illustrate aspects of the specification content provided in this chapter.

A Human Performance Example

This example specifies human performance as a member of an organization.

* However, fuzziness can be used in conjunction with uncertainty. A situation can be modeled as simultaneously fuzzy and uncertain.

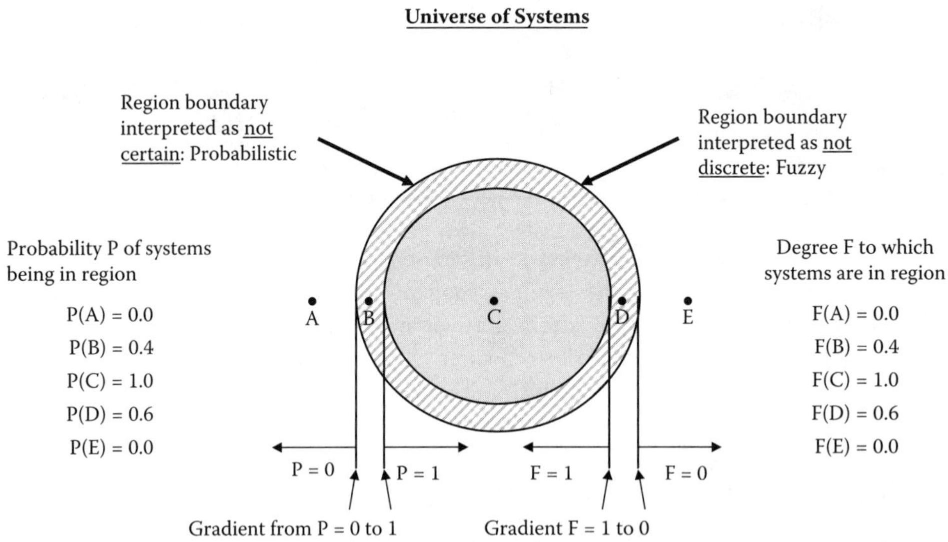

Figure 11.6 Fuzzy specifications use same gradient measure as probabilistic specifications but with different meaning.

We have discussed how defining a role helps narrow the space for specifying human behavior. In fact, a practical assumption when applying the specification approach in this chapter to a human or other agent is that the agent is acting in a specified role. "Person" has an almost unlimited and hard-to-define outcome space. But role behavior is tractable: parent, employee, soldier, factory worker, software agent in XX domain or context, geospatial analyst, researcher, complex systems researcher, complex systems researcher at Santa Fe Institute (SFI), complex systems researcher at SFI in evolutionary dynamics. The example could be something like a factory worker with a well-defined job; but to show the range of the approach in more complex situations, we choose to specify a researcher, which typically is a less circumscribed activity for which it is difficult to specify performance. The specific role for which we will sketch an external specification of expected future behavior is the last one mentioned, that is, complex systems researcher at SFI in evolutionary dynamics.*

The expected behavior could be defined as getting a paper published in evolutionary dynamics. QoS = one paper accepted by a refereed journal in evolutionary dynamics in one year, or one or more publications posted on the SFI public Web site that collectively get > H hits in the next year or > D downloads in the next year with probability of .9. We do not specify what the content of the publication will be, only the domain of evolutionary dynamics (with probability < 1.0, since complex systems researchers are often multi-disciplinary), and some measurable QoS, as suggested above. SFI may in fact not care what the specific contents are, but only in advancing this field and getting recognition and continued funding. So this requirement specification of the researcher "system" is at the appropriate level of generality. The preconditions or

* This example is used here to illustrate specification elements. The example is revisited in Chapter 12 and parsed in a different way, in terms of mapping to the external and internal type models and to traditional artifacts such as a requirements document.

Table 11.4 Possible Researcher Behavior in Each Category

#	Category	Possible behavior
1	Out of range — interpreted as good	Researcher discovers or initiates whole new area of research
2	In range, better than expected	Researcher publishes two or three papers instead of one
3	What is expected	Researcher publishes one paper; satisfies QoS
4	In range, less than expected	Researcher does not write paper, or paper is published but not in a refereed journal
5	Out of range — interpreted as bad	Researcher destroys lab

dependencies include get funding for research, stay at SFI, retain health, etc. Note that getting funding for research may be a precondition if SFI assumes responsibility for funding, or may instead be part of the expected behavior of the researcher if each researcher is expected to find their own funding.

Now consider possible behavior during the next year of the researcher in each of the five behavior categories. See Table 11.4. Note that these behaviors are not all mutually exclusive. The researcher could (1) initiate a new research area, (3) publish a paper, and then (5) get angry and destroy the lab.

Other Sketches

Three brief sketches are given to indicate the applicability range of the specification approach.

Organization

Suppose we want to specify certain characteristics of an organization or enterprise in its current state and its desired state.

External model: Let's assume the precondition is the current actual condition: State 1 (time 1 is now; function or provided service = produces widgets; properties: current profit, size, reputation, #3 in market position, not in 100 best companies to work for, etc.).

Assume post-condition is desired condition: State 2 (time 2 is 18 months from now; same function; 20% more profit, larger or smaller size, better reputation, #1 in market position, in 100 best companies to work for, etc.).

Now the organization just has to redesign itself, that is, define an internal model that will enable it to satisfy the external specification. Example: replace certain components, such as the CEO. That task will not be pursued in this sketch.

Army

Consider an insurgency–counterinsurgency situation.

Internal model: The components include an occupying army, the host nation leaders, host nation army, insurgents, and local population. Define a current internal model of interaction in which the insurgency is causing significant havoc and destruction, has the sympathy of much of the local population, and the host nation army is weak. Then define the goal internal model, where the local population is opposed to the insurgents and supports the host nation leaders and army, and the insurgents are on the decline.

Then define a plan — an actualization process — to move from current model to goal model. An example could be tailoring the U.S. Army counterinsurgency manual (U.S. Army 2006), which is a general counterinsurgency domain model that contains multiple patterns and incorporates CS features of insurgencies.

Everyday Example: Safety Deposit Box

In a safety deposit box situation, a service consumer SC gives a service provider SP item(s) M in condition C at time T1. SP returns the items in the same condition at a later time T2 that will be chosen by SC. The charge for this service is $X/month. The key specification items of the provided service are:

- Precondition: M in condition C at time T1 (C is "as is"; it is not constrained by SP but is rather a benchmark for the invariant)
- Invariant: condition C of M at T2 = condition at T1
- Post-condition: M available to SC at T2
- QoS: respond on demand at unknown T2 with probability P ≥ .99; M will be in condition C at T2 with P ≥ .999
- Cost of service: $X times (Number of months between T1 and T2)

Not included in specification: store M in a safe place; store in dry place; store in temperature range between temp1 and temp2; limit number of employees with master key....

Scaling down the Approach for Software

We claim that the specification approach described in this chapter scales both up and down to any level desired. This sketch shows how it scales down.

Consider a few lines of software code:

```
x = 3;
y = y + 1;
a = x²;
z = sqrt(x² + y²);
r = z - 7;
```

Each line of code or statement can be considered a system, and the types of specification defined in this chapter, specifically an external specification, can be applied to each. This approach was essentially defined 40 years ago by Hoare (1969) in what became known as the Hoare triple: P {Q} R, where P is the precondition, Q is the statement, {Q} represents the execution of the statement, and R is the post condition. One way of characterizing the specification approach of MOSES is as a large extension of Hoare's method to all types of systems at all scales.

In addition, the collective set of software statements listed above can be considered as the internal model of a larger system, where the stuff exchanged among the interacting component statements consists of data and control.

References

Aagedal, J. and E. Ecklund. 2002. Modelling QoS: towards a UML profile. In *Proceedings of UML 2002*, Dresden, Germany, October 2002.

Agrawal, A., M. Amend, M. Das, M. Ford, C. Keller, M. Kloppmann, D. König, F. Leymann, R. Müller, G. Pfau, K. Plösser, R. Rangaswamy, A. Rickayzen, M. Rowley, P. Schmidt, I. Trickovic, A. Yiu, and M. Zeller. 2007a. Web Services Human Task (WS-HumanTask), Version 1.0. June. http://download.boulder.ibm.com/ibonl/pub/software/dw/specs/ws-bpel4people/WS-HumanTask_v1.pdf (Accessed January 11, 2009.)

Agrawal, A., M. Amend, M. Das, M. Ford, C. Keller, M. Kloppmann, D. König, F. Leymann, R. Müller, G. Pfau, G.K. Plösser, R. Rangaswamy, A. Rickayzen, M. Rowley, P. Schmidt, I. Trickovic, A. Yiu, and M. Zeller. 2007b. WS-BPEL Extension for People (BPEL4People), Version 1.0. June. http://download.boulder.ibm.com/ibmdl/pub/software/dw/specs/WS-bpel4people/BPEL4People_v1.pdf (Accessed January 11, 2009.)

Brucker, A. and B. Wolff. 2001. Checking OCL Constraints in Distributed Systems Using J2EE/EJB. Technical Report 157, Albert-Ludwigs-Universität Freiburg, July. http://www.brucker.ch/bibliography/download/2001/tr01.pdf. (Accessed January 11, 2009.)

Ciobanu, G. and D. Lucanu. 2005. A Specification Language for Coordinated Objects. In *Proceedings of the 2005 Conference on Specification and Verification of Component-Based Systems* (Lisbon, Portugal, September 05 - 06, 2005). SAVCBS '05. New York: ACM, 23-30. Available via online ACM SIGSOFT *Software Engineering Notes*, 31(2), March 2006. http://portal.acm.org/citation.cfm?id=1118537.1123062&coll=ACM&dl=ACM&idx=J728&part=newsletter&WantType=Newsletters&title=ACM%20SIGSOFT%20Software%20Engineering%20Notes&CFID=15943256&CFTOKEN=93064975 (Accessed January 11, 2009.)

Dayan, P. and J. Williams. 2006. Putting the computation back into computational modeling. *Pharmacopsychiatry*, 39(Suppl. 1):S50–S51.

De Miguel, M., J. Ruiz, and M. Garcia. 2002. QoS-aware component frameworks. In *Proc. Tenth Int. Workshop on Quality of Service*, 2002.

Gallo, E., M. Siller, and J. Woods. 2007. An ontology for the quality of experience framework. *Proceedings, IEEE Int. Conf. on Systems, Man and Cybernetics.* 7–10 October 2007, p. 1540–1544.

Goodhue, D. and R. Thompson. 1995. Task-technology fit and individual performance. *MIS Quarterly*, 19(2):213–236.

Hemer, D. and P. Lindsay. 2002. Supporting component-based reuse in CARE. In Proceedings of the 25th Australasian conference on Computer science—Volume 4, Melbourne, Victoria, Australia, 2002.

Hoare, C. A. R. 1969. An Axiomatic Basis for Computer Programming. *Communications of the ACM*, 12(1): 576–580, 583.

Holmes, T., Tran, H., Zdun, U., and Dustdar, S. 2008. Modeling Human Aspects of Business Processes — A View-Based, Model-Driven Approach. 4th European Conference on Model Driven Architecture Foundations and Applications, 2008, 246–261. http://www.infosys.tuwien.ac.at/staff/tholmes/publications/human_vbm.pdf. (Accessed January 11, 2009.)

Hybertson, D. 2005. Using a specification approach to facilitate component testing. S. Beydeda and V. Gruhn, Eds, *Testing Commercial-off-the-Shelf Components and Systems*. New York: Springer-Verlag, pp. 213-237.

Jen, E. 2005. *Robust Design: A Repertoire of Biological, Ecological, and Engineering Case Studies*. New York: Oxford University Press.

Kilkki, K. 2008. Quality of Experience in Communications Ecosystem. *J. Universal Computer Sci.*, 14(5): 615–624.

Luckham, D., J. Vera, and S. Meldal. 1995. Three concepts of system architecture. Technical Report CSL–TR–95–674, Computer Systems Lab, Stanford University, July 1995. http://citeseer.comp.nus.edu.sg/cache/papers/cs/1435/http:zSzzSznenya.ms.mff.cuni.czzSzthegroupzSzrel_workzSzLVM.pdf/three-concepts-of-system.pdf. (Accessed January 11, 2009.)

Mehta, N., N. Medvidovic, and S. Phadke. 2000. Towards a taxonomy of software connectors. In *Proc. 22nd Int. Conf. Software Engineering*, Limerick, Ireland, 2000.

Meyer, B. 1997. *Object-Oriented Software Construction, 2nd ed.* Prentice Hall.

Meyer, B. 2003. The grand challenge of trusted components. In *Proc. 25th Int. Conf. on Software Engineering*, May 2003.

OMG (Object Management Group). 2007a. Unified Modeling Language (OMG UML), Superstructure, V2.1.2. OMG Document Number: formal/2007-11-02 Standard document. http://www.omg.org/spec/UML/2.1.2/Superstructure/PDF. (Accessed January 11, 2009.)

OMG (Object Management Group). 2007b. Unified Modeling Language (OMG UML), Infrastructure, V2.1.2. OMG Document Number: formal/2007-11-04 Standard document. http://www.omg.org/spec/UML/2.1.2/Infrastructure/PDF. (Accessed January 11, 2009.)

OMG (Object Management Group). 2007c. Systems Modeling Language (OMG SysML™), V1.0. OMG Document Number: formal/2007-09-01 Standard document. http://www.omg.org/spec/SysML/1.0/PDF. (Accessed January 11, 2009.)

Paige, R. and J. Ostroff. 2001. A proposal for a lightweight rigorous UML-based development method for reliable systems. In *Proc. Workshop on Practical UML-Based Rigorous Development Methods 2001* (co-located with UML 2001), Toronto, Canada, October 2001.

Rajabally, E., S. Snape, S. Whittle, and P. Sen. 2005. Acknowledging Uncertainty in the Provision of Defence Capability: Insights from Literature. INCOSE.

RM-ODP. 1996. Open Distributed Processing — Reference Model Part 2: Foundations, International Standard 107462, ITUT Recommendation X.902; Part 3: Architecture, International Standard 107463, ITUT Recommendation X.903, 15 Sept 1996.

Schall, D., H. Truong, and S. Dustdar. 2008. On Unifying Human and Software Services for Ad-hoc and Process-centric Collaboration. *IEEE Internet Computing* 12(3):62–68.

Selic, B., G. Gullekson, and P. Ward. 1994. *Real-Time Object-Oriented Modeling*, New York: John Wiley & Sons.

SEP (Stanford Encyclopedia of Philosophy). 2003. Modal Logic.

Shapiro, S., Y. Lespérance, and H. Levesque. 2002. The Cognitive Agents Specification Language and Verification Environment for Multiagent Systems. *AAMAS'02,* 15–19 July, Bologna, Italy.

Szyperski, C. 1998. *Component Software: Beyond Object-Oriented Programming*. Addison-Wesley.

Urting, D., Y. Berbers, S. Van Baelen, T. Holvoet, Y. Vandewoude, and P. Rigole. 2002. A tool for component based design of embedded software. In *Proc. 40th Int. Conf. on Tools Pacific: Objects for Internet, Mobile and Embedded Applications*, Vol. 10, February 2002.

U.S. Army. 2006. *Counterinsurgency*. FM 3-24/MCWP 3-33.5. Washington, DC: Headquarters, Department of the Army. http://www.fas.org/irp/doddir/army/fm3-24.pdf. (Accessed January 11, 2009.)

Wang, G., A. Chen, C. Wang, C. Fung, and S. Uczekaj. 2004. Integrated quality of service (qos) management in service-oriented enterprise architectures. *Proc. 8th IEEE Int. Enterprise Distributed Object Computing Conf.* p. 21–32.

Wikipedia. 2008. Normal distribution and scales. http://en.wikipedia.org/wiki/Image:Normal_distribution_and_scales.gif. (Accessed January 11, 2009.)

Zaremski, A. and J. Wing. 1997. Specification matching of software components. *ACM Trans. Software Eng. Methodol.*, 6(4), October 1997.

Zhang, J. and B.H.C. Cheng. 2006. Using temporal logic to specify adaptive program semantics. *J. Systems and Software,* 79(10):1361–1369.

Chapter 12

Mapping and Unification

This chapter discusses the mapping of traditional artifacts and concepts to areas of MOSES described in previous chapters, as well as providing a summary of the unification represented in MOSES.

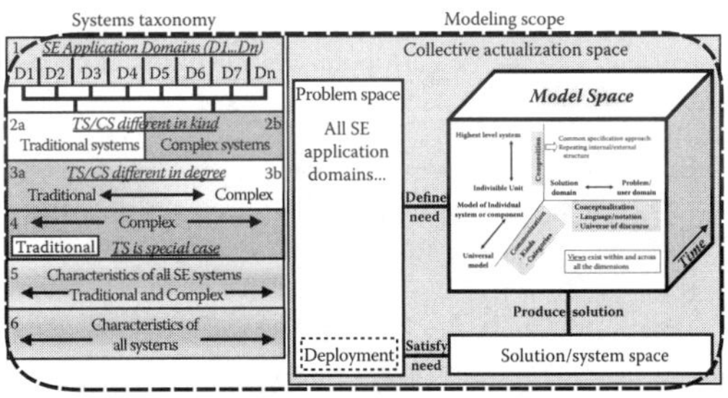

Obvious Mappings

Several of the mappings in this chapter refer back to views in Figure 6.1. To make the connection easier, we duplicate that figure here as Figure 12.1. The previous chapters in this book did not say much about the mapping of traditional artifacts to the external/internal model structure. When they have mentioned the relationship, it is along the lines of the mapping given here.

In the context of the models shown in Figure 12.1, the obvious mappings include:

- A Concept of Operations (CONOPS) document associated with a system S1 maps to an internal model of the interactions of S1 with the elements of its environment. This is view V4 in Figure 12.1.
- A requirements document of S1 maps to an external model of S1, either as a context-general model or as a context-specific model, both of which are indicated as V1 in Figure 12.1. In many cases, the requirements document defines the context-general model, and the context-specific model is defined as part of the CONOPS.
- An architecture description of S1 maps to an internal model of S1, indicated as V3 in Figure 12.1.
- The requirements of subsystems such as S12 in Figure 12.1, indicated as V2, can be described analogously to the requirements of system S1. For example, a requirements document can map to a context-general external model of S12 (not shown in the figure), and the context-specific external model of S12 can be mapped to part of the architecture description of S1 (V2).

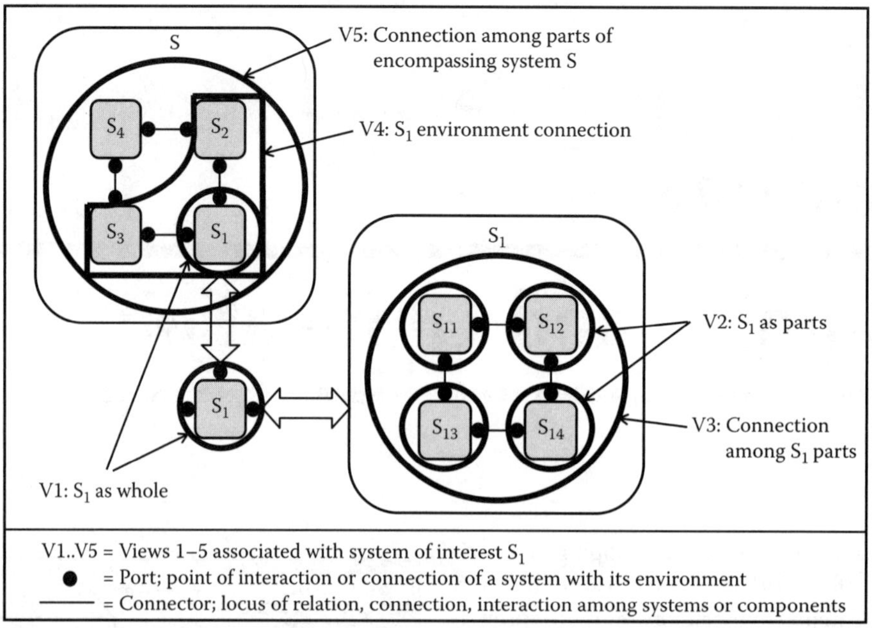

Figure 12.1 Common views associated with a system of interest.

The same requirements/architecture (design) intertwining pattern repeats all the way down the composition hierarchy of S1. But this simple mapping, although clean and elegant, is not sufficient to capture either the real world of TSE requirements documents, or the expanded world of complex systems. The sections that follow develop a more comprehensive approach.

Exploring an Environmental View of Requirements

This section presents more realistic and comprehensive mappings of system requirements to the intertwining internal/external model structure. To allow exploring of the mapping, we decouple traditional requirements documents from external/internal models and then show what the mapping is. Our analysis also includes dependencies because that is an important element of the mapping, but is often implicit in traditional documents and specifications. The purpose of this section is to improve clarity by showing a realistic mapping of requirements to models, and by making the associated dependencies explicit.

Examples

We present four examples of requirements in the context of the intertwining structure. A brief analysis of the mapping pattern illustrated by these examples is given in the next section. We start with a simple traditional system: a car. This is followed by two examples involving people, and by a discussion of the same pattern applied to quality metrics.

Car Example

We normally think of the requirement of a car — what a car is required to do — as providing transportation for a few people from location A to location B. The assumption is that the car and people are at location A, and the people desire to be at location B. But let's examine this more closely. Let's separate it into actual car behavior, what effect the car behavior has in its direct interaction with the environment, and what its indirect interaction is with the user. For each of these, we also spell out the dependencies.

- *Behavior of car:* wheels turn right or left or straight; wheels rotate clockwise or counterclockwise; wheels increase or decrease rate of rotation. *What it depends on for this behavior*: fuel (or other energy source, such as battery); air (assuming an internal combustion engine); driver
- *Direct interaction effect:* car moves relative to a surface such as a road — right, left, straight, forward, backward; constant speed, accelerate, decelerate; tires do not move relative to road (friction, grip); tires move relative to road (rotate, move as a whole). *What it depends on for direct interaction*: dependencies for behavior plus existence of a surface suitable for a car, including supporting friction; force pulling car and surface against each other (within some acceptable range — strong enough for friction and grip to emerge; but not so strong that the car is smashed against the surface); sources of energy renewal (e.g., gas stations or battery recharging stations) are positioned along the road periodically.
- *Indirect interaction effect:* transport one to five people from A to B. *What it depends on for indirect interaction:* Dependencies for direct interaction effect plus: Surface extends continuously from A to B; 1–5 passengers are on board. Those who want the car to provide them with a transportation service must get into the car.

Evolutionary Researcher Example

Recall the specification of behavior of the Santa Fe researcher in evolutionary dynamics in Chapter 11. We modify the example slightly to show its relation to the present mapping. Suppose the specification is:

> *General:* have an impact on the complex systems community.
> *Specific:* publish a paper on evolutionary dynamics in a refereed journal this year, whose quality and import is such that over the next 6 years, the paper is cited by at least ten papers in this and other professional journals and conferences, or in CS books.

- *Behavior of the system:* reads, talks, listens, manipulates research equipment, analyzes, writes paper *What it depends on for this behavior*: adequate health, research funding, office, equipment, knowledge of the domain, access to literature and professional colleagues in the domain, access to research subjects.
- *Direct interaction effect:* submits paper to refereed journal, paper is accepted and published. *What it depends on for direct interaction:* behavior dependencies plus existence and availability of refereed journal with infrastructure for reviewing, accepting, and publishing papers; acceptance of paper by the journal.
- *Indirect interaction effect:* have an impact on the CS community, measured by published paper receiving at least ten citations by papers in professional journals and conferences

or in CS books. *What it depends on for indirect interaction:* dependencies for direct interaction effect plus a community of evolutionary dynamics and broader CS professionals exists and reads this journal and publishes papers and books; members of this community are motivated to cite this paper in their publications — for example, they recognize the importance of these research results and build on them in their research.

Note that what is in each category or level depends to some degree on what the goal is. In this case, if the goal had been to simply get the paper published in a refereed journal, then the direct effect would be to write and submit the paper, and the indirect interaction effect would be to have the paper accepted and published.

Employee Intervention Example

Recall the example in Chapter 3 where Mitch planned to fire Ed but Ellen intervened and convinced Mitch not to fire Ed. We characterize the initial actions in that example. The system of interest in this case is Ellen.

- *Behavior of the system:* Ellen speaks to Mitch. *What it depends on for this behavior:* ability to speak.
- *Direct interaction effect:* Mitch changes his mind and decides not to fire Ed. *What it depends on for direct interaction:* behavior dependencies plus the ability of Mitch to hear and understand Ellen's argument. Another likely dependency is that Ellen has established a degree of credibility with Mitch.
- *Indirect interaction effect:* Ed is not fired. *What it depends on for indirect interaction:* Mitch refrains from taking the planned action to fire Ed.

The employee intervention example illustrates that the direct or indirect effects of a system behavior can be the absence, not just the presence, of an action or event. The employee intervention example as described in Chapter 3 also indicates that the interpretation of effects can change significantly across observer and across time, such that the behavior is considered a service or a disservice.

Quality Metrics

A similar pattern can be illustrated with metrics selected from the specification items defined in Chapter 11. This example involves a model or view of quality metrics that can be defined for each of the three levels:

- *Behavior of system:* quality here is often called system quality or engineering metrics. Examples include response time and reliability of the system itself, and accuracy of system output.
- *Direct interaction effect:* a representative metric for this is quality of service (QoS) because it includes metrics that are understood by and visible to both service provider and consumer.
- *Indirect interaction effect*: an appropriate metric here is quality of experience (QoE) because it is a measure of what the user actually experiences, the indirect interaction effect of what a system does.

Analysis

Kujawski and Miller (2007) argue that requirements of a system need to go beyond engineering properties such as response time, accuracy, and reliability. They must also include operational outcomes or effects. As the application discussed in their paper was military counterterrorism systems, they described effects such as number of opposing weapons neutralized or reduction in the casualties of our forces from these weapons.

The examples given above show that the pattern applies more generally. All four examples show that our intuitive concept of requirements is spread across three levels — from the system itself to direct interaction, to indirect interaction with the user. The examples also emphasize dependencies, which vary with the levels. Traditionally, no clear distinction has been made among these levels or their dependencies, which sometimes leads to poor communication and misunderstanding. The use of the external and internal models of the system and environment help sort both the requirements and associated dependencies into the three levels and distinguish between them. The dependencies accumulate from the system behavior level, which has the fewest dependencies, to the indirect interaction level, which has the most.

The mapping of these requirements levels to MOSES internal and external models is fairly straightforward. The system behavior maps to an external model of the system. The direct interaction with the environment maps to an internal model of the interaction of the system with its environment, which is also the province of a traditional CONOPS. The indirect interaction with the user maps to an external model of the user, described as a consumer of the indirect effect.

The external specification (including the interface specification) in Chapter 11 included some items that can go beyond the behavior of the system itself, specifically QoS and QoE. These items can directly support the broader scope of a requirements view that extends into the direct and indirect effects.

However, there is a more specific and subtle mapping of traditional requirements to external models, and that involves dependencies. The external specification is divided into two categories: what it provides and what it requires — that is, its dependencies. Traditional requirements focus only on what the system provides. However, the requirements can be satisfied only if the dependencies are satisfied. That is why the MOSES external specification emphasizes what a system depends on as well as what it provides. The upshot is this: to be compliant with MOSES, any requirements view, no matter how far into indirect effects it extends, must identify dependencies that have been established to support the services it provides. In addition, it is good practice to provide a map of dependencies to provided services, in terms of the three categories, as we have shown in our examples: system behavior, direct effect, and indirect effect.

The link between services offered and dependencies, along with the MOSES advice to use probabilistic specifications, together lead to the notion of conditional probabilities for each level of requirements being satisfied. The shortest chain of conditional probabilities, which has the fewest dependencies, is reflected in the system behavior; and the longest chain, with the most dependencies, is associated with the indirect interactions. We will not define the conditional probabilities of the examples here but they could be laid out along the lines of the conditional probability example in Chapter 11 (section entitled "Example Using Conditional Probabilities"). Such an approach lends itself to assessing risks, including the risk of specific dependencies not being satisfied.

The thrust of the Kujawski and Miller (2007) paper was on methods of assessing risks of various outcomes. Their probabilistic approach has some features in common with the above approach but we go further in identifying and distinguishing dependencies at the different levels. The indirect effects in our approach correlate at least partially with the concept of real-world effects of a service as defined in the OASIS SOA reference model (OASIS 2006).

Corresponding terms that can be used for behavior of system, direct interaction, and indirect interaction effect are behavior, role, and impact, respectively.

Opposite End of the Telescope: User Viewpoint

Suppose we consider a top-down sequence of the three levels, starting with the indirect interaction effect level. This is a common process when identifying a significant new need. The generic sequence of levels is:

- The indirect interaction effect is first defined in terms of a problem, a user operational need, a gap between current and desired state. This level has the largest potential solution space; no specific solution is identified yet. Multiple alternative solutions may exist.
- The direct interaction effect is then defined in terms of operational requirements. A solution/system is identified. The description is in terms of required interaction of the system with the user or, more generally, the environment. This would map approximately to an internal model of the environment or CONOPS.
- Finally, the actual behavior of the system is defined, that is, what the system can do if its direct dependencies are satisfied.

The car example will be developed from this viewpoint. The sequence is as follows:*

1. Current situation: person P is at location A.
2. Desired situation: P is at location B.
3. Requirements R1: a solution * shall change the current situation into the desired situation, that is, shall transport P from A to B.†
4. Some possible solution approaches for R1:
   a. P walks from A to B.
   b. P rides bicycle from A to B.
   c. P takes public transportation such as a subway or bus from A to B.
   d. P flies in an airplane from A to B.
   e. P takes a boat from A to B.
   f. P takes a taxi from A to B.
   g. P rides in a personal car from A to B.
5. A trade study is done that considers variables such as distance from A to B, time constraints on getting from A to B, availability of potential solutions, cost, reliability, and frequency (is this a one-time need, or will this need recur daily or weekly?).
6. Based on the analysis, (g) is selected as the solution approach S1 for R1, which can now be stated more specifically as R1.1: a personal car shall transport P from A to B in total elapsed time $\leq T$, assuming the dependencies are satisfied. The set of dependencies here corresponds to the indirect interaction effect dependencies given previously in the car example.

---

* Dependencies are mentioned in this list, but are not individually repeated. The set of dependencies was itemized in the earlier car example description.

† Recall from the correspondence set discussion in Chapter 4 that "solution *" means the solution is unidentified or unconstrained

7. A further trade study is done to determine whether to borrow, build, lease, or buy a car. The selected approach is to buy a car. Now R1.1 can be specialized further into R1.1.1*: a purchased car shall transport P from A to B in total elapsed time ≤ T, assuming the given dependencies are satisfied.
   a. Dependencies: same as in (6) plus one: a car that satisfies R1.1 is available for purchase.
8. Now we derive a requirement R1.1.1.1 that constrains what the car must be able to do on the road: buy a car that satisfies these constraints with probability ≥ .99, assuming the dependencies are satisfied. The set of dependencies here corresponds to the direct interaction effect dependencies given previously in the car example.
   a. Can turn at street corners, from a street into a driveway, and into a parking space in a parking lot, and make a U-turn on a 30-foot-wide street.
   b. Can go forward or backward on a road.
   c. Can provide sufficient force on wheels to achieve acceleration ≥ A on road with grade up to 10% incline when coefficient of friction ≥ F.
   d. Can provide sufficient force on wheels to achieve deceleration ≥ D on road with grade up to 10% decline when coefficient of friction ≥ F.
   e. Can generate and sustain sufficient speed to drive the speed limit. (Assumption: this capability enables the car, for example, to traverse the route from A to B in total elapsed time ≤ T.)
   f. Has operating range ≥ 400 miles (e.g., range between stops needed to fill fuel tank or recharge batteries).
9. Now we derive a requirement R1.1.1.1.1 that constrains what the car must be able to do on its own, which is essentially to turn its wheels: buy a car that satisfies these constraints with probability ≥ .99, assuming the dependencies are satisfied. The set of dependencies here corresponds to the car behavior dependencies given previously in the car example.
   a. Has turning radius ≤ R.
   b. Can cause at least two of its wheels to revolve in either a clockwise or counterclockwise direction, with this constraint: all the wheels the car causes to rotate must rotate in the same direction at a given time; that is, the car is prohibited from causing one wheel to rotate clockwise and another wheel to simultaneously rotate counterclockwise.
   c. Can apply force FA on wheels to achieve acceleration.
   d. Can apply force FD on wheels to achieve deceleration.
   e. Fuel tank has capacity C (in gallons) and expected mileage is R (in number of wheel revolutions per gallon), such that C × R ≥ W wheel revolutions; or battery charge is sufficient to last ≥ W wheel revolutions. (W is calculated to yield 400 miles if the car tires mounted on the wheels are in contact with a road while they are revolving.)

Item (3) defines the indirect interaction effect as R1 (transport P from A to B). It allows a large number of potential solutions, as the partial list in item (4) shows. The indirect interaction effect is illustrated with increasing specificity as R1.1 in item (6) (personal car) and R1.1.1 in item (7) (purchased personal car). R1.1.1.1 in item (8) defines the direct interaction behavior of the car on a road. Finally, R1.1.1.1.1 in item (9) defines the system behavior: turn its wheels.

---

* Why did we not just redefine R1 at each point, instead of giving it a different more specialized identity (R1.1, etc.) at each point? The answer is that we want to preserve the more general models, as well as trade study information, because circumstances may change in the future that cause us to go back up to a more general requirement and find a different solution. We want to retain all options and all analyses.

Thus, the car can only turn its wheels forward, backward, or side to side. If it is to move on a road, or transport people, multiple additional dependencies must be satisfied. This means that an external model of a car can only describe behavior of its wheels and tires. For example, an external model can describe the number of tire rotations per battery charge or per tank of fuel, but it cannot describe miles per gallon. That metric exists in the internal model of the interaction of a car with its environment. However, a requirements model can include miles per gallon and, in fact, any of the items from R1 to R1.1.1.1.1, and even beyond this to items internal to the car, as we discuss later.

There is another important mapping illustrated in the above example, namely, user requirements versus traditional engineering spec. Traditionally, we think of users defining what they want in their terms, and then engineers performing systems analysis and requirements analysis, including trade studies, to produce an engineering specification of what the system must do. That sequence is one view of the above example: the user requirements are defined in items (1), (2), and –(3); the trade alternatives and trade studies are presented in items (4) through (7); and the engineering specification is summarized in item (9). Traditionally, item (8) would also be included in the engineering specification.

We must acknowledge that the normal process of a person buying a car would not involve all the foregoing analysis or all the distinctions. A car is a commodity whose basic functions and dependencies are assumed and almost always left implicit. We actually focus on aesthetics such as color and style, and items such as upgrading the CD player. This example has focused strictly on the engineering aspects of the functional basics of a car. The point is to use the structure provided in the model space, and in this context especially the composition dimension, to help sort out and clarify the contents of a traditional artifact such as a requirements document and bring dependencies more strongly into the picture. The technique is overkill for buying or even engineering a car. But if something as mundane and well-understood as a car has this much texture and this many distinctions, then more complex systems or domains that are not well understood might benefit from the technique.

Indirect Effect Requirements and Correspondence Set

The traditional requirements document is virtually always coupled with a system or category of systems, and with an environment or category of environments. This means that the correspondence set {M, MP, W, S} has been defined at the conception of the requirements document, even if implicitly, where M is the requirements document, the mode MP is "shall", W is the intended environment, and S is the intended system — which often does not exist when this correspondence set is conceived.

However, requirements in the form of ultimate user needs and effects are usually defined in terms of the user, not in terms of a system satisfying those needs or achieving those effects. The user need is to be transported from A to B. How that is achieved is not part of the user effects model. What does correspondence mean in this situation, and what do requirements mean?

We now describe the correspondence sets for the transportation and evolutionary researcher examples, and use those to answer the foregoing questions.

1. *Transportation example.* User need/effect: user U needs to get from location A to location B.
   - M = Precondition: U is at A time T1. Postcondition: U is at B at time T2, where $A \neq B$, and $0 < T2 - T1 \leq P$, and P is a time period (such as 5 minutes or 2 hours or 2 days). It might be that A and B are not single fixed points; they may represent sets of locations, and a single element from each may be selected each day or each hour.

- MP = shall (satisfy M)
- W = world in which U, A, and B exist at T1.
- Case 1. S = * = no solution has been identified
- Case 2. S = S1 = an identified system, such as a car.

2. *Evolutionary researcher example.* User U = evolutionary dynamics community. User effect: U publishes at least ten publications that reference the researcher's paper in the next 6 years.
   - M = Precondition: paper is published at time T1; no one in U has cited it. Postcondition: at least ten U publications have cited the paper at time T2, where T2 – T1 ≤ 6 years.
   - MP = shall (satisfy M).
   - W = world in which U and the paper exist during the period from T1 to T2, and U produces publications during the period from T1 to T2.
   - Case 1. S = *.
   - Case 2. S = S1 = an identified system, such as a researcher capable of producing a paper that achieves this effect; specifically the Santa Fe researcher.

These correspondence sets look like "normal" correspondence sets but they differ in at least one way. The model M in a correspondence set usually is seen as a model of the system S in the correspondence set. But in this case, the model M is a model of the effect on the user of an implied solution. As stated previously in the analysis section, M is therefore an external model of the user, not an external model of the solution or system.

The requirements are: a solution (whether identified or not) shall do something to cause or influence the user effects to happen — given that the proper dependencies of the solution to achieve these effects are satisfied. In the case where a solution S1 has been identified, all user effects can be allocated as requirements of S1. But as stated previously, if the decision is to include user effects in a system requirements document, the document must be unambiguous and clear about the dependencies of S1 on the user and other elements of the environment in order to achieve the effects.

CS Actualization

It turns out that the SE instincts that separate the requirements of a system and extend them into direct or indirect interaction effects with the environment serve us well when we consider engineering complex systems. Due to autonomy, such systems require more environment shaping than direct system engineering, so the indirect interaction is appropriate for that. The example of the evolutionary researcher given above illustrates the approach. We do not tell a person or other autonomous agent what behavior they should exhibit; we tell them what effect we would like them to achieve (or they decide what effect they would like to achieve).

Other Engineering Mappings

Exploring Requirements by Looking Inward

A traditional requirements document can include not only requirements associated with the environment of the system, but also requirements and constraints on the internal design or architecture of the system. Common examples include a requirement that a given standard must be used in the system; or the system is constrained to using a certain technology; or in the case of software, all the software must be written in a certain language.

While this is accepted and to some extent legitimate practice in the context of TSE, it needs to be revisited in the broader context of the future SE and especially for CS. The danger in over-constraining CS internals is that it tends to interfere with autonomous activities like self-organization and the generation of emergent properties and behavior.

Resulting Requirements Mapping Spectrum

We have looked at two separate continuums regarding how a requirements spec extends an external spec. We looked extensively at the outward mapping, and briefly at the inward mapping.

- Outward: one continuum is an extension into the internal model of the *system environment,* including the external model of the user, to define constraints and effects. One extreme end of this spectrum is a pure external system spec, which includes only system behavior dependencies, and specifies only what goes in and comes out of the system boundaries/ports. The other extreme defines the complete environment, the indirect interaction effects on the user.
- Inward: the other continuum is the previously acknowledged extension into the *internal model of the system,* to place constraints on it. One extreme end of this spectrum is a pure external spec, which places no internal constraints. The other extreme is where the entire design or internal model is specified in the external model, and the internal of each part, and so on all the way down to the bottom of the composition hierarchy.

Elements across this entire spectrum can be incorporated into a requirements document. However, as mentioned above, extending outward is in many cases better practice than is extending inward, especially for complex systems.

Mapping Interaction Artifacts

The CONOPS document maps fairly cleanly to the environment relation model V4 in Figure 12.1, although it could mandate other constraints on the system architecture.

The architecture description maps simplistically to the internal model of S1, which is V3 in Figure 12.1. Note that this internal model, as all internal models, includes the external models of its parts, indicated in Figure 12.1 as V2. This is where it stops. However, the realistic mapping of an architecture can drive down all the way to the bottom of the composition stack by imposing constraints on all elements of the system, analogous to the inward looking extension of the requirements document.

Significance of the Overall Engineering Mapping

The engineering mapping examples and analysis presented in this section have shown that MOSES clarifies certain aspects of traditional artifacts, and places some constraints on them but also allows considerable flexibility in them. Several significant features of the mapping can be identified. Each is presented and briefly discussed in the following subsections.

The Mapping Makes Dependencies Clear

We stated that the "provides" and "requires" structure of an external/interface specification goes beyond traditional requirements documents to specify dependencies. Requirements that comply with MOSES must make the dependencies clear in the specification, whether they are the minimal dependencies of system behavior or the maximum dependencies of indirect effects.

Knowledge of these dependencies and how they are allocated are essential for adequate testing. It is relatively simple to define tests for system behavior because the dependencies are minimal. But dependencies of indirect effects can be substantial, and if they are not understood, testing can be very complicated and unfocused, and the results hard to interpret.

Compliance is actually based on *established* dependencies. This refers to the fact that in some views, no dependencies have been established. In the R1 requirements view of the car example in the section entitled "Opposite End of the Telescope: User Viewpoint," no solution is identified, only the indirect effect. No dependencies are established in this view. When solutions are introduced, as in (4 a) through (4 g), each of them introduces its own set of dependencies. The trade-off analysis considers both what each solution provides and what its dependencies are.

All Artifacts Are Views of the Model Space Structure

Because traditional artifacts can range across the composition spectrum, both up and down from the point of interest, what mechanism can we use to model them? The answer is *view*. All artifacts, including CONOPS, requirements, architecture, and design, are views of the intertwining composition structure that can extend well beyond a nominal internal or external model. The mapping also makes the target of the view clear; it is not always the system of interest. In addition, the external-internal structure of MOSES composition makes clear each of the elements of the view (artifact) and how they relate to each other.

Note that the treatment of traditional artifacts as views is reinforced by the fact that the artifacts can also vary across other model space dimensions. For example, any of the artifacts can vary in specificity, which is supported in the commonization dimension.

The latitude given the artifacts raises potential overlaps and conflicts between the documents or views. Good practice is to define boundaries between them so no overlaps exist. But if overlaps are allowed, they must be adjudicated to be consistent, or better yet, to decide for each rule or constraint which view is the owner, and the others reference it. The owner is responsible for configuration management of the rule. For example, if the requirements document defines environment effects, which is traditionally the province of the CONOPS, an agreement should be made on which is the owner of this rule. Even more overlap can occur on the inward side if both the requirements view and the architecture view want to constrain all the way down the composition hierarchy. For example, both may mandate that a certain standard be used at all levels. A decision should be made as to whether the requirements view or the architecture view owns this rule.

SE General Problem Solving

Some requirements are stated as a model of a user rather than as a model of a system. A systems engineer must define and assess possible solutions, and determine whether the solution should involve an SE system. Table 12.1 shows the pattern. The analysis of (a) includes a variety of possible solutions, many of them not SE systems. The car example in the section entitled "Opposite End of the Telescope: User Viewpoint" listed a variety of solutions. Suppose the systems engineer

Table 12.1 Pre-Systems Problem Solving

Traditional Artifact	*Level*	*MOSES Model*
a. Need or gap	Indirect effect	External model of user (but not a requirement of the user)
b. CONOPS	Direct effect	Internal model of interaction of system with its environment
c. Requirements specification	System behavior	External model of system

assessed the possible solutions, and based on the situation recommended (4a) — walk. If this solution were adopted, it would probably not be considered an SE system. Deciding whether an SE system is needed as the solution to a problem is part of SE, and it includes the model type in row (a) in Table 12.1. In row (a), no system or solution is identified; in row (b) and (c), a system is identified as the solution.

The point is that part of the work of systems engineers is general problem solving, as opposed to engineering or actualizing a system *per se*. The general problem-solving work, however, does occur in the collective actualization space.

Autonomous Systems without External Models

In cases where an autonomous system decides to achieve the indirect effect or the direct effect, no external model of this system is needed. The system is free to decide which external behavior will achieve the desired effect. Requirements are stated in terms of the direct or indirect effect models.

This may be extended to some degree to more traditional systems. For example, the government may decide to define direct effects on the environment and allow a contractor to be the autonomous agent that decides what the system behavior (external specification) needs to be to achieve those effects. In those cases, the government needs to make clear the desired effects and the environmental constraints, and the contractor needs to make clear the dependencies of the solution.

Adaptation and Versatility

The analysis thus far is only part of the story. For a given system that is realized, we define the effects and the system performs in a way that achieves those effects. First, deployment of a system involves moving, copying, or instantiating the system from the solution space into a deployed state in one or multiple problem space environments (such as a factory producing multiple Boeing 777 aircraft to send to different airlines, or a software system being distributed to multiple users). Part of the deployment may be configuring, customizing, or adapting to the specific environment during installation (and continuing thereafter). The solution space system and the associated deployed system(s) correspond to Martin's (2004) intervention system and deployed system, respectively.

Second, when the system is deployed, it not only enters the problem space, but it also becomes part of the problem space and thereby changes the problem space. We now have to consider that the system and its environment are mutually adapting entities within the problem space. As soon as the system is successful in achieving its effects, the environment changes. Those changes often mean that the deployed system needs to change in some way. And so on in a repeating cycle. To do this means that the deployed system cannot be developed in the old regime of satisfying a fixed set of requirements. The actualization process must anticipate the changes the system might induce in its environment and the changes the new environment requires of the system, and the resulting system must be able to deal with those changes in an ongoing process.

In the context of this repeating cycle in which system and environment change and adapt to each other in a series of interactions or moves, the original system in general needs two features built in: (1) it must have an inherent capability to change its behavior, (2) it must be sufficiently adaptable to change (either be changed by the systems engineer or autonomously change itself) so that it can respond to changes the environment presents to it. These two features reflect, respectively, change levels L1 (range or versatility of behavior) and L2 (adaptability of system) as defined in Chapter 9.

TSE has been doing this in certain arenas, notably in the security area. Ongoing mutual adaptation of security mechanisms and people trying to break the mechanisms has been a long-standing pattern. The back-and-forth adaptation between Internet hackers and anti-virus security specialists has become ritualized to the extent that we now get virus definition file updates almost every day to protect against new viruses that appear. In the area of CS, these mutual adaptation patterns are even more dominant. The development of complex adaptive systems concepts has resulted in models of these patterns across a broad range of systems and environments.

MOSES provides several foundational elements that support the features identified above, including:

- The repeating and intertwining structure of internal and external specifications provides a clear picture of the distinction between system behavior, direct effects, and indirect effects.
- The inherent expectation of continual change, and models of change, supports adaptability and versatility.
- The modularity and decoupling aspects of MOSES also support adaptability.
- The flexibility of general models enables the bounding and modeling of the adaptation space of system and environment, as well as system versatility.
- The inclusion of social and psychological models that depict motivation and goal-directed behavior help inform the content of the general models of the adaptation space, in effect yielding better models of the phase space and trajectories of the mutual game that is played between system and environment.

Definitions of Traditional Artifacts

Given the material presented thus far in the book, we are in a position to define traditional artifacts in MOSES terms. We have given brief definitions in several places but now we can give more complete definitions. Concept of operations, requirements document, architecture, and design are defined. Recall that, in MOSES, each of these is regarded as a model, and more specifically as a view that cuts across the model space.

A *concept of operations* or CONOPS exists in a context in which a system S has been identified and an environment E has been identified as the environment of S. E consists of all entities that are related to S or interact with S. Furthermore, the CONOPS is regarded as the CONOPS of S in E; that is, a correspondence has been established that associates the system S with the CONOPS model in the world E, perhaps with a modality such as "shall" or "desired." The viewpoint in which the CONOPS view is defined is the locus of an internal model of the interaction of S with E. Normally, the interactions among elements of E (i.e., interactions not involving S) are excluded from the CONOPS but such interactions are not prohibited from appearing in the CONOPS. That is, in Figure 12.1, V4 would normally be the CONOPS, but V5 could also be the CONOPS.

The contents of the CONOPS view include all information in the internal specification table in Chapter 11 selected to be made explicit in the model, where the components are S and the E entities, and the connectors are the vehicles of interaction between S and the E entities. Because part of the internal specification is external specifications of each component and connector, those specifications are also part of the CONOPS, to some level of detail. This means that the CONOPS will include an external model of S and of each element of E. In addition, the CONOPS view may specify additional rules and constraints that would normally be included in the internal model of S, or even further down the composition hierarchy of S. Because this practice tends to break modularity, it is not generally encouraged but it is sometimes appropriate.

A *requirements document,* or requirements view, usually exists in a context in which a system S has been identified, and sometimes an environment E has been identified, at least in a general way, as the environment of S. Furthermore, the requirements view R is regarded as the requirements of S; that is, a correspondence has been established that associates the system S with the requirements model R with modality "shall" (or other prescriptive modality), perhaps in the world E. The viewpoint in which R is defined is the locus of an external model of S. If R assumes E, it is a context-specific external model; if R assumes no specific environment, it is a context-general external model. The contents of R can vary considerably in terms of S, E, and the internals of S. As discussed in this chapter, R can define the effects on elements of E resulting from S, or the behavior of S itself. R can also constrain the internals of S down multiple levels inside the S composition hierarchy. If R defines the effects on elements of E, then it includes external models of those elements. Thus, R can include a range of models both external and internal, and the contents include the elements in the corresponding tables in Chapter 11 that are chosen to be explicit.

An *architecture description,* or architecture view, exists either as a standalone model, or in a correspondence context in which a system S has been identified or a class of systems C has been identified. In the latter case, the architecture A is regarded as the architecture of S or of C; that is, a correspondence has been established that associates S or C with A, perhaps with a modality such as "shall." The viewpoint in which the A view is defined is the locus of an internal model of S or C. The contents of the A model includes all information in the internal specification table in Chapter 11 selected to be made explicit in the model, where the components and connectors are parts of S or C. Because part of the internal specification is external specifications of each component and connector, those specifications are also part of the A view, to some level of detail. In addition, the A view may specify additional rules and constraints that would normally be included in the internal models of parts of S, or even further down the composition hierarchy of S. Typically, the A view will consist of multiple architecture views, as discussed in Chapter 10. That is, the A view will be a composite view or a multi-view model. Note also that S or C may be any category of SE system. For example, an architecture of a system such as a commercial aircraft is an internal model of the aircraft. An enterprise architecture is an internal model of the enterprise or organization. Both consist of multiple views.

The locus of a *design* of a system S is also an internal model of S. So how does this differ from an architecture of S in MOSES? Formally, MOSES does not distinguish between architecture and design; both are internal models. In this perspective, the definition or description of architecture given above applies to design as well. However, architecture and design usually differ by convention. Architecture is often taken to mean a general internal model while design is a more specific internal model. For example, architecture may define the model parameters, while design defines the specific values of the parameters. Architecture, by intent or by default, often describes a class of systems, while a design tends to be more system specific.

Mapping Enterprise Management Elements

Traditional SE as generally conceived includes not just engineering, but also management. Partly for that reason, the extension and unification goals of MOSES include both systems engineering and enterprise engineering/management elements. This section reflects the enterprise portion of Table 1.3 in Chapter 1. It has two boxes: enterprise engineering/management (EEM) and enterprise autonomy (EA).

Mapping traditional management to EEM and EA reveals two significant differences. The first is the addition of an engineering approach, including model-oriented management. This is primarily reflected in EEM. The second is the addition of autonomous behavior of enterprise components, such as self-organization or self-management. This is primarily reflected in EA.

Enterprise Engineering/Management (EEM)

Traditional management is mostly an art that depends heavily on the experience and instincts of each manager. What EEM brings to the picture is some science-based enterprise engineering, including model-oriented management. What does that mean? We explain with an example: organizational structure.

Suppose we are inventing or reinventing an organization. One of the decisions to be made is the "org chart," which reflects the organizational structure. Traditionally, the head of the organization or an appointed committee would define the organizational structure based on personal preferences or simply use the only structure they know.

Now suppose we regard the organization as an enterprise system and the org chart as an important view of the architecture of this system. Further, suppose that a body of knowledge exists in the model space, such as organizational theory or management science, confirmed through collective experience, that indicates which organizational structures work best for different types of organizations (such as R&D, government agencies, commercial companies, manufacturing companies, etc.; or maybe finer-grained categories). If that body of knowledge is applied and possibly tailored to this particular organization, then we claim that this organizational structure decision has moved from the practice of traditional management to the practice of EEM.

The scope of models in the knowledge space goes beyond enterprise structures to include planning, resource management, budgeting, cost analysis, schedules, and staffing models. All of these types can exist for, and be applied to, organizations and projects as well as other enterprise types.

Enterprise Autonomy (EA)

Members of an enterprise often have their own ideas about enterprise decisions and activities. Continuing the organization structure example from the EEM discussion, members of the organization participate in decisions about the structure and also form their own autonomous structures. These activities represent the EA box in Table 1.3.

Other models of enterprises in the model knowledge space are drawn from the CS community. These models portray self-organizing and self-managing enterprises. The effect these models have on EEM is to cause enterprise engineers and managers to recognize the role of EA in shaping the organization structure, and hence to share with EA the realization of the organization structure and to encourage EA participation in the activity.

The proportion of the organization structure determined by EEM versus EA is a variable that ranges across a spectrum from all EEM (traditional top-down management) to all EA (completely self-organizing). Taken together, this EEM/EA picture represents the MOSES approach to incorporating and extending traditional management into the larger collective actualization space of the future SE.

Summary of MOSES Features and Unification

Areas of MOSES Unification

This section discusses areas where MOSES promotes and enables unification of the expanded SE of the future. To some extent, this list reprises areas that were introduced in Chapter 1 and discussed in the various intervening chapters. Table 12.2 summarizes elements of MOSES and what areas of SE each element unifies. Each shaded cell indicates which areas are unified. These are now discussed.

A glance at Table 12.2 reveals that most of the elements of MOSES support the unification of TSE and CSE. This is not accidental. As the book has emphasized throughout, a primary driver of MOSES is to extend SE to include complex systems, and to unify the engineering of traditional and complex systems.

Model orientation, the model space, similarities and differences, and the expanded systems taxonomy unify TSE and CSE by bringing them into a common taxonomy framework that is

Table 12.2 Areas of SE Unified by MOSES

	What SE Areas It Unifies				
MOSES Element or Concept	*TSE-CSE*	*Disciplines*	*Management-Engineering*	*Development-Operations*	*Composite SE BoK*
Model orientation	■	■	■		■
Whole systems modeling		■	■		
Model space	■	■	■		■
Self-similarity patterns	■	■	■		■
Common specification approach	■		■	■	■
Agents	■				
Use of systems science	■	■			■
Similarities and differences	■	■	■	■	■
Expanded systems taxonomy	■				
Collective actualization	■			■	
Opportunistic process	■				
Support for change	■			■	

based on all artifacts being considered as models, and that incorporates into the models similarities and differences of TS and CS. In addition, model orientation and the model space unify the scope of SE, including all associated and supporting disciplines, management and engineering, and the composite SE body of knowledge, by considering the artifacts of all these disciplines as models, and incorporating all general models in the knowledge space part of the model space.

Self-similarity patterns are part of the model space but are important enough to call out separately. They support unifying all areas that the model space unifies because the variables and dimensions of the model space are self-similar.

Whole systems modeling helps unify the associated disciplines by including all spaces in Figure 1.2 — problem space, model space, solution space, and actualization space. The common specification approach supports whole systems modeling through its unification of TSE and CSE, disciplines, management and engineering, development and operations, and composite SE BoK. The same external/internal specification approach and information types are used for all these areas.

The remaining areas are summarized as follows:

- The concept of agents unifies TSE and CSE by treating human, software, and other types of agents the same way.
- The use of SS helps understand what is common across all systems, both TS and CS, and all system-related disciplines, and all SE BoK, which helps unify all these areas.
- Similarities and differences is a very broad element that helps unify all areas of SE by factoring commonalities out of the differences.
- Collective actualization is a broader term that unifies TSE and CSE by covering the engineering or control of TSE, as well as the influence of CSE, and includes management.
- Opportunistic process and the emphasis on supporting change help unify TSE and CSE and also strengthen a weak area of TSE.

Summary of Key MOSES Features

The key features of MOSES are:

- Defines and supports an expanded and unified taxonomy of systems
- Defines a unified and simplified specification approach that requires only two basic types: internal and external
- Expands the concept of engineering to the broader concept of actualization; incorporates management as part of actualization
- Embraces the CS concepts of tension, duality, paradox, balance
- Disciplines: uses classic universal modeling concepts
- Model spectrum reveals and exploits self-similarity of model space dimensions, thus supporting scalability
- Foundation is well grounded in SS and other sciences
- Captures the best patterns and practices and principles of TSE
- Does not disrupt current best practice
- Explicitly allows partially implicit specifications, via the distinction between model and specification
- Is applicable to multiple technologies but does not depend on any of them
- Does not need to invent the solution pieces; they already exist in the broader SE and supporting disciplines

- Fully supports the blurring of lines that occurs between artificial and natural systems
- Supports the hybrid systems that systems engineers actually address
- Enables increasing SE scope of more ambitious arenas

References

Kujawski, E. and G. Miller. 2007. Quantitative risk-based analysis for military counterterrorism systems. *INCOSE J. Systems Eng.*, 10(4):273–289.

Martin, J. 2004. The seven Samurai of systems engineering: dealing with the complexity of 7 interrelated systems. Presented at the *14th Annu. Int. Symp. Int. Council on Systems Engineering (INCOSE)*. Toulouse, France. 21–24 June.

OASIS. 2006. Reference Model for Service Oriented Architecture 1.0. OASIS Standard. 12 October. http://docs.oasis-open.org/soa-rm/v1.0/. (Accessed January 11, 2009.)

Chapter 13

SE Actualization Processes

The title of this chapter reflects the broader scope of the future SE and model orientation. The composite MOSES map in Chapter 1 depicted a space called the collective actualization space. The structure and elements of this space are defined in Table 1.3 in Chapter 1. In MOSES and the expanded SE, two changes occur. First, model orientation unifies engineering and management into a common modeling approach. Second, the expansion to CS leads to reinterpretation of the complexion of engineering, management, and modeling from an orientation of definitive control and certainty to one of influence, shaping, and actualization in the context of acknowledged uncertainty and autonomous agents. Therefore, the term "collective actualization" in this book and this chapter is this reinterpreted form of engineering, management, modeling, and autonomy. The scope of actualization includes the full taxonomy of produced systems, both TS and CS, and enterprises, such as organizations and projects.

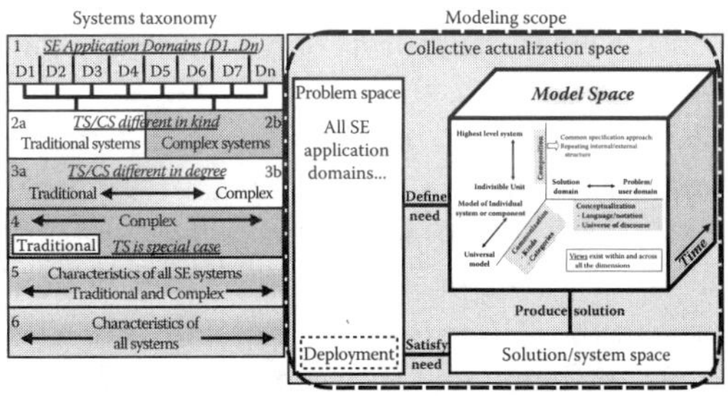

Because MOSES is positioned as a science of SE, as opposed to a method of SE, it does not address how to do SE, or even how to do model-oriented SE (MOSE). It is intended to be a foundation on which a variety of SE methods can be developed, including methods that are model oriented and methods that address complex and traditional systems. Nevertheless, this chapter on actualization processes is included in the book in an attempt to alleviate some of the difficulties caused by current SE processes, and to point to more natural processes and a more balanced focus.

Based on this rationale, the sections of the chapter discuss, respectively, TSE methods and variances with MOSES, CSE methods, model-oriented methods, and characteristics of MOSES unified processes, including collective actualization.

Traditional Processes

This section discusses TSE processes, including some prominent general life-cycle models, standards, and a few related sources. It then summarizes some of the weaknesses of the TSE processes.

General Models

Several process models came out of the software engineering community. The classic process model that has influenced both software engineering and SE for four decades is the waterfall model (Royce 1970). Royce's recommended model that included an initial full prototype effort (he called it "simulation") has been largely ignored and forgotten. His simplest and most linear model (requirements > analysis > design > code > test), the one he said was "risky and invites failure," was widely adopted.

Boehm (1988) introduced the spiral model as an iterative development (but not deployment) approach that focused on risk reduction through a series of prototypes followed by the final product. Each spiral reassesses risk and goes through a quick development activity that produces a prototype that reduces a risk and improves understanding of the final product.

Forsberg and Mooz (1991) introduced the Vee (or "V") model of systems engineering, focused on the technical aspects of a project. Its name derives from the shape of the graphic, which is shown in simplified form in Figure 13.1. It is correlated with two dimensions of the model space: composition and time. It starts with user needs in the upper left, decomposes from system to segment to hardware and software configuration items to components and assemblies and units, then implementation (software development and hardware fabrication), and finally composes units into components and so on back up to the full system. The activities coming down the left side of the Vee are modeling and design, while those going up the right side are integration and test. The horizontal levels on opposite sides of the Vee are linked in the sense that they are at the same composition level and they represent the flip sides of design and integration/test at that level. In terms of services provided and required (dependencies) in MOSES specifications, the Vee

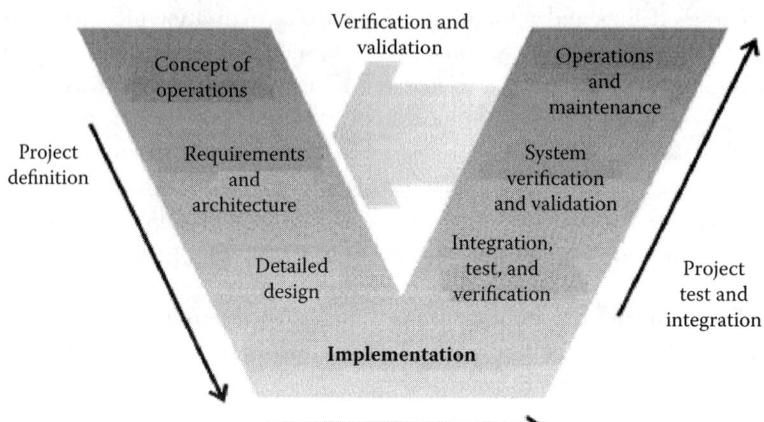

Figure 13.1 SE Vee model. (*Source:* Clarus Concept of Operations, Federal Highway Administration, 2005.)

model illuminates one sequence constraint. There is no constraint in MOSES on the sequence of defining services provided but there is one on defining services required. That constraint is that the full set of dependencies of a given level of system in the composition hierarchy cannot be defined until the dependencies at the next lower level are defined. This constrains the sequence of defining (or at least completing the definition of) dependencies to go from the bottom of the Vee up the right side.

Why does this constraint exist in MOSES? The reason is that design decisions can affect external dependencies. For example, it is possible to define the services that a car provides without knowing the internal design, such as the type of engine. But to fully specify the dependencies of the car, we need to know, for example, what type of engine it has or will have, and whether it is automated. If the car has an internal combustion engine, it depends on air and fuel, while if it is battery powered, it does not depend on either air or fuel, but does depend on a way to recharge the batteries. If it is not automated, it depends on a driver; if it is automated, it does not depend on a driver. This constraint is why Chapter 12, in the discussion of specifying dependencies, said that the *established* dependencies must be specified.

Nguyen (2006) observed that most SE methods assume a blank slate, engineering a system from scratch, rather than modifying an existing system. He discussed adapting the Vee model to address system modification.

Another approach is iterative and incremental development (IID). This concept has been around for several decades but has become more visible in the past decade. A history of this approach is given in Larman and Basili (2003). Although much of its use has been in software engineering, one of the earliest examples was its successful use on the North American X-15 rocket-powered aircraft in the 1950s, which illustrates its more general SE applicability.

Cockburn (2008) has described the use of incremental and iterative development in the context of agile software development. He has also incorporated and adapted the SE Vee model using multiple smaller V's with multiple deliveries to fit the agile paradigm.

Standards

IEEE and ISO

There are two relevant standards from the IEEE and ISO/IEC. IEEE 1220 (IEEE 2005b) is the IEEE Standard for Application and Management of the Systems Engineering Process. ISO/IEC 15288 (ISO/IEC 2008), which was adopted as IEEE 15288 (IEEE 2005a), is the Systems Engineering — System Life Cycle Processes standard.

ISO/IEC 15288 has a modular design. It defines a set of process modules but does not require any specific sequence. However, it does require the establishment of a sequence in which the components are the process modules. It defines four groups of processes: (1) enterprise (the enterprise and management environment), (2) project, (3) technical, and (4) agreement (acquisition and supply).

INCOSE: Handbook and Guide to BoK

The International Council on Systems Engineering (INCOSE) has developed a handbook (INCOSE 2007) that also is patterned after ISO/IEC 15288. It addresses the four groups of processes from ISO/IEC 15288, and also enabling SE activities such as risk and opportunity management, additional life-cycle activities such as configuration management and resource management, and specialty engineering activities such as logistics and human factors.

The INCOSE guide to the SE body of knowledge (INCOSE 2004) provides discussion of and pointers to many of the processes and standards identified in this section, including the INCOSE Handbook, the Vee SE model, Plowman's synthesis (see below), and the U.S. Department of Defense life cycle.

U.S. Department of Defense (DoD) Acquisition

The general approach of the DoD to systems engineering is described in DoD (2001). There is also an extensive set of standards, regulations, and instructions that define the systems acquisition process, called the DoD 5000 series. Much of this information is described in DAG (2006).

Related to the DoD acquisition process is their capabilities-based assessment (CBA) process as described in the Joint Capabilities Integration and Development System (JCIDS 2007). The key CBA concept of note for MOSES is the up-front analysis of a problem that does not assume that a machine or a materiel system will be the solution. The analysis examines alternative solutions, which might include training, change in doctrine, or personnel, as well as or instead of materiel systems. This is an important element in MOSES, discussed in Chapter 12 as the general problem-solving aspect of SE.

Other Sources

Many other general and specific SE processes and methodologies are described in a variety of sources, including commonly used books such as Kossiakoff and Sweet (2003) and Sage and Armstrong (2000). We briefly mention one example: Hitchins' five-layer model.

Hitchins (2008) has defined a five-layer SE process model. The layers are:

- *Layer 5:* Socio-economic, the stuff of regulation and government control.
- *Layer 4:* Industrial Systems Engineering, or engineering of complete supply chains/circles. Many industries constitute a socio-economic system.
- *Layer 3:* Business Systems Engineering — many businesses constitute an industry. At this layer, systems engineering seeks to optimize performance somewhat independent of other businesses.
- *Layer 2:* Project or System layer. Many projects constitute a business.
- *Layer 1:* Product layer. Many products constitute a system. The tangible artifact layer. Many engineers and their institutions consider this to be the only "real" systems engineering.

The layers form a composition hierarchy that extend beyond the normal scope of TSE and include at least part of the new expanded scope of SE. However, the described processes and characteristics of the layers emphasize the differences between them, and not so much the commonalities, in the manner of TSE. MOSES would emphasize the commonalities, as well as the differences.

Problems with Traditional Processes

The following characteristics of TSE processes are observed as problems from the perspective of MOSES:

- Focus on process
- Problem solving: engineering process is "disciplined" to be linear and unnatural
- Does not deal with change: heavyweight; over-constrained; change is a negative

- Myth of the blank slate
- Myth of requirements driven

Each is briefly described below.

Focus on Process

This issue or weakness comes from a perspective that is broader than the scope of this chapter. A comparison of the primary documents of the SE discipline — including the most used textbooks, the body of knowledge, and the standards — with the primary documents of a mature discipline such as chemical engineering, shows the weakness as discussed in Chapter 5. Mature engineering disciplines have models of systems, while SE has models of development processes. The belief that models of systems are more helpful to systems engineers than models of development processes is a major reason why MOSES is defined in terms of a model space and general models.

Problem Solving: Engineering Process Is "Disciplined" to be Linear and Unnatural

The basic TSE processes as widely practiced are still variations of the waterfall model, to a large degree. The obvious weakness is the linear sequential nature of the waterfall. The sequence is from stating the requirements to designing and building the solution. Why is this a problem? Because the nature of problem solving in general, and engineering a system in particular, does not follow a linear sequence. First, the natural process of design is not linear; it is more like chaos (in a technical sense), as we discuss in the section entitled "A Unified Actualization Process." Second, the forces that shape producing a system do not occur in sequence; they are usually concurrent. See the sidebar in this chapter on the history of process.

Does Not Adequately Deal with Change

Possibly the most significant weakness of TSE methods is the difficulty of dealing with change. This is partially related to the rigid application of the waterfall model. The processes have tended to become heavyweight, ponderous, over-constrained, and over-controlled. If a process is not working very well, the reason is often seen as not having enough discipline, or enough detail in the process, or enough processes. So we add more discipline, more oversight, more detail, and more processes, and the result is that everything slows down even more. A corollary to this is that change is seen as a negative — it represents a failure of some sort. We want to avoid all change if possible, and if it is not possible, then make change in a very carefully controlled manner. Change is not seen as an opportunity. Unfortunately, the world does not stop changing during the long, slow development (or operation) of a system, so change must be dealt with.

Myth of the Blank Slate

This weakness is an assumption that we are engineering a new system from scratch, or from a blank slate. Over the decades of SE practice (and software engineering practice), it has become the exception rather than the rule that we create whole new systems; in practice, we much more often modify existing ones. And even when we do create new systems, their requirements and designs are not often entirely new — they are usually variations of existing ones. This is to be expected in any discipline or application domain. When a domain is new, many or most of the systems built are new designs; but as the domain gains some maturity, more of the design space of this domain has been explored and systems built. Each new project that comes along, each new system desired, is more likely to be similar to one or more systems that have already been built. Classes of similar systems now have common architecture models and other general models that can be specialized for this system, instead of having to start from scratch. The problem in SE, and in software engineering, is that our processes have not caught up with our practice in this regard. Our processes continue to be oriented toward engineering a whole new unique system from scratch.

Myth of Requirements Driven

A common assumption is that the primary — perhaps even the only — force that determines what a system does is the needs of the user, and these are in some sense *a priori* or preordained. That is, a requirements-driven process is assumed. However, the reality is that in most cases, the needs of the user are only a limited driver of what a system does. The largest driver is what technology exists or is within reach. George Washington could have written down all the requirements he wanted for aerial bombing in the Revolutionary War, and offered a large amount of money, but it would not have done him any good because nobody knew how to build an airplane. Technology at any point in time defines a space of possibilities of system behavior, and users are constrained to define their requirements within that space. The boundary of that space is fuzzy. The fuzziness represents the fact that the needs of users can push the state of the art modestly beyond its current space, but not much beyond. So, requirements of systems are pegged either to existing technology or slightly beyond it.

In addition to available technology and user needs, the third significant driver of the capabilities of a system is resource constraints, which include cost, schedule, and staffing, as well as operational resources. Cost is the dominant resource constraint. The order of the three big drivers from most to least consequential is probably available technology, resource constraints, and finally user needs. That is, if technology defines a space of possibilities, resource constraints further reduce that to a more limited space, and users define or negotiate their requirements within that limited space, although resource constraints are sometimes also negotiable.

All three drivers can change over time, even within one system. In general, the technology changes the least, while the user needs and resources can change significantly.

The foregoing discussion of drivers illustrates why a linear requirements driven process is not realistic for two reasons: (1) the dominant effect of available technology, and (2) the continuing change of all drivers. The upshot of the linear process evaluation is that a linear process is not the natural way that system designers work, and it is not a realistic way for an SE project to proceed.

Offbeat History of Process

Context

This sidebar presents a view of process based on an examination of how process has been treated in several SE-related disciplines. The view, and the discussion that follows it, explain the role that process plays in MOSES.

The View

Historical practice in several disciplines shapes the view presented here. The two primary historical examples are traditional architecture and science, especially the scientific method.

Architecture. An evolution of the design process in the traditional architecture field in the 1960s and 1970s, especially in Britain, is described by Jones (1963), Hillier, Musgrove, and O'Sullivan (1972), Rittel (1972), and Broadbent (1979).

In the 1960s, the people who studied the methods of architectural design decided that the process needed to become more systematic and more quantified. Architects needed to clear their mind of preconceptions. They needed a rationalized process of analysis then synthesis, with more emphasis on user requirements, and more open-ended designs and options as opposed to a standard set of built-in requirements. This approach, called Analysis-Synthesis-Evaluation, has been designated a "First Generation" design method. (The way architectural design was done during the thousands of years prior to this period does not seem to deserve any name or generation number; perhaps we could call it Generation 0. Apparently, process was not a focus. In his ten books on architecture written in the first century BCE, for example, Vitruvius (1914) focused on designs and not so much on the process of creating them.)

The result of this first-generation process, according to Hillier, Musgrove, and O'Sullivan (1972), was deterioration in user acceptability of buildings that were built. The problem seemed to be that designs did not incorporate the built-in knowledge of requirements they previously relied on and instead they relied too heavily on specified user requirements.

A second generation then emerged, containing elements of political consideration; it was a rebellion against the notion that an expert designer was more knowledgeable than other stakeholders about how to design the system. The approach was more like design by committee, in which it was politically incorrect to think that anyone knew more than anyone else about design. The designer acted as coordinator. Also included were some legitimate ideas such as an argumentative approach in which each idea was to be open to discussion and challenge, rather than accepted as a pronouncement from an expert at face value. This approach resulted in buildings that were not quite usable because amateur designers on the committee wanted exotic features that were not workable, but the architect was not in a position to overrule them, and could not persuade them of the folly of their ideas until after the building was built.

The third-generation process that followed owes much to the philosophy of Karl Popper (1965), and is called the conjecture-refutation model. It has these characteristics: it returns to the notion of an expert designer proposing a design in the form of a conjecture, which is then examined and opened up to refutation. The process also portrays definition of the problem and development of the solution(s) as simultaneous activities that support each other. The argument is that the old sequence view (define problem, design, implement) is inappropriate because problem and solution emerge together. The third generation seems to be the application of problem solving to design. That is, it seems to be an explication of how architecture was done before all these "generations" began, with perhaps the added value of explicit recognition of refutation and debate on proposed designs.

Science. Medawar (1969), Weinberg (1994), and Bauer (1994) discuss the process of science and the scientific method. The basic argument of Medawar and Bauer is that scientists have for the most part ignored the discussion about process and are able to be successful despite ignoring it. They are not really aware of what process they use, and are unable to accurately describe it if they are asked to do so. Furthermore, the people who have tried to define the scientific method, such as philosophers and methodologists, describe a process that is not what scientists actually follow. The physicist Weinberg (1994) relates that many physicists puzzle over two phenomena: the unreasonable effectiveness of mathematics, and the unreasonable ineffectiveness of philosophy, in supporting the scientific enterprise.

The general pattern that seems to apply in architecture, science, mathematics, and traditional engineering is that practitioners do not pay much attention to process, but methodologists and philosophers do. We would say that practitioners in all these disciplines apply the human problem-solving process. It is a natural, messy, trial-and-error, opportunistic, iterative, intuitive process that is part creative insight, part rational reasoning, part hard grinding work, and part social interaction with its attendant emotional and non-rational elements. The results, when successful, give the illusion of resulting from a clean, straightforward process. Philosophers, however, try to capture the illusion and prescribe it as the process to follow. That process is unnatural, clean, sequential, rationalized, and objective, and does not square with the practice of human problem solving.

Applying the View

We have mentioned in several places in the book that TSE (and software engineering) has focused on process. We have also mentioned that the heritage of MOSES lies more in the orientation of traditional engineering handbooks and less in the process focus of software and systems engineering. More specifically, the heritage is a combination of the conventional designs of traditional engineering handbooks and the complex systems emphasis on ultimate effects, versatility, and adaptability. The structure and emphasis of the overall book reflects this heritage. The heart of the book, Chapters 5 through 11, is directed to models and the model space — the actual designs etc.

The process chapter is last, and from the view presented above, one might wonder why we include it at all. It is included for two related reasons. First, the extended and unified SE of the future as depicted in this book has some implications for the SE process, and it is important to spell out those implications. Perhaps the most important general implication is to reduce the emphasis on process. Other implications are spelled out in this chapter. Second, the point of presenting this view is not to say that process does not matter. The point is that the history of process in SE-related disciplines, and the over-reliance on process in recent decades in software and systems engineering, both indicate that the future SE needs to find a more appropriate balance between product and process, or in MOSES terms, between models and modeling or actualization. Part of the needed balance is the recognition of CS and actualization processes.

Candidate CSE Expansion Processes

This section briefly introduces several candidate processes for engineering complex systems. The scope of the section does not include traditional or hybrid systems — only complex systems.

Garbage Can Model

If one takes an organization to be a complex organic human-based system, then it is reasonable to consider the engineering of an organization to be similar in many ways to organizational decision making, properly conceived. By properly conceived, we mean an organizational decision method or model that regards organizations as organic systems, not as mechanistic systems. On that basis, the garbage can model (GCM) of organizational choice (Cohen, March, and Olsen 1972) is a candidate method at least for engineering organizations, and possibly for a broader class of CS.

The GCM was designed to model "organized anarchies," defined as organizations or decision situations that have three properties: Its goals or preferences are inconsistent and ill-defined; its processes are not understood by its members; and participation — both audience and decision makers — in making choices is highly fluid. Four streams flow through the organization, and their interaction generates decision situations. The four streams are problems, solutions, participants, and choice opportunities. Decision-making styles are defined as resolution (problem is resolved), oversight (choice is made quickly), and flight (no problem is resolved).

The authors simulated these streams using a variety of flow rates and other conditions. Among the results obtained was that the oversight and flight decision styles were more common than resolution.

Fioretti and Lomi (2008) developed an agent-based model that simulates the GCM and reproduces many of its properties. They also state that the GCM has been highly visible and influential in organizational theory since it was introduced, and remains so.

Goal-Oriented Methods

Goal orientation in requirements engineering has an extensive history (Van Lamsweerde 2001) but the primary reason for including the topic in this section is that, recently, methods and languages for goal-oriented modeling of agents and adaptive systems have emerged. Brown et al. (2006) describe the use of the Zhang and Cheng (2006) adaptation extension to LTL that was mentioned in Chapter 11, to develop goal-oriented adaptation requirements using the KAOS methodology. KAOS provides a specification language for capturing requirements in terms of goals, constraints, actions, and agents (Darimont et al. 1997). It supports refinement of requirements and assignment as agent responsibilities (Nakagawa, Karube, and Honiden 2006).

CAS Organizational Framework

Axelrod and Cohen (2000) have defined a complex adaptive systems (CAS) framework for organizational actions and decisions that adapts models such as biological evolution to enterprises. An enterprise is conceptualized as a population of more or less autonomous agents of various types. The key processes are variation, interaction, and selection. Variation involves mechanisms for copying (with error) or recombining ideas or information. The goal is to achieve a balance between variety and uniformity. Interaction patterns form among the agents that influence their behavior and their strategies. A variety of selection mechanisms act to increase or decrease certain behaviors and strategies. Selection does not always lead to improvement but when it does, it is called adaptation.

One of the instructive features of the Axelrod and Cohen framework is the adaptations they made to evolution for the enterprise context. One is that they are interested in shaping evolutionary processes rather than just observing them. Another is that the individuals involved are assumed to be intelligent, with language and culture, instead of relying on genetic heritage; that is, they are modeled as social entities rather than biological entities. A third is that the measures of success have greater variety than the ability to have offspring. Their framework emphasizes bottom-up management, and thus corresponds substantially to the enterprise autonomy (EA) element of MOSES collective actualization.

CASE

White (2008) has defined a methodology called Complex Adaptive Systems Engineering (CASE) designed to support complex systems engineering as we have defined it in this book, with a focus on an organizational setting. The methodology lays out a set of eight activities:

1. *Create climate for change:* in an organizational context, this involves convincing management to shift from hierarchical control to a more networked model.
2. *Architect a strategy:* determine how to engineer the environment such that it will influence the system to evolve and adapt in the desired direction.
3. *Target outcome spaces:* in MOSES terms, this is a description of indirect effects on the user or the environment to be achieved by the system.
4. *Reward results:* define and give rewards to the system for achieving the targeted outcomes.
5. *Formulate decision-making heuristics:* define indicators to help managers improve their decision-making abilities in the context of the system and the environment.
6. *Stimulate natural processes:* exploit the natural initiative and problem-solving ability of the system developers to achieve system innovation and integration.
7. *Develop in operational environs:* improve and adapt the system while it is operating.
8. *Assess, learn, and re-plan:* learn from this cycle and plan the next one.

The core process is a cycle of activities 1 – 2 – 3 – 4 – 8 and back to 1. The first two activities engineer the environment, while activities 3 and 4 engineer the system itself. Between or in parallel with activities 3 and 4, activities 5, 6, and 7 can be performed.

These activities reflect the MOSES CSE themes of influencing the environment and exploiting the autonomy and initiative of CS such as people; the reduced control over CS, the indirect environment effects of a system, the simultaneous operation and development of systems, and the adaptation due to learning and feedback.

Joint Cognitive SE Process

Hollnagel and Woods (2005) describe the cognitive systems engineering approach to socio-technical systems. A joint cognitive system (JCS) is a system that includes both humans and machines, which MOSES calls hybrid systems. The focus of cognitive SE is not on the human and machine as separate interacting components, but rather on the external functions of JCS as a single system. The human-machine relation is viewed as co-agency rather than interaction.

Agile Processes

Agile processes are an important aspect of addressing some of the TSE weaknesses. This approach has been used mostly in software engineering, represented in methods such as XP (extreme programming; Beck and Andres 2005) and Scrum (Schwaber 2004). The major thrust is a lightweight approach that emphasizes small and fast, as in small teams, small deliveries, and rapid development and frequent deliveries.

General discussions of the approach include those by Cockburn (2006) and Larman (2004). Proponents of the approach have formed an Agile Alliance (Agile 2008). They adhere to the agile manifesto (Manifesto 2008), which states the following values:

- Individuals and interactions over processes and tools
- Working software over comprehensive documentation
- Customer collaboration over contract negotiation
- Responding to change over following a plan

The basic tenets of agile processes are a piece of the overall unified MOSES process characteristics.

Enterprise Engineering

The concept of enterprise engineering appeared briefly in 1995 and 1996 (see, e.g., Liles and Presley, 1996) but it never got traction. Recently a similar but expanded idea has reappeared in the context of complex systems, and is sometimes called enterprise systems engineering.

Swarz and DeRosa (2006) define a framework for enterprise systems engineering processes. They describe enterprise as a different kind of entity that goes beyond a business entity or a large system; it is a collection of systems that includes people, processes, and technology; has imprecise boundaries; and is characterized by emergent behavior and environment dependencies. Their framework addresses not just these systems but also the larger environment in which they exist, and the factors shaping the change or evolution of the systems. They define five processes for addressing such systems: (1) Technology Planning, (2) Capability-Based Engineering Analysis, (3) Enterprise Architecture, (4) Strategic Technical Planning, and (5) Enterprise Analysis and Assessment.

Rebovich (2006; 2007) defines enterprise in a similar manner and addresses some of the same themes. He brings in concepts of evolutionary biology and social systems, and his focus is on the notions of actualization and influence of enterprises as opposed to the classical specify and build approach of TSE. The limited control of enterprises as complex systems necessitates moving toward shaping and improving outcomes rather than engineering specific results. The approach is intended to complement classical SE by shaping the environment of those systems.

Bricolage and Muddling Through

Three concepts discussed in the book are now brought together to form a candidate CSE process that can apply to the creation of whole new technology areas, not just the engineering of individual systems. The concept of continual change and evolution was discussed in Chapter 9 and is an underlying theme of this chapter. Autonomy and self-actualization, which bring in a bottom-up flavor, were introduced in Chapter 1 and are a major theme of this chapter. The third concept is collaboration, also introduced in Chapter 1.

A significant portion of SE effort focuses on large systems that tend to grow and evolve over an extended period of time. We have suggested that evolution and autonomy, as well as collaboration, will play a larger and more visible role in these types of systems in the future SE.

But do these processes also apply to the development of a whole new technology area? We mention a case study and two related processes that imply the answer is yes. The case study is a comparison of the approaches to the development of wind turbine technology used in Denmark and the United States from the mid-1970s into the 1990s, described in Garud and Karnøe (2003). They characterize the Danish approach as bricolage, meaning resourcefulness and improvisation of multiple collaborating agents and stakeholders in co-shaping the trajectory or path of the technology; they summarize this as "emergent co-shaping" (2003, 284). The American approach is characterized as breakthrough, meaning agents attempt to produce sudden dramatic outcomes. The American breakthrough approach is consistent with traditional entrepreneurship literature that associates success and failure of initiatives with individuals. The Danish bricolage approach is consistent with a school of thought (and literature) referred to as social construction of technological systems (SCOT), where the outcome results from a process that is more distributed across agents along the technology path.

The Danes did not pursue a design-intensive research approach but instead took designs and materials that were available and started producing simple prototypes. They learned not just from their individual experiments and prototypes, but also the various firms working this area communicated with and learned from each other, and formed an association to further the collaboration. They also had extensive involvement with users, and their scale-up of the design was based on learning from the prototypes and the users.

The Americans, on the other hand, focused on developing the best design and on defining a knowledge base of basic engineering science in wind power, and they de-emphasized field testing. The learning process was based more on a technology-push model of research results from the labs to industry developers. The policy makers also had a breakthrough mentality, attempting to jump-start the technology, and then later abruptly removing subsidies.

We would summarize the Danish/bricolage approach as growth and opportunistic learning in a user environment, and the American/breakthrough approach as determining the optimum answer in a research environment. The result was that, despite an almost order of magnitude advantage of the United States over Denmark in R&D funding and subsidies, the Danes became the world leaders in wind power, including market share, while the U.S. effort did not achieve viability.

A related process is one that developed in the 1950s and is more associated with enterprises, especially organizations and societies. This process, called "muddling through" or incrementalism by Lindblom (1959; 1979), is also closely associated with the bounded rationality of Herbert Simon in his study of organizations. Both argue against the proposition that decision makers in organizations and societies can (and therefore should) make decisions using a "scientific" approach based on complete information and optimization of all factors and trade-offs involved. Instead, the available information and optimization (rationality) are both limited. Simon used the term "satisficing," meaning satisfactory, in contrast with the goal of optimizing.

These closely related processes indicate that for complex systems, and for the majority of SE systems, trying to get everything "right" and complete at the start, and relying on a breakthrough, is not realistic. A more viable approach is to start small and learn, grow, scale, adapt, and evolve. That is a more natural problem-solving process.

Modeling Approaches

We mentioned several modeling approaches in Chapter 4 where the view was that they were part of the rationalized genealogy of MOSES. We mention them here in a different view: they are also potential applications of the MOSES concepts. That is, they are, for the most part, not a science of model-oriented systems engineering but rather a method for applying it. To the extent that they are consistent with MOSES as defined in this book, they can be considered MOSES applied methods. In the context of this chapter, this category of methods focuses less on the distinction between TSE and CSE and more on the underlying foundation of modeling, which enables both TSE and CSE. In that sense, they represent a segue to the section entitled "A Unified Actualization Process," which spells out the MOSES unified processes.

Object Process Methodology

Dori has developed a method called Object Process Methodology (OPM) available as a book (Dori 2002) and a Web site (OPM 2008). OPM supports modeling systems that include humans (agents), machines, and information. It includes a language called the Object-Process Language (OPL), which has a formal basis but is dual-purpose in that it is intended to be both machine readable and human readable. It uses a constrained subset of English to achieve the latter. It can be viewed as an alternative to UML or SysML.

The content of OPM includes areas that map directly to several MOSES model space dimensions: dynamics (maps to Time); structure and aggregation (Composition); generalization and instantiation (Commonization); systems theory and modeling natural and artificial systems (Systems science).

OPM is a method that is largely consistent with MOSES and is a potential application of it as a MOSE method.

Service-Oriented Modeling Framework

The Service-Oriented Modeling Framework (SOMF) defined by Bell (2008) was cited in Chapter 10 as an example of an SOA framework. This chapter views SOMF as an example of a modeling approach that follows the SOA style. It addresses both business and technical modeling, and encourages generalization of service concepts for more general use. It defines three modeling environments: (1) conceptual environment, (2) analysis environment, and (3) logical environment. It also emphasizes the ability to deal with "metamorphosis" or the continually changing business environment. Modeling notations are defined for analysis, business integration, design, composition styles, and architecture.

Others

Other model-oriented and model-based approaches described in Chapter 4 include the mathematically oriented, model-based SE of Wymore (1993), which is being revisited at INCOSE as a foundation of the MBSE effort there. Estefan (2008) mentions several MBSE-related approaches, including OPM and model-based architecting and engineering (MBASE) of Boehm and Port (2008). Other significant efforts at OMG and INCOSE are also relevant here. On the CSE side, Bocarra (2004) has been mentioned in various places as a source of CS models. On the TSE side, Sage and Armstrong (2000) could reasonably be characterized as a model-based TSE book.

A Unified Actualization Process

This section discusses aspects of a unified collective actualization process that is consistent with MOSES and the expanded scope of the future SE. The section does not lay out a complete process or set of processes but it does identify characteristics, process types, and example processes. The first subsection lists a set of essential characteristics and depicts a set of basic process types in the actualization space. The remaining subsections then discuss views and example processes that reflect these characteristics and types.

Characteristics and Types

Essential characteristics of the collective actualization process include the following:

1. Allow natural opportunistic problem solving in the actualization space.
2. Expect, manage, and exploit change at various levels.
3. Recognize and differentiate elements that are volatile versus stable; use the stable to anchor and guide the volatile; one important source of stability is domain knowledge, captured in the general models/knowledge space of the model space; another source is the general models that reflect the vision of the system being engineered.
4. Frequent feedback and learning: hindsight is usually better than foresight.
5. Modest planning: recognize that plans are needed but they always change, so time spent on plan details is wasted.
6. Recognize and support a range of control from controlled engineering of traditional machines to setting goals based on indirect effect on the environment to indirect influence via environment shaping.
7. The set of actualization processes described in this chapter apply to both radical design and normal design. Radical design is more like research or R&D; normal design is more like tailoring. They emphasize different parts of the processes.

Figure 13.2 shows a view of the actualization space that emphasizes process types and de-emphasizes subspaces, especially the problem space and solution space, which are not visible in this view. The process types include creative process, repeatable process, and autonomous process. The figure also contrasts MOSES with traditional emphasis of selected disciplines that include software engineering, SE, and traditional engineering disciplines such as chemical engineering and mechanical engineering. It should be noted that both creative and repeatable processes are part of MOSES as well as part of software and systems engineering. The difference is one of emphasis: MOSES, following traditional engineering handbooks, relies more on existing models and knowledge (the model space), while software and systems engineering traditionally rely more on repeatable processes. Additional items to be understood from Figure 13.2 include:

- To varying degrees, all three processes make use of existing models and knowledge.
- In terms of collective actualization elements from Chapter 1, all three process boxes in Figure 13.2 apply to both enterprise and produced system in Table 1.3. Creative and repeatable processes apply to EEM and SE; autonomous process applies to EA and SA.
- In terms of change levels from Chapter 9, all three process boxes apply to Level 2 or 3 change of an enterprise or produced system.

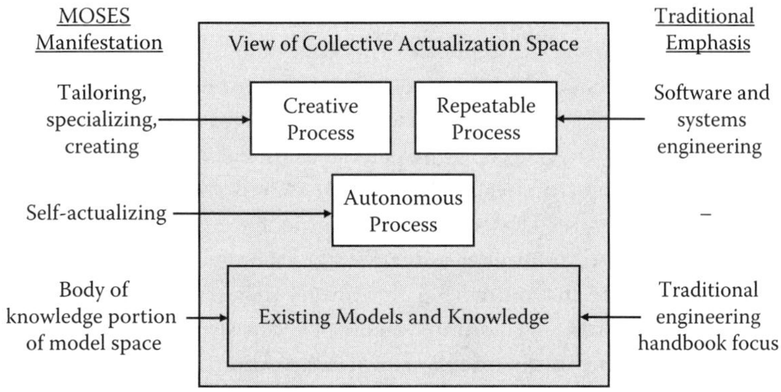

Figure 13.2 Process view of actualization space.

- Examples of each box:
  - Creative: specializing a design pattern; creating a radical new design.
  - Repeatable: configuration management; a factory or production line process.
  - Autonomous: self-organization.
  - Existing: conventional design or design pattern.

Opportunistic Process View

The first step in addressing existing processes is to investigate the actual process that competent engineers use in engineering systems, as opposed to the process they are told to use. Figure 13.3 shows the results of experiments in tracking the activities of experienced designers (Conklin 2006). The actual process is the jagged line, going back and forth between problem and solution, between

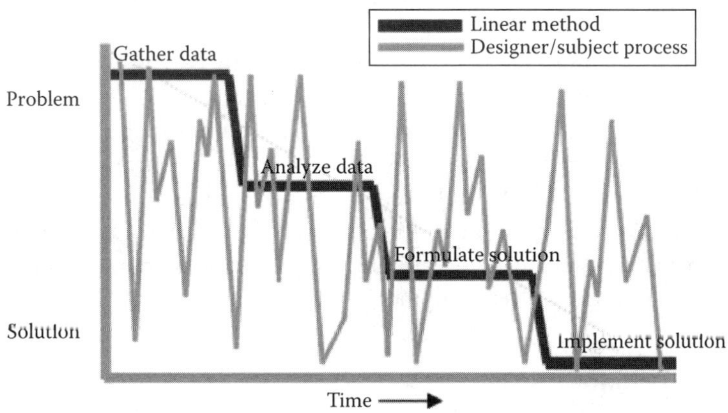

Figure 13.3 Pattern of cognitive activity of one designer: the "jagged" line. (*Source*: Conklin, J. *Dialogue Mapping*: *Building Shared Understanding of Wicked Problems, Chichester,* England: John Wiley. 2006. Used with permission.)

top and bottom. The straighter line is the waterfall model. It is an unnatural act for an engineer to try to engineer in a linear waterfall sequence. This conclusion is based not only on the result of this one study, but also on considerable experience that is consistent with this study. Conklin calls it opportunity-driven process. The opportunistic behavior pattern of designers (up and down, in terms of top level and bottom level) is the same pattern as the behavior of certain chaotic dynamical systems. An example, taken from Martelli (1999, 31–32), is the difference between two orbits of the dynamical system $x_{n+1} = 4x_n(1 - x_n)$. The two orbits have initial values x0 = 0.3 and y0 = 0.3000001. A plot of the iteration number n against the absolute difference $|x_n - y_n|$ shows a pattern similar to the jagged line in Figure 13.3. It appears that the real design process is chaotic instead of linear and "disciplined." This individual opportunism is also consistent with the collaborative opportunism evident in the bricolage process mentioned earlier in the chapter.

Why is this important? In general, the natural human problem-solving process is not linear; it is opportunistic, and moves back and forth to different parts of the problem being solved and to different aspects of the solution. The constraint of a linear step-by-step engineering process is unnatural and counterproductive. In engineering, the resulting models and *relations* between models, and between models and systems, are important. The sequence in which the models were defined is not in itself important. This is not exactly a new idea; see Parnas and Clements (1986). However, its importance has not been understood by the top-down linear engineering process movement of TSE.

Another view of the natural process is the puzzle metaphor. Putting together a jigsaw puzzle or a crossword puzzle involves jumping around to wherever we can make progress next.

This raises a question. If engineering and actualization activities are unconstrained and allowed to jump around in a chaotic fashion, how are projects to be managed? The preliminary answer is in two parts:

1. They are not unconstrained. Sequence constraints are removed but not the real constraints, which are the content of the models and their relations, as mentioned above.
2. The opportunistic process is one view of the overall actualization process. Other views are depicted in the subsections that follow.

Examples of (1): In the development of a given system, remove the constraint that a complete set of requirements must be specified and agreed on before we start thinking about architecture. Instead, constrain the architecture and the requirements to be consistent; this must be true whether the requirements were defined first, or the architecture was defined first, or they were defined in parallel. Remove the constraint that an architecture, which defines interacting components, be specified before each component is specified and built. Instead, constrain the relation to be that the components match the architecture. This must be true whether the architecture is defined first, as in top down TSE; or the architecture and the components were specified in parallel; or the components were specified and built before the architecture was defined, as allowed in system-of-systems engineering, reverse-engineering, component-based development, or SOA.

Change Process View

The levels of change were identified and described in Chapter 9. In this subsection we briefly note how an actualization process that is consistent with MOSES would use these levels.

Level 1 is the variation in system behavior: an approach to anticipating change at this level is to design systems to be as versatile as possible, to have a broad range or repertoire of behavior, and possess the ability to play different roles. However, this must be balanced with considerations of simplicity and focus.

Level 2 is changes in the system itself: The approach to anticipating change at this level is to design systems to be both adaptable and robust. Modularity and encapsulation are two classic enabling principles that still apply in the new SE. A source of material on robust design of complex systems in the expanded space of the new SE — including biology, ecology, and engineering — is the collection of case studies in Jen (2005). A related actualization discussion of controlling chaos is in Erdi (2008).

Level 3 is variations and changes in models. One primary process on a given project is tailoring general models into specific models for this system, that is, specializing from the knowledge space to the project workspace. Within the workspace, the specific models may also change throughout the lifetime of the system.

Collective Actualization

This section describes characteristics of the collective actualization process, as depicted in Table 1.3, accounting for the views described thus far. The ideas in this section are a blend of concepts from material in previous chapters, and from CSE and modeling methods in this chapter, along with ideas from Boehm and Turner (2004), Boehm and Lane (2007), Schein (2004), and some of the Cutter Consortium literature on agile methods including Spann (2006).

The Challenge

A significant tension exists in SE, including management, between the complexity and size of many SE projects and the need to adapt quickly to change. Complexity and size require significant funding, staffing, organizing, and planning to manage in a way that will produce the desired system. On the other hand, change in project variables requires the agility to respond to the change.

Methods exist separately for both of these needs. Traditional plan-driven SE methods are targeted to large complex projects. They emphasize discipline, planning, control, compliance, stability, heavyweight processes, extended time phases, documentation, and contracts. A major quality indicator is compliance to process.

Agile methods are targeted to small simpler projects. They emphasize responsiveness to change, interaction and tacit shared knowledge, short cycles, lightweight processes, and iteration. The major quality indicator is customer satisfaction.

The challenge is that SE projects do not necessarily align with either one of these two extremes. The increasing complexity and adaptability requirements of projects indicate that plan-driven methods and agile methods will need to be combined in some fashion on more projects (Boehm and Turner 2004). Given the significant differences between them, how can this be accomplished?

Concepts of the Actualization Process

Balance and Focus

A key part of the answer from Boehm and Turner is to develop a balance of the two approaches. They provide some examples of projects that have worked toward this balance. Projects need the enterprise context of planning, funding, and resource management. The plan-driven approach provides this context, and supports scale-up to larger projects. On the other hand, the same projects must produce results in a timely fashion and must be responsive to change. The agile approach can support this aspect.

Boehm and Turner presented several conclusions on this issue from their study. We discuss three of them. One conclusion is that combined methods should be tailored up from an agile starting point rather than tailored down from a full plan-driven method. That way, any additions need to be justified. Tailoring down tends to not happen in practice for a number of reasons. We would support this recommendation by noting that the traditional tailoring approach with government projects is more often the opposite: a tailoring-down method. For example, architecture frameworks are given to contractors with the possibility of tailoring them down. In practice, the result is often no tailoring because the government manager wants to be conservative and ask for everything, and the contractor is not inclined to disagree.

A second conclusion of Boehm and Turner is a corollary of the first: some balanced methods are beginning to emerge, and most of these are methods built up from agile methods. A more recent balanced method is the Incremental Commitment Model (ICM) defined by Boehm and Lane (2007). The ICM builds on several previous models including agile, concurrent engineering, and spiral. Some of the key ICM features we incorporate into the MOSES collective actualization process include the incremental, evolutionary, and iterative system definition and development and stakeholder commitment. ICM maps to the life-span stages of formation and maturation. Formation is the ICM stage I incremental definition plus the first step in stage II incremental development, which produces the first operational increment. From that point on is a succession of scheduled periods in which development of the next release is proceeding concurrently with planning and architecting future capabilities.

A third conclusion is that while methods are important, major gains are more likely to be achieved by more focus on people and expectations management. We would characterize this as putting more focus on outcomes and less focus on process. The pendulum over the past few decades in both SE and software engineering processes has moved from the outcomes side to the process side. This trend has two results. First, the end goal (producing a working system) loses some of its urgency on a project, and second, the process, as the object of attention, gradually becomes larger. Together these results conspire to slow everything down. We need to regain balance, urgency, and focus on the end goal. The discussion in Chapter 12 on outcomes and ultimate effects is a basis for this aspect of the actualization process.

In discussing agile adaptive management, Spann (2006) describes the traditional management view of the world as constrained. This view leads managers to push people hard and to focus on control. By contrast, agile adaptive management views the world as full of options. The primary feature is the view that people are innovative because they are drawn to the purpose and potential outcomes of a project, not because the manager pushes them to it. Schein (2004, 159) notes that people at different levels of management have different time horizons and planning cycles. Higher-level managers have a longer cycle — several years out — while low level-managers and workers have shorter cycles, such as weeks. These differences can be exploited, as we discuss below.

Radical and Normal Design

Radical design is required in a situation where a new type of system is needed that is not well understood. General models do not exist in the knowledge space for this type of system. It represents a high level of risk. The focus is on the opportunistic process, research, and experimentation.

Normal design is used in situations where the general problem and system type are understood, that is, where general models do exist in the knowledge space. The focus in this case is more on tailoring and specializing the general models for this specific case, but within that framework the opportunistic process also applies.

Maintaining and Verifying Fidelity of Models

Verifying of models needs not only to be an independent activity but also an ongoing activity. Because of continual change of system and environment, we cannot assume that because a model had good fidelity some time in the past, it has good fidelity now or will have in the future.

The maintenance of models is also an issue. General models in the knowledge space are presumably maintained and conditioned by the community at large, via mass collaboration and use. Specific models in the project space are maintained by the project.

Part of the conditioning of models is replication. Replication or alignment of models — and testing equivalence of models — is a way of accreting a body of knowledge. Axtell et al. (1996) defined a process for this. Chen et al. (2004) discussed aligning computational models to see if they can produce equivalent results.

Applying the Concepts

Considerations

Using concepts from the book, we build on the above ideas with a few other actualization process considerations. First, we can use the commonization and time dimensions in dealing with change. We can find stability by defining models (including both management models such as schedules, and engineering models such as requirements) at a higher level of generality, and add detail for near-term models. This approach fits naturally with the different horizons and focus of different levels of management. High level managers see the general models that are more stable and can project further into the future. Low level managers see the specific models that are more volatile and will be used in the near-term horizon.

Second, referring back to Table 1.3, we can exploit the self organization, self governance, and autonomy of the EA and SA side of the actualization process. This is consistent with the agile adaptive management discussed above, and facilitates the natural innovative behavior of people.

Third, we can exploit the concept of views in the tension between plan-driven methods and agile methods. While the balance between plan-driven and agile recommended by Boehm and Turner is a reasonable solution to the problem, we suggest that views give even more leverage of both the plan-driven and agile aspects. That is, the separation of views allows a project to apply the plan-driven approach to the more stable and long-term aspects such as resource and budget planning, research, and general architecture, as well as benefiting from interaction with the larger enterprise context, and to apply the agile approach to short-term activities, especially releases of capability in the iterative development of a system and reaction to rapid changes. The integration of views allows a two-way transfer and feedback of learning between the two project views. Long-term direction and evolution continually guides the short-term activities, while the changes occurring in the short-term activities feed into and influence the long-term direction.

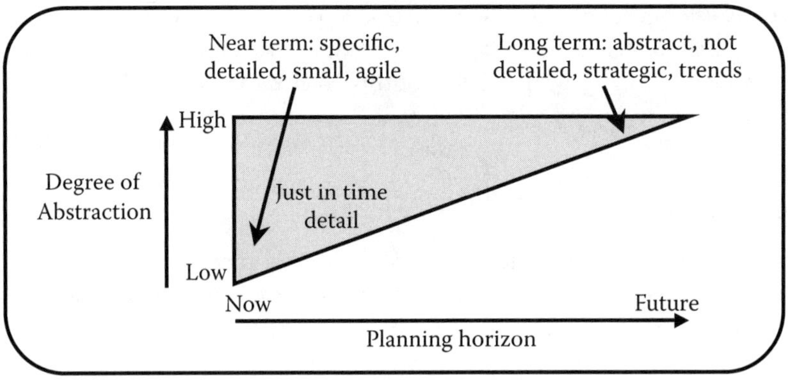

Figure 13.4 Unifying planning and agility.

Visualizations

Based on these considerations, we present some visual perspectives of how contrasting views can address both the planning needs and agile needs and bring them together into a workable framework. The approach combines different levels of abstraction (more or less detail) with different planning horizons (shorter or longer terms). Figure 13.4* illustrates a gradient of the planning horizon spectrum from now to the future. The longer term future is the most abstract and general; now is the most specific and concrete.

Figure 13.5 shows a planning process at a point in time that ranges from development and planning details for the current period or year to successively fewer details, that is, higher levels of abstraction, for future years. An arbitrary set of four levels of detail are defined, ranging from Level 1 (the most detailed) to Level 4 (the most abstract). Level 1 details are planned for current-year management and development (2009 in the figure). Level 2 plans are defined for the next year (2010), and so on up to Level 4 plans for the furthest out years. Research and experimentation are planned for as far out as necessary. The choice of one-year planning periods is also somewhat arbitrary; it could be any appropriate length such as 6-month or 3-month periods.

Figure 13.6 shows how the Figure 13.5 process is repeated each year, in a rolling wave that fills in more details for the next year and pushes the level of abstraction gradient out one more year on the planning horizon. For example, if 'now' is 2009, Level 1 plans are defined for current year (2009), Level 2 for next year (2010), etc. If 'now' is 2010, Level 1 plans are defined for current year (2010), Level 2 for next year (2011), etc.

Figure 13.7 shows a release view of the rolling wave. At the bottom is long term planning and research, and at the top are the releases, which may be per year or a shorter period as mentioned above. Each release is a small complete rapid system development cycle that fills a gap between long term vision and current state. The long term and short term models and activities both inform each other. The long term or general models guide the near term release development, and the learning from developing releases, as well as the user feedback from those releases, can modify the long-term direction. The other external forces shown in Figure 13.7, including environment

* Some of the concepts and figures in this approach were developed in collaboration with David Craig, a MITRE colleague.

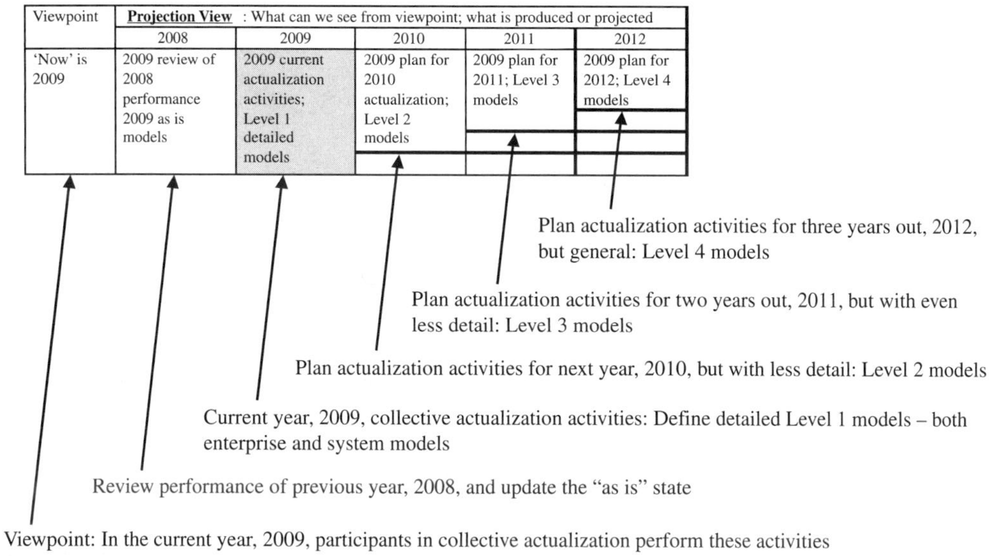

Figure 13.5 Short and long term actualization.

Viewpoint	Projection View : What can we see from viewpoint; what is produced or projected							
	2008	2009	2010	2011	2012	2013	2014	2015
'Now' is 2009	2009 review of 2008 performance; 2009 as is models	2009 current actualization activities; Level 1 detailed models	2009 plan for 2010 actualization; Level 2 models	2009 plan for 2011; Level 3 models	2009 plan for 2012; Level 4 models			
'Now' is 2010		2010 review of 2009 performance; 2010 as is models	2010 current actualization activities; Level 1 detailed models	2010 plan for 2011 actualization; Level 2 models	2010 plan for 2012; Level 3 models	2010 plan for 2013; Level 4 models		
'Now' is 2011			2011 review of 2010 performance; 2011 as is models	2011 current actualization activities; Level 1 detailed models	2011 plan for 2012 actualization; Level 2 models	2011 plan for 2013; Level 3 models	2011 plan for 2014; Level 4 models	
'Now' is 2012				2012 review of 2011 performance; 2012 as is models	2012 current actualization activities; Level 1 detailed models	2012 plan for 2013 actualization; Level 2 models	2012 plan for 2014; Level 3 models	2012 plan for 2015; Level 4 models

Figure 13.6 Rolling wave model.

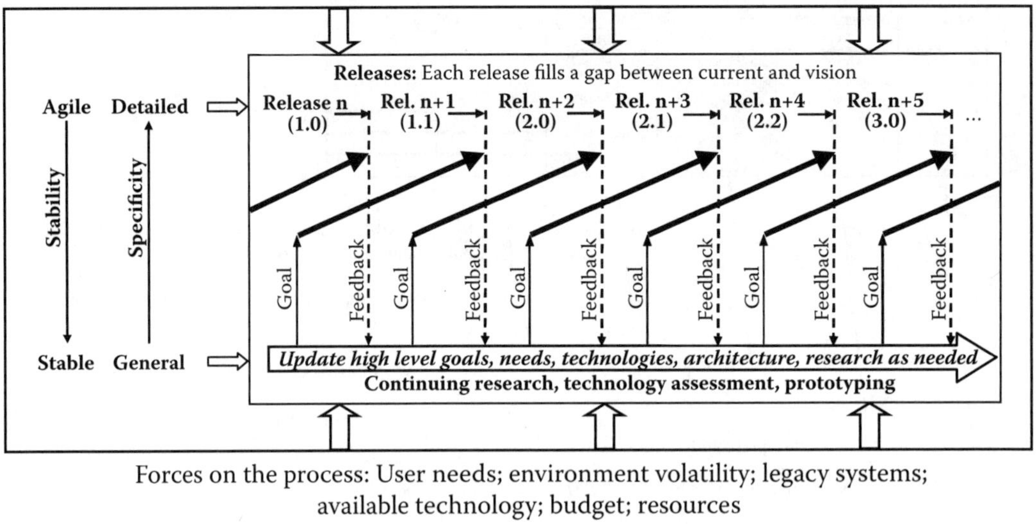

Figure 13.7 Release view of rolling wave.

volatility, budget and resource limits, and available technology, can induce change in both long term and short term models and goals.

References

Agile. 2008. Agile Alliance. http://www.agilealliance.org/. (Accessed January 11, 2009.)

Axelrod, R. and M. Cohen. 2000. *Harnessing Complexity: Organizational Implications of a Scientific Frontier.* New York: Basic Books.

Axtell, R., R. Axelrod, J.M. Epstein, and M.D. Cohen. 1996. Aligning simulation models: A case of study and results. *Computational Mathematical Organization Theory,* 1(2):123–141. http://www.econ.iastate.edu/tesfatsi/DockingSimModels.pdf. (Accessed January 11, 2009.)

Bauer, H. 1994. *Scientific Literacy and the Myth of the Scientific Method.* Urbana, IL: University of Illinois Press.

Beck, K. and C. Andres. 2005. *Extreme Programming Explained: Embrace Change, second edition.* Addison-Wesley.

Bell, M. 2008. *Service-Oriented Modeling: Service Analysis, Design, and Architecture.* Hoboken, NJ: John Wiley & Sons.

Boccara, N. 2004. *Modeling Complex Systems.* New York: Springer-Verlag.

Boehm, B. 1988. A spiral model of software development and enhancement. *Computer,* May 1988, p. 61–72.

Boehm, B. and J. Lane. 2007. Using the incremental commitment model to integrate system acquisition, systems engineering, and software engineering. *CrossTalk: J. Defense Software Eng.,* October, p. 4–9.

Boehm, B. and D. Port. 2008. Model-Based Architecting and Engineering (MBASE). http://sunset.usc.edu/research/MBASE/. (Accessed January 11, 2009.)

Boehm, B. and R. Turner. 2004. *Balancing Agility and Discipline: A Guide for the Perplexed.* Boston, MA: Addison-Wesley.

Broadbent, G. 1979. The development of design methods - a review. *Design Methods Theories,* 13(1):41–45. (Reprinted in Cross, 1984.)

Brown, G., B. Cheng, H. Goldsby, and J. Zhang. 2006. Goal-oriented specification of adaptation semantics in adaptive systems. In *Proc. of the ICSE 2006 Workshop on Software Engineering for Adaptive and Self-Managing Systems (SEAMS),* Shanghai, China.

Chen, L.C., K.M. Carley, B. Karrinsky, T. Tummino, E. Casman, D. Fridsma, and A. Yahja. 2004. *Aligning Simulation Models of Biological Attacks.* http://www.casos.cs.cmu.edu/publications/papers/chen_2004_aligningbiological.pdf (Accessed January 11, 2009.)

Cockburn, A. 2006. *Agile Software Development: The Cooperative Game, 2nd edition.* Boston, MA: Addison-Wesley Professional.

Cockburn, A. 2008. Using both incremental and iterative development. *Crosstalk: J. Defense Software Eng.*, May, p. 27–30.

Cohen, M., J. March, and J. Olsen. 1972. A garbage can model of organizational choice. *Administrative Sci. Quart.*, 17(1)1–25.

Conklin, J. 2006. *Dialogue Mapping: Building Shared Understanding of Wicked Problems*, Chichester, England: Wiley.

Cross, N. (Ed.) 1984. *Developments in Design Methodology.* Chichester, England: John Wiley.

DAG. 2006. *Defense Acquisition Guidebook.* https://akss.dau.mil/dag/. (Accessed 26 June 2008.)

Darimont, R., E. Delor, P. Massonet, and A. van Lamsweerde. 1997. GRAIL/KAOS: an environment for goal-driven requirements engineering. In *Proc. 19th Int. Conf. Software Eng.*, 17–23 May 1997, p. 612–613.

DoD (Department of Defense). 2001. *Systems Engineering Fundamentals.* Fort Belvoir, VA: Defense Acquisition University Press.

Dori, D. 2002. *Object-Process Methodology.* Berlin: Springer-Verlag.

Erdi, P. 2008. *Complexity Explained.* Berlin: Springer-Verlag.

Estefan, J. 2008. Survey of Model-Based Systems Engineering (MBSE) Methodologies, Jet Propulsion Laboratory, California Institute of Technology, Pasadena, CA.

Fioretti, G. and A. Lomi. 2008. An agent-based representation of the garbage can model of organizational choice. *J. Artificial Societies and Social Simulation*, 11(1):1. http://jasss.soc.surrey.ac.uk/11/1/1.html. (Accessed January 11, 2009.)

Forsberg, K. and H. Mooz. 1991. The relationship of system engineering to the project cycle. *Joint Conference sponsored by the National Council of System Engineers and the Am. Soc. Eng. Management,* Chatanooga, TN, October 1991.

Garud, R. and P. Karnøe. 2003. Bricolage versus breakthrough: distributed and embedded agency in technology entrepreneurship. *Res. Policy* 32:277–300.

Hillier, B., J. Musgrove, and P. O'Sullivan. 1972. Knowledge and design. In *Environmental Design: Research and Practice*, W.J. Mitchell, Ed. Los Angeles: UCLA, p. 29-3-1 through 29-3-14. (Reprinted in Cross, 1984.)

Hitchins, D. 2008. World Class Systems Engineering — The 5-Layer Model. http://www.hitchins.net/5layer.html. (Accessed January 11, 2009.)

Hollnagel, E. and D. Woods. 2005. *Joint Cognitive Systems: Foundations of Cognitive Systems Engineering.* Boca Raton, FL: CRC Press.

IEEE. 2005a. Adoption of ISO/IEC 15288:2002. Systems Engineering — System Life Cycle Processes. IEEE 15288-2004. New York: IEEE Computer Society.

IEEE. 2005b. IEEE Standard for Application and Management of the Systems Engineering Process, IEEE Std 1220-2005, New York: IEEE Computer Society.

INCOSE. 2004. Guide to the Systems Engineering Body of Knowledge — G2SEBoK. International Council on Systems Engineering. http://g2sebok.incose.org/. (Accessed January 11, 2009.)

INCOSE. 2007. *INCOSE Systems Engineering Handbook*, v3.1. Seattle, WA: INCOSE (International Council on Systems Engineering).

ISO/IEC. 2008. Systems and Software Engineering — System Life Cycle Processes. ISO/IEC 15288:2008, 2nd ed. 18 March.

JCIDS. 2007. Operation of the Joint Capabilities Integration and Development System. CJCSM 3170.01C. DoD Joint Chiefs of Staff. May.

Jen, E. 2005. *Robust Design: A Repertoire of Biological, Ecological, and Engineering Case Studies.* New York: Oxford University Press.

Jones, J. 1963. A method of systematic design. In *Conf. Design Methods*, J.C. Jones and D.G. Thornley, Eds. New York: Macmillan Company. (Reprinted in Cross, 1984.)

Kossiakoff, A., and W.N. Sweet. 2003. *Systems Engineering: Principles and Practice*. Hoboken, NJ: John Wiley & Sons.

Larman, C. 2004. *Agile and Iterative Development: A Manager's Guide*. Boston, MA: Pearson Education.

Larman, C. and V. Basili. 2003. Iterative and incremental development: a brief history. *Computer*, 36(6):47–56.

Liles, D. and A. Presley. 1996. Enterprise modeling within an enterprise engineering framework. *Proc. 1996 Winter Simulation Conf.*, p. 993–999.

Lindblom, C. 1959. The science of "muddling through." *Public Administration Rev.*, 19(2):79–88. http://www.jstor.org/stable/973677. (Accessed January 11, 2009.)

Lindblom, C. 1979. Still muddling, not yet through. *Public Administration Rev.*, 39(6):517–526. http://www.jstor.org/stable/976178. (Accessed January 11, 2009.)

Manifesto. 2008. Agile Manifesto. http://www.agilemanifesto.org/. (Accessed January 11, 2009.)

Martelli, M. 1999. *Introduction to Discrete Dynamic Systems and Chaos*. New York: John Wiley & Sons.

Medawar, P. 1969. *Induction and Intuition in Scientific Thought*. Philadelphia: American Philosophical Society.

Nakagawa, H., T. Karube, and S. Honiden. 2006. Analysis of multi-agent systems based on KAOS modeling, *Proc. 28th Int. Conf. Software Eng.*, 20–28 May 2006, Shanghai, China.

Nguyen, L. 2006. Adapting the Vee model to accomplish systems engineering on change projects. NDIA. In *9th Annu. Systems Eng. Conf.: Focusing on Improving Performance of Defense Systems Programs*, San Diego, CA, 24–26 October 2006. http://www.dtic.mil/ndia/2006systems/Thursday/ngut.pdf. (Accessed January 11, 2009.)

OPM. 2008. Object-Process Methodology. http://www.objectprocess.org/. See also http://www.er.byu.edu/er2003/slides/ER2003PT1Dori.pdf. (Accessed January 11, 2009.)

Parnas, D. and P. Clements. 1986. A rational design process: how and why to fake it. *IEEE Trans. Software Eng.*, 12(2):251–257.

Popper, K. 1965. *Conjectures and Refutations: The Growth of Scientific Knowledge, 2nd ed.* New York: Basic Books.

Rebovich, G. 2006. Systems Thinking for the Enterprise: New and Emerging Perspectives. *Proc. 2006 IEEE/SMC Int. Conf. System of Systems Eng.*, Los Angeles, CA, April 2006.

Rebovich, G. 2007. Engineering the enterprise. *1st Annu. IEEE Systems Conf.*, Waikiki Beach, Honolulu, HI, 9–12 April 2007.

Rittel, H. 1972. Son of Rittelthink: Horst Rittel on the state of the art in design methods. In *The DMG 5th Anniversary Report: DMG Occasional Paper No. 1*, p. 5–10. (Reprinted in Cross, 1984, as "Second Generation Design Methods.")

Royce, W. 1970. Managing the Development of Large Software Systems, *Proc. 9th. Int. Conf. Software Eng.*, IEEE Computer Society, 1987, p. 328–338. (Originally published in *Proc. IEEE WESCON, 26:1–9.*)

Sage, A.P. and J.E. Armstrong, Jr. 2000. *Introduction to Systems Engineering*. New York: John Wiley & Sons.

Schein, E.. 2004. *Organizational Culture and Leadership. 3rd ed.* San Francisco: Jossey-Bass.

Schwaber, K. 2004. *Agile Project Management with Scrum*. Redmond, WA: Microsoft Press.

Spann, D. 2006. *Agile-Adaptive Management Model: Building the Bridge from Innovation to Business Value*. Cutter Consortium Executive Report, Vol. 9, No. 1.

Swarz, R. and J. DeRosa. 2006. A Framework for Enterprise Systems Engineering Processes. ICSSEA 2006.

Van Lamsweerde, A. 2001. Goal-oriented requirements engineering: a guided tour. Invited minitutorial. In *Proc. RE'01 — Int. Joint Conf. Requirements Eng.*, Toronto, IEEE, August, p. 249–263. http://citeseer.ist.psu.edu/vanlamsweerde01goaloriented.html (Accessed January 11, 2009.)

Vitruvius Pollio, M. 1914. *The Ten Books on Architecture*. Morris Hicky Morgan (translator). Cambridge, MA: Harvard University Press. Original: Latin, 1st Century BCE. Available as Project Gutenberg EBook #20239, Dec. 2006. http://www.gutenberg.org/files/20239/20239-h/29239-h.htm. (Accessed January 11, 2009.)

Weinberg, S. 1994. *Dreams of a Final Theory*. New York: Vintage Books.

White, B. 2008. On Complex Adaptive System Engineering (CASE). Paper presented at *8th Understanding Complex Systems Symp.*, University of Illinois at Urbana-Champaign, 12–15 May 2008.

Wikipedia. 2008. Systems Engineering Process. http://en.wikipedia.org/wiki/Image:Systems_Engineering_Process_II.gif. (Accessed January 11, 2009.)

Wymore, W. 1993. *Model-Based Systems Engineering*. Boca Raton, FL: CRC Press.

Zhang, J. and B.H.C. Cheng. 2006. Using temporal logic to specify adaptive program semantics. *J. Systems and Software* 79(10):1361–1369.

Acronyms

AADL	Architecture Analysis and Design Language
AAS	Advanced Automation System (FAA)
ACM	Association for Computing Machinery
ACSL	Advanced Continuous Simulation Language
ADL	Architecture Description Language
ADML	Architecture Description Markup Language
AML	Agent Modeling Language
AMS	Analog and Mixed Signal
AOD	Audio on Demand
ARC	Australian Research Council
ASSL	Autonomic System Specification Language
ATM	Automated Teller Machine
BDI	Belief, Desire, Intention
BFO	Basic Formal Ontology
BoK	Body of Knowledge
BPMN	Business Process Modeling Notation
BSA	Business Services Architecture
BWW	Bunge-Wand-Weber
C4ISR	Command, Control, Communications, Computers, Intelligence, Surveillance, and Reconnaissance
CAS	Complex Adaptive Systems
CASE	Complex Adaptive Systems Engineering
CASL	Cognitive Agents Specification Language
CBA	Capabilities-Based Assessment
CBDI	Component-Based Development and Integration
CEO	Chief Executive Officer
CESF	Center for Engineering Systems Fundamentals
CIMMS	Center for Integrative Multiscale Modeling and Simulation
CIO	Chief Information Officer
CMS	CROSI Mapping System
CONOPS	Concept of Operations
CoS	Cost of Service

COTS	Commercial Off-the-Shelf
CPU	Central Processing Unit
CROSI	Capturing, Representing, and Operationalising Semantic Integration
CS	Complex Systems
CSCS	(University of Michigan) Center for the Study of Complex Systems
CSE	Complex Systems Engineering
CSS	Complex Systems Science
CSSL	Continuous System Simulation Language
DARPA	Defense Advanced Research Projects Agency
DBC	Design by Contract
DL	Description Logics
DLRM	Digital Library Reference Model
DoD	U.S. Department of Defense
DoDAF	DoD Architecture Framework
DOLCE	Descriptive Ontology for Linguistic and Cognitive Engineering
DRIVER	Digital Repository Infrastructure Vision for European Research
EA	Enterprise Architecture
ECSAM	Embedded Computer System Analysis and Modeling
EE	Enterprise Engineering
ESA	European Space Agency
ESL	ESA Simulation Language
FEA	Federal Enterprise Architecture
FOC	Full Operational Capability
GCM	Garbage Can Model
GFO	General Formal Ontology
GME	Generic Modeling Environment
GML	Geographic Markup Language
GOTSE	Good Old Traditional Systems Engineering
GPS	Global Positioning System
GUMO	General User Model Ontology
HCI	Human Computer Interface
HOT	Highly Optimized Tolerance
HPS	Human-Provided Services
IBM	International Business Machines
IEC	International Electrotechnical Commission
IEEE	Institute of Electrical and Electronic Engineers
IID	Iterative and Incremental Development
INCOSE	International Council on Systems Engineering
IOC	Initial Operational Capability
ISFO	Integrated System of Foundational Ontologies
ISO	International Organization for Standardization
ISSS	International Society for Systems Sciences
JCIDS	Joint Capabilities Integration and Development System
JCS	Joint Cognitive System
KAOS	Knowledge Acquisition in autOmated Specification
KARO	Knowledge, Abilities, Results, Opportunities
KIF	Knowledge Interchange Format

KR	Knowledge Representation
KWIC	Keyword in Context
LTL	Linear Temporal Logic
MA	Management Autonomy
MARTE	Modeling and Analysis of Real-Time and Embedded
MathML	Mathematical Markup Language
MBASE	Model-Based (Systems) Architecting and Software Engineering
MBSE	Model-Based Systems Engineering
MDA	Model-Driven Architecture
MDSOC	MultiDimensional Separation of Concerns
ME	Management Engineering
MIT	Massachusetts Institute of Technology
MO	Model Orientation
MoDAF	Ministry of Defense Architecture Framework
MOF	Meta-Object Facility
MOSE	Model-Oriented Systems Engineering
MOSES	Model-Oriented Systems Engineering Science
MSC	Message Sequence Charts
MTBF	Mean Time Between Failure
NAF	NATO Architecture Framework
NATO	North Atlantic Treaty Organization
NRC	National Research Council
NSL	Neural Simulation Language
OASIS	Organization for the Advancement of Structured Information Standards
OBO	Open Biological Ontology
OCL	Object Constraint Language
OMB	Office of Management and Budget
OMG	Object Management Group
OPL	Object Process Language
OPM	Object Process Methodology
OR	Operations Research
OWL	Web Ontology Language
PBSE	Patterns-Based Systems Engineering
PDF	Probability Distribution Function
PIM	Platform Independent Model
POSIX	Portable Operating System Interface for Unix
QoE	Quality of Experience
QoS	Quality of Service
R&D	Research & Development
RDF	Resource Description Framework
RM-ODP	Reference Model for Open Distributed Processing
SA	System Autonomy
SAE	Society of Automotive Engineers
SDL	Specification and Description Language
SE	Systems Engineering
SES	Systems Engineering Science
SFI	Santa Fe Institute

SGML	Standard Generalized Markup Language
SGSR	Society for General Systems Research
SML	Service Modeling Language
SOA	Service-Oriented Architecture
SOAP	Simple Object Access Protocol
SOC	Self-Organized Criticality
SOI	System of Interest
SOMA	Service-Oriented Modeling and Architecture
SOMF	Service-Oriented Modeling Framework
SoSE	System of Systems Engineering
SS	Systems Science
SUMO	Suggested Upper Merged Ontology
SVG	Scalable Vector Graphics
SysML	System Modeling Language
TOGAF	The Open Group Architecture Framework
TS	Traditional Systems
TSE	Traditional Systems Engineering
TTF	Task-Technology Fit
UD	Universe of Discourse
UDDI	Universal Description, Discovery, and Integration
UEML	Unified Enterprise Modeling Language
UML	Unified Modeling Language
UPDM	Unified Profile for DoDAF and MoDAF
VHDL	VHSIC Hardware Description Language
VIN	Vehicle Identification Number
VM	Virtual Machine
WS-BPEL	Web Services Business Process Execution Language
WSDL	Web Services Description Language
XMI	XML Metadata Interchange
XML	Extensible Markup Language
XPDL	XML Process Definition Language
XSL	Extensible Stylesheet Language
YAWL	Yet Another Workflow Language

Index

f indicates figures; t indicates tables

E

F

J

K

Q

R

T

U

V